Praise for
THE SKEPTICS' GUIDE TO THE UNIVERSE

"In his 1995 book *The Demon-Haunted World*, the great Carl Sagan prophesied a descent into superstition and ignorance. Well, that world has arrived. Fortunately, Steve Novella and his coauthors are here to help us navigate it with critical thinking and scientifically appropriate skepticism, along the way exposing the anti-science and pseudoscience so prevalent in our public discourse today."

—Michael Mann, distinguished professor, Penn State University, and coauthor of *The Madhouse Effect: How Climate Change Denial Is Threatening Our Planet, Destroying Our Politics, and Driving Us Crazy*

"Steve and the gang have done it again. *The Skeptics' Guide to the Universe* is the best and most popular podcast on science and skepticism out there (I am honored to have been their first guest!). And now this book is the best, soon to be the most popular, guide to what's really real, so far as we can tell."

—Massimo Pigliucci, K. D. Irani Professor of Philosophy at the City College of New York and author of *Nonsense on Stilts: How to Tell Science from Bunk*

"There's nothing more riveting, nor more frightening, than the economy-size ability of the citizenry to embrace unscientific explanations for puzzling events. We hardly blink when ordinary folk are seduced by easy-to-grasp—if wrong—explanations. But we *should* be alarmed. The science may be pseudo, but the grievous consequences are real. Steven Novella and his colleagues offer a fascinating collection of the many contemporary phenomena now ascribed to mysterious or even conspiratorial causes. For those who want to know, not merely believe, this book is an essential read. It's also a great one."

—Seth Shostak, senior astronomer, SETI Institute

"Using examples ranging from Monty Python to Monty Hall, *The Skeptics' Guide to the Universe* offers the first ever page-turner that teaches you how to think clearly."

—Paul A. Offit, MD, author of *Bad Advice: Or Why Celebrities, Politicians, and Activists Aren't Your Best Source of Health Information*

"If everyone in the world were to read this book, we might just arrest humankind's depressing slide into truthlessness. Someone should put *The Skeptics' Guide* on the vaccination schedule."

—Tim Minchin, musician/comedian

"Our lack of critical thinking skills and embarrassing scientific illiteracy are among the most critical issues of our time. We are drowning in a sea of myths, lies, deceptions, fakes, and superstitions. We are inundated and surrounded by pseudoscience and illogic. We have lost our way. The timing couldn't be better for the publication of a brilliant new book, *The Skeptics' Guide to the Universe*. It's a friendly, comprehensive, clear, and direct guide to these pressing issues and guess what…it's a hell of a lot of fun to read."

—Dean Edell, MD, host of the *Dr. Dean Edell Show* and author of *Eat, Drink & Be Merry*

"Reads like a dream: casual, occasionally irreverent, full of great anecdotes and humor but ultimately a well-researched and authoritative account of the vagaries of human folly and foolishness."

—Bruce Hood, author of *SuperSense*, *The Self Illusion*, and *Possessed*

"If there's one thing we all need in these confusing and stressful times, it is clear thinking and accurate appraisals of the information flooding our senses. Thankfully, this book gives to the reader precisely the tools needed to bring this about. *The Skeptics' Guide to the Universe* is an essential resource for a lifetime of critical thinking and analysis."

—Michael Whelan, award-winning artist and author of *The Art of Michael Whelan* and *Beyond Science Fiction*

The SKEPTICS' GUIDE to the UNIVERSE

How to Know What's Really Real in a World Increasingly Full of Fake

DR. STEVEN NOVELLA

with **BOB NOVELLA, CARA SANTA MARIA, JAY NOVELLA, AND EVAN BERNSTEIN**

GRAND CENTRAL
PUBLISHING

NEW YORK   BOSTON

A number of the topics discussed in this book are related, in one way or another, to disease, medicine, or health. Such information is not intended to replace the advice of the reader's own physician or other medical professional. You should consult a licensed and qualified medical professional in matters relating to your health, especially if you have existing medical conditions, and before starting, stopping, or changing any regimen that has been prescribed by your doctor or health professional, or the dose of any medication you are taking. Individual readers are solely responsible for their own health care decisions.

Grand Central Publishing
Hachette Book Group
1290 Avenue of the Americas, New York, NY 10104
grandcentralpublishing.com
twitter.com/grandcentralpub

First Edition: October 2018

Grand Central Publishing is a division of Hachette Book Group, Inc. The Grand Central Publishing name and logo is a trademark of Hachette Book Group, Inc.

The publisher is not responsible for websites (or their content) that are not owned by the publisher.

Library of Congress Cataloging-in-Publication Data
Names: Novella, Steven, author. | Novella, Bob, author.
Title: The skeptics' guide to the universe : how to know what's really real in a world increasingly full of fake / Dr. Steven Novella with Bob Novella [and three others].
Description: First edition. | New York, NY : Grand Central Publishing, Hachette Book Group, 2018. | Includes bibliographical references and index.
Identifiers: LCCN 2018013089| ISBN 9781538760536 (hardcover) | ISBN 9781478988496 (audio download) | ISBN 9781538760512 (ebook)
Subjects: LCSH: Science—Popular works. | Errors, Scientific—Popular works.
Classification: LCC Q162 .N68 2018 | DDC 500—dc23
LC record available at https://lccn.loc.gov/2018013089

ISBN: 978-1-5387-6053-6 (hardcover), 978-1-5387-6051-2 (ebook),
978-1-5387-6432-9 (B&N signed edition)

Printed in the United States of America

LSC-H

10 9 8 7 6 5 4 3 2 1

Dedicated to Perry DeAngelis,
A friend and skeptic of some note

Contents

SECTION 2:
Adventures in Skepticism

SECTION 3:
Skepticism and the Media

SECTION 4:
Death by Pseudoscience

SECTION 5:
Changing Yourself and the World

Introduction

There is a theory which states that if ever anyone discovers exactly what the Universe is for and why it is here, it will instantly disappear and be replaced by something even more bizarre and inexplicable. There is another which states that this has already happened.

—Douglas Adams

———————

Spock lied to me.

And it wasn't just Spock (well, actually Leonard Nimoy as the host of the popular but dubious TV show *In Search of...*) who lied but the media, journalists, corporations, politicians, salespeople, and just about every adult and authority figure in my life. Some were deliberately lying, while others were fudging the truth for what they thought was a good purpose. Most, however, were simply mistaken, misinformed, or self-deluded and were spreading what they assumed was the truth but were just the lies others had told them.

We all start as children believing pretty much whatever we're told. The gulf of knowledge and experience between adults and young children is so great that to a child, any adult is perceived as the ultimate authority on any topic. As we mature we realize that not all adults agree with each other, so some of them must be wrong. We become more sophisticated in our choice of authority figures, but we still tend to rely on others to know what to think—on experts, leaders, religious figures, celebrities, talking heads, or just "common knowledge."

Science is another authority. When I was younger I had an intense fascination with all things scientific. Scientists simply had the best stories to tell: how gigantic dinosaurs roamed the earth millions of years ago; how our primitive ancestors made tools out of stone; how the sun, Earth, and moon formed billions of years ago; how a single cell develops into a human; and how life evolved from simple creatures to everything we see today.

My brothers and I watched every TV science documentary we could find, instinctively knowing they were much better than any sitcom or hokey drama. Mixed in with the science documentaries, however, were what I now know to be pseudoscience or fake documentaries. This is where Leonard Nimoy comes in, as the host of the popular series *In Search of...*, which aired from 1977 to 1982.

In each episode Nimoy would narrate how scientists had discovered that aliens built the Nazca lines (the enigmatic drawings etched into the sands of the Nazca desert in Peru) or were on the verge of discovering a large creature living in Loch Ness. He built a compelling case (or so I thought) for the existence of extrasensory perception, Atlantis, and Bigfoot. Just like the hosts of the other science documentaries we would watch, he would show us evidence and interview experts. We ate it up.

At the same time, we were being raised as good Catholics. Looking back, it amazes me how, as a young child, I completely compartmentalized my belief in evolution and Adam and Eve. Abraham, Mary, and Joseph were historical figures to me, taught with the same authority as were the exploits of George Washington. In church, faith was taught as an absolute virtue, so being a good person meant simply believing.

This exciting world of fantastic narratives that society had built in my head—evolution, Genesis, Loch Ness monsters, and many more—was not sustainable. Eventually I had to sort out the apparent conflicts among these various narratives. The claims of evolutionary biology and a literal interpretation of Genesis could not both be true at the same time. This meant questioning and doubting. I had to figure out which narratives to believe and which ones to reject.

This is the essence of skepticism: How do we know what to believe and what to doubt?

Once you begin to ask questions like "How do we actually know anything?" our beliefs start to fall one by one. How would we know if aliens were visiting Earth or if sticking needles in special points on the body could cure disease? This questioning doesn't always end well, as some people will choose to reject science so they can maintain their belief in aliens or miracles. Others keep questioning. This process is never done—it is an endless journey.

The question is, How far will you go, and in which direction will you head? This is ultimately a personal odyssey, but we aren't completely alone. We are part of ever-expanding social rings with which we share our quest for knowledge and understanding. The biggest ring encompasses all of humanity. As a species, we are on a journey of discovery, trying to figure out what to believe and what knowledge is reliable.

Your travels will take you through some interesting and exciting twists and turns. It can be scary. At some point you will confront a belief you really want to be true, that may even be part of your identity, but that does not stand up to close scrutiny. Ultimately this is a journey of self-discovery, and hopefully this book will serve as a guide on that journey. In these pages you will learn about the many ways in which your brain fails, its tendency to prefer nice clean and reassuring stories, the absolute mess that is your memory, and all the preprogrammed biases in your thinking.

You will also learn about the many ways in which society fails, the imperfections in the institutions of science, learning, and journalism. It's intimidating to realize that we live in a world overflowing with misinformation, bias, myths, deception, and flawed knowledge. We are all children and there are no adults. There are no ultimate authority figures; no one has the secret; there is no revealed knowledge; and there is no place to look up the definitive answers to our questions (not even Google).

We are all struggling through this complex universe just like everyone who came before us.

Fully realizing this essence of the human condition can make you cynical, denying all knowledge. That, however, is just another bias, another narrative to help us deal with the chaos. Cynicism is also cheap—it's easy just to doubt or deny everything. But it doesn't get you or society anywhere. Skepticism goes beyond cynicism. While it may start with doubt, that's the beginning, not the end. There isn't any definitive or ultimate knowledge (no Truth with a capital *T*), but we can grind out knowledge about the world that is sufficiently reliable for us to treat it as provisionally true and act upon it. We can, for example, send probes to Pluto and get back gorgeous pictures of a place that was previously entirely mysterious. If our knowledge of the universe wasn't at least to some degree true, then our efforts to see Pluto wouldn't have been rewarded with beautiful images of this distant and frozen world.

So, while we cannot trust the stories we are told, tradition, faith, convenient or reassuring narratives, charismatic figures, or even our own memories, we can slowly and carefully build a process by which to evaluate all claims to truth and knowledge. A big part of that process is science, which systematically tests our ideas against reality, using the most objective data possible. Science is still a messy and flawed process, but it is a process. It has, at least, the capacity for self-correction, to move our beliefs incrementally in the direction of reality. In essence, science is the process of making our best effort to know what's really real.

Combined with science are logic and philosophy, which are simply ways of thinking about things really carefully, to at least make sure there is no internal contradiction. Valid reasoning, careful and systematic observations, recording and counting all data (not just the data we want), and trying to disprove our own conclusions is a lot of work, but it moves us forward on our journey. Skeptics combine knowledge of science, philosophy, human psychology, and all the flaws and biases of human thinking to evaluate all claims and beliefs, especially their own.

I took a major step in this direction when I watched the series *Cosmos*, cowritten and hosted by Carl Sagan. It was perhaps the first time I saw a science documentary that didn't just state facts but explored how we know what we know. Sagan investigated what we thought in the past, and how new discoveries changed our thinking. In one episode he dedicated a segment to questioning the alleged evidence for alien visitation. That segment was a revelation. If there was ever a moment when I tipped over into becoming a skeptic, that was it. Here was a respected scientist carefully arguing with logic and evidence why the case for UFOs was not compelling. In fact it was utter crap. That meant that all those alien visitation documentaries (including half of the *In Search of…* episodes) were not just wrong, they were nonsense. I felt betrayed.

Those documentaries weren't real science, they were what Sagan called "a cheap imitation." They were pseudoscience. Wow—what other claims were as flimsy as the evidence for UFOs? The dominoes started to fall.

My journey took another step when I encountered creationism, the denial of evolution. This blew me away. By then I'd read dozens of books and articles on evolution, which was one of my favorite scientific topics, and I was well versed in the mountains of evidence scientists had accumulated showing that life on Earth has evolved.

Now I was reading or listening to people who denied all this evidence, denied that evolution had happened at all, or at least how scientists claim it had happened. I couldn't help but slow down and take a close look, like viewing a horrific accident on the highway. The arguments of creationists ranged from silly to sophisticated, but ultimately they were all flawed. I tried my best to understand where they went wrong. At first I naively believed that if I could just explain to creationists (when I had the opportunity, during random in-person encounters) the flaws in their reasoning or the factual errors in their premises, I could change their minds. While this isn't impossible, it proved far more difficult than I had imagined.

Something was going on, something I didn't understand. I went to college as a science major struggling with all of this and my rapidly waning faith. I was well on my way on a skeptical journey that continues to this day, three decades later.

This obsession with understanding how we know what we know and how human belief goes off the rails has become my life's work. I love the process and findings of science and working out how to explain all of this to others. When I discovered that there was an active skeptical movement of people with the same set of obsessions, I never looked back.

One day my close friend and fellow skeptic Perry DeAngelis noticed that while there were dozens of skeptical groups around the country, there were none in our home state of Connecticut. We should start our own group, he suggested, and I knew instantly we were on the right path.

We were joined by our mutual friend Evan Bernstein and my two brothers, Bob and Jay. Those are the same brothers who watched *In Search of...* with me as teenagers. We helped each other through those early years spent transitioning from believing all manner of nonsense to being skeptics. I still remember (well, as much as our flawed memories can record anything) the day I told Bob that I no longer thought ESP (extrasensory perception) was real. He was shocked, but a thirty-minute conversation was all it took to lay out the gaping flaws in the evidence.

Together, in 1996, the five of us created the New England Skeptical Society. We were content (sort of) to print a newsletter and host local lectures. We would have loved to be more involved, but we all had day jobs. We did go on investigations, attend UFO conventions, and confront proponents of all sorts of pseudoscience. Those were the years when we "earned our bones." We also had the opportunity to meet scientists, philosophers, magicians, and other skeptical activists.

Then social media hit with full force. In 2005, when podcasting was a new thing, I had the idea of starting a weekly science and critical thinking podcast. We settled on the name *The Skeptics' Guide to the Universe* (inspired by the Hitchhiker's Guide to the Galaxy series of books we all love)—SGU for short. As of writing this book, we are still going strong.

Along the way we picked up a fellow traveler, Rebecca Watson, who cohosted the show for nine years. We also lost Perry, who died in 2007 from complications of scleroderma. We were joined by a new Skeptical Rogue (as we call ourselves) in 2015, Cara Santa Maria.

Every week the five of us try to make sense of what seems like an increasingly crazy world. The internet and social media have been a mixed blessing, allowing us to teach science and critical thinking to a far wider audience but also facilitating the spread of misinformation like never before.

In the last two decades, while we've been fighting for science and reason, it seems like the stakes have only gotten higher. My own profession—medicine—has been thoroughly infiltrated by the pseudoscience of so-called "alternative" medicine. The very process of science is under attack. There are entire movements dedicated to denying and opposing the hard-won fruits of scientific discovery. And now it seems that truth and facts themselves are easily tossed aside as an inconvenience. There are those who think, with good reason, that humanity is stepping back from the Enlightenment to huddle in echo chambers of comfortable, or at least familiar, beliefs.

This is a generational struggle, one that will likely never end.

Like in the universe of Douglas Adams, we have created this helpful and reassuring guide, which will prove invaluable on your skeptical journey, because the world is actively trying to deceive you and fill you with stories and lies. The forces of ignorance, conspiracy thinking, anti-intellectualism, and science denial are as powerful as ever.

It's true that we have to struggle with these incredibly flawed meat machines in our skulls. But we are standing on the shoulders of giants.

A lot of smart people over a very long time have thought and argued carefully about the nature of reality and our ability to understand it. We have some powerful tools, like science and philosophy. We do know stuff, and we have ways of making sense of it all.

Don't panic. This whole notion of thinking for yourself and questioning everything is actually quite fun and empowering. We can do this together.

SECTION 1

Core Concepts Every Skeptic Should Know

The truth may be puzzling. It may take some work to grapple with. It may be counterintuitive. It may contradict deeply held prejudices. It may not be consonant with what we desperately want to be true. But our preferences do not determine what's true.   —Carl Sagan

Before you start on any journey, you need the right equipment—so we'll grab the necessary tools, some trail mix, and a good jacket, and put on a pair of comfortable shoes. In this section of our guide we'll load you up with skeptical gear (like a fashionable utility belt), the things that we decided were the most important for the journey, things that at least will give you options to deal with the obstacles or pitfalls you might encounter on your way.

These tools—your core concepts of scientific skepticism—can be broken down into four categories. The first set of skills comprise what I like to call "neuropsychological humility." This category includes knowledge of all the ways in which your brain function is limited or flawed. The primary tool with which you probe and understand the universe is your brain, and we should all have a better understanding of how it works.

The second category of skeptical "gear" is called metacognition—thinking about thinking. Metacognition is an exploration of all the ways in which your thinking is biased. This has obvious overlap with the first category but focuses more on critical thinking skills than on the hardware of the brain. It would be nice if humans were perfectly logical beings, like a certain pointy-eared fictional character mentioned in the introduction, but we're not. We are emotional, semi-rational creatures, plagued with a host of biases, mental shortcuts, and errors in thinking.

The third type of skeptical equipment has to do with science—how it works, the nature of pseudoscience and denialism, and how science can go wrong.

The fourth category of core concepts takes you on some historical journeys, reviewing iconic examples of pseudoscience and deception as cautionary tales.

All of this skeptical gearing-up is to prepare you for your adventure and get you started (or help you on your way if you have already begun). This isn't some weekend getaway—the journey is the rest of your life.

Are you ready to become part of an epic quest, one that has taken us from huddling in dark caves to stepping foot on the moon? (Yes, we really did that.) Like all adventures, this one is foremost a journey of self-discovery. The monsters you will slay and the challenges you will face are mostly constructs of your own mind. But if you can master them, the rewards are indeed epic.

1. Skeptics' Guide Entry: Scientific Skepticism

Section: Core Concepts
See also: Critical Thinking

Scientific skepticism, a term first popularized by Carl Sagan, is an overall approach to knowledge that prefers beliefs and conclusions that are reliable and valid to ones that are comforting or convenient, and scientific skeptics therefore rigorously and openly apply the methods of science and reason to all empirical claims, especially their own. A scientific skeptic provisionally accepts a claim only in proportion to its support from valid logic and a fair and thorough assessment of available evidence. A skeptic also studies the pitfalls of human reason and the mechanisms of deception so as to avoid being deceived by others or themselves. Skepticism values method over any particular conclusion.

> We ignore public understanding of science at our peril.
>
> —Eugenie Scott

You are reading the *Skeptics' Guide* because we advocate the overall worldview known as scientific skepticism. There can be a lot of confusion about what it means to be a skeptic, however. What do we do and what do we believe?

Being a skeptic means doubting, but philosophical skepticism is distinct from scientific skepticism and is not what we advocate.

Philosophical skepticism is essentially a position of permanent doubt: Can we actually know anything? And what is the nature of knowledge itself? Before science revolutionized our thinking, philosophical skepticism was a reasonable position. Most knowledge was based on authority and tradition, so wiping the slate clean by doubting everything was probably a step in the right direction. René Descartes famously said that all we really know is "I think therefore I am." His idea was to eliminate everything that passed for knowledge up to that point, to start fresh and then try to see what he could reason out from first principles (self-evident starting points).

Lucky for us, we don't live in a prescientific age. There are hundreds of years of carefully worked-out knowledge leading to where we are today. Philosophers focus on how to think in a clear, precise, unambiguous, and internally consistent way. Science works within a philosophy of methodological naturalism (all effects have natural causes) and uses a refined set of methods to check our theories against reality.

We still can't know anything to 100 percent metaphysical certitude, but we can know things. We can methodically build a set of knowledge that is internally consistent and logically valid, not only consistent with what we can observe about reality but that actually helps us predict how the universe will behave.

That's why "scientific" skeptics are not philosophical skeptics, professing that no knowledge is possible. We are also not cynics, who doubt as a social posture or just have a generally negative attitude about humanity. We are not contrarians who reflexively oppose all mainstream opinions. The term "skeptic" has also been hijacked by deniers who want to be viewed as genuine skeptics (asking the hard and uncomfortable questions) but are really just pursuing an agenda of denial for ideological reasons.

We are scientific skeptics because we start with doubt, but we then carefully try to separate what we can and do know from fantasy, wishful thinking, bias, and tradition.

I believe that modern scientific skepticism has several facets, not only a worldview but also a body of knowledge and an area of expertise. Here are the tools and methods scientific skeptics use to help them parse reality:

Respect for Knowledge and Truth—Skeptics value reality and what is true. We therefore endeavor to be as reality-based as possible in our beliefs and opinions. This means subjecting all claims to a well-founded process of evaluation.

Skeptics believe that the world is knowable because it follows certain rules, or laws of nature. The only legitimate method for knowing anything empirical about the universe follows this naturalistic assumption. In other words, within the realm of the empirical (factual knowledge based on evidence), you don't get to invoke magic or the supernatural.

Promotion of Science—Science is the only set of legitimate methods for investigating and understanding the natural world. Science is therefore a powerful tool—and one of the best developments of human civilization. Those of us who choose to be activists promote the role of science in our society, public understanding of the findings and methods of science, and high-quality science education. This means protecting the integrity of science and education from ideological intrusion or antiscientific attacks. It also involves supporting high-quality science, which requires examining the process, culture, and institutions of science for flaws, biases, weaknesses, and fraud.

Promotion of Reason and Critical Thinking—Science works hand in hand with logic and philosophy, and therefore skeptics also promote better understanding of these fields and critical thinking skills.

Science vs. Pseudoscience—Skeptics are the first, and often the last, line of defense against incursions by pseudoscience. In this role, we seek to identify and elucidate the borders between legitimate science and pseudoscience, to expose pseudoscience for what it is, and to promote knowledge of how to tell the difference. If I had to say briefly what our core area of expertise is, it's pseudoscience. When countering common but false beliefs, it's not enough to understand the relevant science, you also have to know how science goes wrong, how people form and maintain false beliefs, and how they promote those beliefs. Such expertise is generally lacking within mainstream academia, and that's where skeptics come in.

Ideological Freedom / Free Inquiry—Science and reason can only flourish in a free society in which no ideology (religious or otherwise) is imposed upon individuals or the process of science and free inquiry.

Neuropsychological Humility—Being a functional skeptic requires knowledge of the various ways in which we deceive ourselves, the limits and flaws of human perception and memory, the inherent biases and fallacies in cognition, and the methods that can help mitigate all these flaws and biases.

Consumer Protection—Skeptics endeavor to protect themselves and others from fraud and deception. We do this by exposing fraud and educating the public and policy makers to recognize deceptive or misleading claims or practices.

In addition, skeptics tend to be a source of institutional memory. The same scams and false beliefs tend to come around again and again: Seemingly, every generation has to make the same mistakes for themselves. It's useful to have those who study the history of scams, errors, and pseudoscience to help recognize and deflect such things when they inevitably reappear.

While activist skeptics tend to be science communicators, we aren't *just* that. We have a particular expertise and niche—knowledge of pseudoscience, the mechanisms of deception, and how to counter misinformation. A generation ago, science communicators felt that all they had to do to counter belief in pseudoscience and myths was to teach science. We now know this is sadly not the case.

For example, a 2017 study by John Cook et al., confirming prior research, showed that exposing people to misinformation about the scientific consensus on global warming had a polarizing effect based on political ideology. People who already accepted the consensus did so even more, and those who rejected the consensus held more firmly to their rejection. Correcting that misinformation had almost no effect on reducing this polarization—facts were simply not enough to change people's minds.

However, if you started by explaining to the subjects how fake experts can be used to falsely challenge the scientific consensus, the polarizing effect of this misinformation was completely neutralized.

This is exactly why we choose to promote science partly by exposing pseudoscience, and not just the false information of bad science but the deceptive (sometimes self-deceptive) tactics that pseudoscientists use. It's not enough to just teach people science, you have to teach them how science works and how to think in a valid way. This book is meant to be one giant inoculation against bad science, deception, and faulty thinking.

No one is arguing that these are the most important issues in the world. Often we're criticized for tackling one type of belief over another, but that's what we call the fallacy of relative privation, the notion that what you are doing is not valuable because there is a more important issue out there that needs attention. "Don't bother doing anything until we cure childhood cancer," for example.

I call this "Gilligan's Island logic." Sure, if you are one of the few people stranded on a deserted island, you should tackle the most pressing survival problems first. But on a planet with over seven billion people, that makes no sense. People should feel free to take on whatever issues are important to them, or where they feel their talents and inclinations lie.

So when you see the word "skeptic" in this book, it means an advocate for science and critical thinking. Perhaps after reading the book you will consider yourself a skeptic too.

NEUROPSYCHOLOGICAL HUMILITY AND
MECHANISMS OF DECEPTION

Around the turn of the twentieth century, the invention of the airplane—a heavier-than-air flying machine—was widely anticipated. In the decades leading up to the Wright brothers at Kitty Hawk in 1903, and even for a time afterward, there was a rash of sightings of such flying machines.

Rumors of sightings would break out and engulf communities. Local newspapers would report how prominent citizens (police, the mayor) saw the ships and could even make out the pilot at the controls. Many people swore they heard the distant roar of the engine. Others reported finding debris dropped by the airships or even being taken for a ride in one. Often it was assumed that famous inventors like Thomas Edison were involved.

We now know that all these incidents were hoaxes and mass delusions. There were no airships. Drawings of the alleged aeroplanes by eyewitnesses resemble quaint notions of contraptions with flapping wings, not the planes that were eventually developed.

What these and many other similar incidents reflect is the constructed and unreliable nature of perception, memory, and belief. They are the products of expectation, cultural influence, and psychology.

When the great airship hoax was eventually revealed to be just that, many people who'd been taken in were angry. I suspect, however, that they were mainly angry with themselves for being so gullible. But as we'll see, they were just being human.

2. Skeptics' Guide Entry: Memory Fallibility and False Memory Syndrome

Section: Neuropsychological Humility
See also: Perception, Self-Deception

There are numerous ways in which our memories are unreliable. In fact, entirely false memories can be easily fabricated, sometimes by misguided therapists.

> Cognitive psychology tells us that the unaided human mind is vulnerable to many fallacies and illusions because of its reliance on its memory for vivid anecdotes rather than systematic statistics. —Steven Pinker

Our real-time experience of the world is fleeting. The moment after we experience something, it becomes a memory—first a short-term memory, some of which is later consolidated into long-term memory. Everything we personally know about the world is memory.

But how reliable is human memory? Think of a vivid childhood remembrance, the kind of memory that you recall often and that represents an important part of your history and identity. I hate to break this to you (actually, no I don't), but that memory is probably mostly, or even entirely, fake.

You may find that notion disturbing, and right now you are thinking, "No way. I clearly remember that trip to the toy store when I was ten. No way is it fake." But, I'm sorry to say, a century of memory research is not on your side.

Our memories aren't accurate or passive recordings of the past. We don't have squishy camcorders in our skulls. Memories are constructed from imperfect perceptions filtered through our beliefs and biases, and then over time they morph and merge. Our memories serve more to support our beliefs than to inform them. In a way, they are an evolving story we tell ourselves.

Most people have had the experience of being in a heated conversation, and then once the dust settles and everyone tries to resolve the discrepancies, it becomes clear that each person has a different memory of the conversation that just happened. You may be tempted in these situations to assume your memory is accurate and everyone else is losing their mind. Of course, that's what everyone else is thinking about you.

You may also have had the experience of reminiscing about an event long ago with someone who was there with you. On one episode of the SGU, Bob, Jay, and I were recalling a family trip to Pompeii and Herculaneum from our childhood. Bob and Jay clearly remembered an encounter with a strange tour guide who took perverse pleasure in pointing out every ancient depiction of a penis to our mother. My memory is that our mother told us of this encounter from her previous trip to Pompeii, but we never met the tour guide ourselves. Bob and Jay were shocked to learn that their memories were mostly fabricated (although they stubbornly insist it's my memory that's flawed).

Researchers have learned much about the process of constructing our memories. As we will discuss in chapter 3, "Fallibility of Perception," what we perceive minute by minute is also an active constructive process. The various sensory streams are all filtered for the important stuff, assumptions are made about what we are likely perceiving, and everything is compared in real time to the rest and to our existing models and assumptions about the world. Preference is given to continuity, making sure everything syncs up.

That perceptual stream then becomes a memory. Recent research suggests that short-term and long-term memories may form at the same time, but the short-term memory dominates initially and then fades quickly. Over a few days, long-term memory may consolidate and dominate. The actual relationship between short-term and long-term memory is likely more complex than this, and research is ongoing, but those are the basics.

There are various kinds of long-term memory, and while they overlap considerably, they have different "neuroanatomical correlates" (involve different networks in the brain).

There is declarative memory (also called explicit memory), which is factual knowledge stored in long-term memory and consciously recalled. This is what most people think of when they think of memory. Procedural memory (also called implicit memory) is more automatic and involves learning how to do motor tasks such as throwing basketballs or writing in script.

Declarative memory is further divided. Episodic memory is memory of events and our own experiences. Episodic memories form what we call autobiographical memory, which is memory of what has happened in our own lives, usually remembered in the first person. There is also a specific type of autobiographical memory known as a "flashbulb" memory, which is an unusually vivid memory of an emotionally significant event. Remembering where you were and what you were doing when you learned about the attacks on 9/11 or other similar events is a flashbulb memory.

Semantic memory is factual memory about the world, not specific to your own experiences. Semantic memory also has different components, which are stored separately. We have memory for facts, a separate memory for the "truth status" of each fact (is it true or false?), and a still separate memory for the source of the fact. This is why it is common to remember that you heard something somewhere but you can't remember where—and may not even remember if it's true or not.

Memories are Malleable

Memories are flawed from the moment we construct them, but they're also not stable over time. Each time we recall a memory we are actually reconstructing and updating it. Memory, like perception, emphasizes internal consistency. We alter memories to fit our internal narratives about reality, and as those narratives change, we update our memories to fit them.

The life of a memory is a chaotic adventure with many twists and turns. Here are some of the ways we mold our memories along the road:

Fusion—We can fuse the details of different memories, mixing them up or even combining two separate memories into one. Consider that hypothetical trip to the toy store when you were a child. Perhaps you remember a scary encounter with a stranger in the store. But that encounter with a stranger may have happened at another time and place, and for some reason you merged the experience with your memory of visiting the toy store. It's also possible that the scary encounter happened in a movie, or to someone else, and you combined it with your own memory.

Confabulation—Put simply, we make shit up. This is a completely automatic and subconscious process. Again, our brains want to construct a continuous and consistent memory, so if there are any missing pieces it just makes them up to fill in the gaps.

People with dementia tend to confabulate more, because their failing memory and cognition creates more gaps for them to fill. In severe cases it's possible to get patients with dementia to confabulate entire events that never happened, and they need little encouragement.

Details are invented in order to emphasize the emotional significance of an event. In fact, the overall thematic and emotional significance of a memory is stored separately from the supporting details.

In 2008, when asked about a trip to Bosnia a few years earlier, Hillary Clinton recalled:

I certainly do remember that trip to Bosnia, and as Togo said, there was a saying around the White House that if a place was too small, too poor, or too dangerous, the president couldn't go, so send the First Lady. That's where we went. I remember landing under sniper fire. There was supposed to be some kind of a greeting ceremony at the airport, but instead we just ran with our heads down to get into the vehicles to get to our base.

Clinton then fell victim to the age of video. In a video of the event found by the press, Clinton is shown walking casually with her daughter, Chelsea, without any apparent rush or fear. She even stops on the tarmac to greet a young girl. Hillary's account and the video record are in stark contrast.

Clinton was widely criticized for "lying" about the event, but no such accusation is necessary. She remembers the fear and anxiety surrounding her trips to troubled parts of the world, and her brain could have easily filled in the details to support that emotional theme.

In 1995 Elizabeth Loftus and Jacqueline Pickrell performed a study, which has become known as the "Lost in the Mall" study. They interviewed older relatives of twenty-four subjects in order to create a booklet for each subject detailing four stories from their childhood. However, only three of the stories were true; the fourth was a completely made-up story about being lost in the mall.

The subjects were able to recall some details about 68 percent of the true stories. However, 29 percent also remembered the false event either partially or fully. Some subjects could recount the false event in extreme detail and had high confidence in the accuracy of their account.

In a 2010 study by Isabel Lindner et al., researchers found that simply observing another person performing an act can create false memories that we performed that act. They report:

In three experiments, participants observed actions, some of which they had not performed earlier, and took a source-memory test. Action observation robustly produced false memories of self-performance relative to control conditions.

This research follows other research demonstrating that imagining an event is often enough to create the false memory of that event. Imagination activates many of the same brain areas that a true memory would. In essence, a memory of the imagination may over time become indistinguishable from a memory of a real event—and a false memory is born.

This is especially relevant to many UFO abduction therapists who use hypnosis and encourage their clients to imagine themselves being abducted. This "research" is in fact optimized to produce (rather than uncover) false memories of alien abduction.

A 2015 study by Julia Shaw and Stephen Porter found that many adults can be convinced that they committed a nonexistent crime after just three hours of interrogation by police. She reports:

> Our findings show that false memories of committing crime with police contact can be surprisingly easy to generate, and can have all the same kinds of complex details as real memories.

In essence our brains lie to us all the time. They tell us stories constructed from multiple sources, with details added as necessary to fill in any cracks.

Personalize—There is a tendency to shift memories from happening to other people to happening to ourselves. If someone tells us an emotionally gripping tale of an encounter, months or years later we may remember ourselves in the starring role, or at least as a direct witness. (Perhaps that explains Bob and Jay's obviously false memory of Pompeii.)

NBC anchor Brian Williams may also have fallen victim to this phenomenon. Over the years, he's told a story of being in a helicopter that was forced down under fire while flying over Iraq. The problem is, his story has evolved over the years and differs from the memories of others who were there. It seems he has in retrospect personalized a story of another helicopter being forced down and merged it with his own less eventful helicopter trip. Williams was actually fired for, most likely, having the same flawed memory as everyone else.

Contamination—We are social creatures. Part of our social nature is that we place high value on the testimony of others. When people discuss an event together, sharing the details of their individual memories, they are likely to contaminate each other's memory. People will take details offered by someone else and seamlessly incorporate them into their own memories.

That is exactly why courtroom witnesses aren't allowed to meet or talk with each other prior to giving their testimony.

Distortion—Details of a memory can also simply change or be distorted. The details may drift over time, or they may change in a way specifically to support the emotional narrative of the memory.

Memories can be distorted by suggestion. Merely suggesting a detail to someone while they are recalling a memory may cause them to incorporate that detail into their memory. If, for example, they saw a hooded figure and couldn't discern the figure's gender, an interviewer referring to the person as "her" could be enough for the memory to change to one of a hooded woman.

The different elements of memory can also be distorted independently of each other. A 2005 study by Skurnik et al. supports the notion that we remember the "truth status" of facts separately from the facts themselves. It's as if we have a little check box next to memories to indicate if they are actually true. Skurnik first exposed subjects to factual claims and told them that some of the claims were myths. An example might be, "It is a myth that vaccines cause autism." After three days, they surveyed the subjects and found that 27 percent of the young adults in the study misremembered false statements as true. The figure was 40 percent for older adults. Apparently, we don't pay as much attention to that little check box as we should.

Confidence and Accuracy

Perhaps one of the most iconic memory studies was by Neisser and Harsch in 1992, in which they looked at flashbulb memories related to the *Challenger* explosion. They gave 106 students in an introductory psychology class a questionnaire asking them to recall how they heard about the *Challenger* explosion, which had happened within the last twenty-four hours. They were given seven specific questions about what they were doing and how they felt at the time. Two and a half years later, the same students were given a follow-up questionnaire. In this survey they were also asked to rate their confidence in the accuracy of their memory on a scale from 1 to 5.

Of the seven details they previously recorded, on average the students could remember only 2.95 of them. A quarter of the students scored zero out of seven, and half scored two or less. In fact, only a quarter of the students even remembered taking the survey previously. Despite their terrible recall of the event, the average score of confidence in the accuracy of their memories was a whopping 4.17 out of 5. Other studies have also shown the lack of any relationship between confidence in a memory and its accuracy. We tend to think that vividness and confidence predict accuracy, but they don't.

The clear lesson here is that we all need to be humble when it comes to the accuracy of our own memories. As we will see, failure to appreciate the true nature of memory can create great mischief.

False Memory Syndrome

Belief systems and myths have incredible cultural inertia, and they're difficult to eradicate completely. That's why belief in astrology, while held by a minority, persists.

Professions, however, should be different. A healing profession should be held to a certain minimum standard of care, and that standard should be based upon something real, which means that scientific evidence needs to be brought to bear. Professionals, especially medical professionals, are not excused for persisting in false beliefs that have long been discredited.

The 1980s saw the peak of an idea that was never based in science: the notion that people can suppress memories of traumatic events and those repressed memories can manifest as seemingly unconnected mental health issues, such as anxiety or eating disorders. The idea was popularized by the book *The Courage to Heal* (1988), in which the authors took the position that their therapy clients, especially women, who had such issues should be encouraged to recover memories of abuse, and if such memories could be dredged up, then they were real.

The concept of repressed memories led in part to the satanic panic of the 1980s, and many of those subjected to recovering techniques "remembered" not only being abused but also being part of satanic ritual abuse.

Recovered memory syndrome was a massive failure on the part of the mental health profession. The ideas, which were extraordinary, were never empirically demonstrated. Further, basic questions were insufficiently asked: Is there any empirical evidence to support the incredible events emerging from therapy, for example? Is it possible that the recovered memories are an artifact of therapy and are not real?

Now, with three decades of hindsight, we can say a few things with a high degree of confidence. Recovered memory syndrome is mostly, if not entirely, a fiction. People generally do not repress memories of extreme trauma (the existence of rare exceptions remains controversial). Further, as Elizabeth Loftus pointed out, memories are constructed and malleable things. Also, independent investigations by the FBI, other law enforcement agencies, and scholars never found any evidence of the satanic ritual abuse, murders, and other atrocities emerging in recovered memory sessions. The events simply never happened.

What emerged from the entire sad episode was an increased understanding of what is now called false memory syndrome, the construction of entirely fake memories. This is accomplished through guided imagery, hypnosis, suggestion, and group pressure. These techniques violate one of the basic rules of investigation: Never lead someone by putting words in their mouth. This is especially important if the person is vulnerable and confused.

While there is some legitimate controversy over whether it is even possible to repress such memories and accurately recall them later, there is no question that the massive repressed-memory industry of the 1980s and '90s was not evidence-based. It was essentially an industry creating false memory syndrome. The fact that a controversial idea was put into practice so widely, despite the risks to patients and their families, indicates a systemic lack of self-regulation within the mental health profession.

More disappointing is the fact that recovered memory therapy is still ongoing today. Even if you wish to adhere to the minority opinion among memory experts that repressed memory is possible, the evidence doesn't justify the practice of recalling such "memories." The potential for harm is more than sufficient grounds to suspend the practice pending further evidence.

In 1990, George Franklin was convicted of murder after just one day of deliberation by the jury. The conviction was based entirely on the "recovered" memories of his daughter, who did not remember the alleged murder until twenty years after it occurred. Her memories of witnessing her father rape and murder her eight-year-old friend came back to her over time and under therapy, after being "repressed" for two decades. There was no physical evidence to support the accusation. After serving about five years of a life sentence, Franklin's conviction was overturned, and he was not retried, in part because of the misuse of "recovered memory" and hypnosis in the case. The jury was apparently impressed by the vividness and confidence of Franklin's daughter's memory, enough to give him life without parole.

They shouldn't have been.

Even when you aren't sitting on a jury, it's important to remember the true nature of memory. Whenever you find yourself saying, "I clearly remember..." stop! No, you don't. You have a constructed memory that is likely fused, contaminated, confabulated, personalized, and distorted. And each time you recall that memory you reconstruct it, changing it further.

When it comes to memory, be skeptical and humble. Unless the memory is of a trip to Pompeii, and then, of course, your brothers are wrong.

3. Skeptics' Guide Entry: Fallibility of Perception

Section: Neuropsychological Humility
See also: Illusion

The act of perception is a complex, highly filtered, and active constructive process by your brain. We do not passively perceive external stimuli like a camera. This constructive process introduces many possibilities for illusion and misperception.

> Our beliefs do not sit passively in our brains waiting to be confirmed or contradicted by incoming information. Instead, they play a key role in shaping how we see the world.
> —Richard Wiseman

We have all marveled at optical illusions. One second you're looking at a picture of snow scattered over rocks, and then suddenly you're looking at a picture of a dalmatian. Or perhaps you're looking at clearly wavy lines and then are informed that they are actually straight and parallel. You may have trouble believing it, so you get out a straight edge, and damn—those wavy lines aren't wavy, they're straight. How can that be?

We're a bit freaked out by really good optical illusions because they force us to directly confront a reality we tend to ignore as we go through our daily lives: What we think we see is not objective; it is a process of our brains, and that process can be fooled.

Perception Is Constructed

The bottom line is this: Your real-time perceptions are not a passive recording of the outside world. Rather, they are an active construction of your brain. This means that there is an imperfect relationship between outside reality and the model of that reality crafted by your brain. Obviously, the model works well enough for us to interact with that reality, and that's actually the idea. Constructed perception is not optimized for accuracy but rather for functionality.

First, only a minute fraction of information from the outside world even makes it to the portions of your brain that construct your perception. Much of it is missed because your organs of perception aren't perfect and there are numerous trade-offs. For example, only the tiny fovea of our eyes records images in any detail. This means that you have the potential to have 20/20 vision only in the area of a postage stamp held at arm's length. That small patch of vision, densely packed with light-sensing cells, carries about half the information from your retina. The other half is spread out over the rest of your retina, and for that part of your vision you are legally blind. Outside the fovea your vision is essentially a blurry mess.

There's also a blind spot in each eye corresponding to the location on the retina where the optic nerve exits the eye. You don't perceive the blind spots or the blurriness because your brain weaves a seamless image out of this information. After, of course, a lot of processing.

That processing crafts colors out of the information from the cones. It enhances edges, adjusts contrast, and also adjusts for apparent perspective. All of this has only created a two-dimensional image—your brain also needs to process all the cues to distance, size, and relative position, make the best assumptions it can, and then construct a three-dimensional visual model out of this partial and heavily processed information.

Hold on, we're just getting started. Now you have a crafted image, but it doesn't mean anything yet. The next area of the visual cortex assigns meaning to the image—is that a tree or a whale? Okay, it finds a match, then makes further adjustments to the basic processing so that the image it constructed matches better with the thing it thinks it's seeing.

Did you catch that? Visual processing is a two-way street. The basic visual information is processed up the chain as your brain constructs a meaningful image, and then the brain communicates back down the chain to tweak the construction so it fits better. Essentially, if your visual association cortex thinks you are looking at an elephant, it communicates back to the primary visual cortex and says, "Hey, make that look even more like an elephant." It changes what you actually see, not just how you interpret it. This all happens automatically, outside of your awareness.

And we're not done. Now you have a tweaked image of an object, but what meaning does this object have for you? Does it seem to be moving in a non-inertial frame (meaning that its movement does not seem to be just passively following gravity)? If so, then it must have agency. Your brain then sends that visual stream along a different path and assigns emotional significance to the image.

But wait, we need to animate all this. As objects move through your visual field, their images morph, and light plays across their surfaces. You don't want to be confused by this, so our brains smooth it all out over time so we see it as one object moving through our vision. All of this processing also takes time (a few hundred milliseconds). By the time we see that baseball flying at our face, it might have hit us, so our brains project movement a little bit into the future to compensate for the delay in processing time.

If you think that's it, think again, because so far I've only been talking about visual processing. Our brains are simultaneously processing auditory information, sensation from our bodies, vestibular information about gravity orientation and acceleration, and feedback from our muscles to tell how we're moving. Our brains favor continuity and internal consistency over accuracy, so all these streams are compared in real time and further adjustments made so they all fit together nicely. In a way, our brains are constructing a narrative about what is happening, and making that narrative make sense to us.

Also in the stream, however, are our prior knowledge and expectations. We know elephants are big, so when we see a small elephant our brains tend to assume it is big and therefore it must be far away. But it could be small and close up (okay, if it were a miniature elephant).

Glitches often occur when the stimulus is ambiguous or contradictory. We call these glitches illusions. Your brain might construct the information in more than one way. It may even flip back and forth, but each construction will be compelling while it's dominant. For example, there is a spinning girl illusion. The animated image is specifically constructed so that it makes equal sense if the girl is spinning in either direction. But you can't see the girl spinning to the left and the right at the same time; it's one or the other as your brain flips assumptions and resulting constructions.

As an example of how our brains compare different sensory streams and then tweak them, consider something known as the McGurk effect. When you listen to someone speak, you look at how their lips are moving, and if the sounds you hear are ambiguous, you will not only read their lips, your brain will change what you hear so that it matches how the lips appear to be moving. (I'm sorry, did you say I should go pluck myself?)

Your brain also constructs its internal model of your body. You probably take for granted that you are in your body, you own your body, and you can control your body. It might seem like these sensations are inevitable consequences of the obvious fact that you *are* your body. This isn't true, however. All of those bodily sensations are active constructions by the brain, requiring specific circuits dedicated to that purpose. First, we feel like we are in our bodies because our brains compare what we see and what we feel. We literally see ourselves in our own body, and we feel the parts of our body, and when those things sync up, our brain constructs the sensation of being in our body.

This construction can be disrupted, which results in an out-of-body sensation. Some drugs can do this, as can seizures in certain parts of the brain, or even just some easy trickery. Researchers have found that it is actually quite simple to get people to feel as if they are inside a virtual body or a prop body rather than their own. The setup is straightforward: Wear virtual reality goggles that are being fed real-time video from a camera behind you. This way you are seeing the back of yourself. Someone then touches the back of your shoulder. You see the virtual image in front of you being touched while simultaneously feeling the touch. This is enough for your brain to construct the sensation that you occupy the virtual image of yourself rather than your actual body.

This has cool implications for people who love video games. Imagine playing a virtual reality game and feeling as if you actually are your avatar (the virtual representation of you in the game). Awesome.

There is also a module in the brain called the ownership module. This module is part of a circuit that includes somatic information and also emotional processing. Essentially you feel as if your arm is your arm—it belongs to your body and is part of you. Related to this are circuits that make you feel as if you control the various parts of your body. These circuits compare information about your intention to move to visual and feeling sensation about your actual movements. When your movements match your intentions, your brain creates the sensation that you control your body. If this circuit is interrupted, however, the result is something called alien hand syndrome. With this syndrome your body part seems to act on its own, without your control.

The lesson here is that even the most basic components of your existence are actively constructed by your brain. Each component can be disrupted and erased.

How does all this affect critical thinking? Well, just as with memory, be wary of saying, "I know what I saw." Hmm...no, you don't. You have a constructed memory of a constructed perception based on filtered partial sensation and altered by your knowledge and expectations. Psychologists have recognized some specific and dramatic manifestations of the limits of human perception. Mentalist Derren Brown has made a career partly out of turning these psychological demonstrations into magic tricks, because they are amazing.

One such phenomenon is called attentional (or inattentional) blindness. This means that we don't see what we are not paying attention to, even if it happens right in front of our eyes. The now classic demonstration of this comes from a 1999 video by researchers Daniel Simons and Christopher Chabris in which students in white or black T-shirts are passing basketballs. Your task is to count how many times the students in white pass the ball.

If you have your smartphone handy, search for "selective attention test" on YouTube before reading on if you want to avoid spoilers.

Okay, done? Did you see the gorilla? About 40 percent of people do not see the gorilla—they miss something happening right in front of them. I didn't see the gorilla the first time I watched the video, and like most people, I went back to make sure it was really there and I wasn't the victim of trickery. Nope, the only trickery was in my own limited perception.

This research demonstrates the effect of attention on perception. We focus on the students in white and miss everything else, which reflects the fact that attention is a limited resource of our brains. We can only pay attention to so much at once. We can also either make our attention diffuse but sacrifice detail or focus down to catch detail but then miss everything else. Following up on this research, psychologists Janelle Seegmiller, Jason Watson, and David Strayer found that subjects who had better working memory were more likely to see the gorilla than those who scored lower on a test of working memory. This likely reflects a larger budget of attention.

Another follow-up study by Trafton Drew asked radiologists to read a CT scan of the lungs. He placed an obvious picture of a gorilla in a dark part of the scan. It is clearly visible, and yet 82 percent of trained radiologists missed the gorilla. Radiologists, while experts at reading such scans, are still humans. They're not looking for a gorilla. They're looking for tumors and other known pathology, and you don't see what you are not looking for.

It also turns out that research showing inattentional blindness goes back quite a long time. Probably the first studies were published in the *Journal of the Society for Psychical Research* in 1959. They were only inadvertently about inattentional blindness, however. The researcher was interested in ghost sightings, so he donned a sheet and walked through a college campus in one study and across a movie theater stage in another (while a trailer was playing). In the first case no one reported seeing anything unusual, and in the second only about half the audience noticed anything. The author concluded that "real" reports of ghost sightings must be different, that they must contain some psi (extrasensory) component. What he really documented, however, was the first experiment to demonstrate inattentional blindness. The ghost fared about as well as the gorilla—maybe a ghost gorilla would do better?

Inattentional blindness obviously has many implications for our day-to-day lives, not the least of which is when driving. Increasingly, studies show the dangers of distracted driving. In another example, a 2012 study simply asked subjects to name the location of the fire extinguisher closest to their office. Only thirteen out of fifty-four subjects (24 percent) were able to correctly name the location, even if they'd been working at the same place for years. After reading this study, I tried to recall the nearest fire extinguisher to my office and learned that I didn't know there was one literally two feet from my office door.

The researchers point out that this could be due to not remembering as well as not noticing the fire extinguishers. In a follow-up study one month later, however, all of the original fifty-four subjects could remember the location. Drawing their attention to the fire extinguisher had a lasting effect on their memory. This, unfortunately, justifies why flight attendants have to review that boring safety procedure every single time you get on a plane. The reminders do help. But that doesn't make you want to sit through them.

There is a related phenomenon called change blindness, which reflects our inability to notice changes in detail. We do tend to notice changes that occur in front of us, but not when they occur outside our direct view. So, if you're looking at a complex picture or a painting with many details, and one of those details changes in front of you, you will likely notice it. If, however, the picture blinks (you look away and back, or an image on a computer screen literally blinks off and back on) and the change occurs in the blink (referred to as a visual disruption), you're unlikely to notice it. Even big changes can be missed in this way.

Another iconic video demonstrating this effect is the color (or "colour" since it's British) changing card trick by our friend Richard Wiseman, which can be found by searching "colour changing card trick" on YouTube. Watch it and be amazed. You can also just Google "change blindness" and be rewarded with all sorts of fun.

Change blindness can be dramatic. In many psychological experiments it's been shown that you may not notice you are talking to a different person than the one you were just talking to. Dan Levin and Dan Simons have done a series of experiments in which an experimenter walks up to a stranger on the street to ask them for directions. Another experimenter then walks between them with a door, and after it passes by, there is a different person talking to the stranger (previously hiding behind the door and swapped with the original person). Only about half the subjects notice the change.

The same experiment has been done while talking to a person behind a counter. The experimenter bends down to retrieve a form and then a different person stands up, with only about half the subjects noticing the switch.

There are factors that affect how many people notice, however. We are more likely to notice the change if the person is from our social group. We are also more likely to notice if there is a change in gender, race, or age group. What this suggests is that we tend to classify people into broad categories—I am speaking to an old white guy—and that is all the detail we notice. As long as the person swapped in belongs to the same general category, the change may not register. Think about the implications of this research for eyewitness testimony—did the witness really see the accused at the crime scene? How can we trust such testimony when half of subjects won't notice when one person is swapped for another?

With all these limitations to our perception, it may seem amazing that we can even function. However, day-to-day our perception works fine. We encounter glitches all the time, like mishearing what someone said, and we laugh those experiences off. They're usually not that consequential. Occasionally, however, we may find ourselves in an unusual situation, or something unexpected may happen. We may be in poor viewing conditions, or we may be sleepy. At these times our perceptions become even less reliable.

We see something sudden and brief out of the corner of our eyes, and our brain constructs an image of what we think we saw. That is the image we remember. It's just a story our brain told us, but we remember that we actually saw it. We may see three bright lights in the sky and our brain connects the dots so that now we see a huge object that isn't there. We may see something small, close to us, moving slowly, but our brains assume it's large and therefore far away and moving fast. Then, once we think we saw a spacecraft, our brain backfills the details. That's how a blade of grass turns into a flying saucer.

There are many iconic historical examples of these kinds of perceptual mistakes. On July 17, 1996, TWA flight 800 exploded over Long Island Sound. Many eyewitnesses swore they saw a missile fly up and hit the plane. However, we know from reconstructions of the wreckage and other lines of evidence that there was no missile. People saw a sudden, unusual, and unexpected event and their brains filled in details to make sense of it as best they could.

Of course, the moment you perceive something, it becomes a memory, and we already discussed how unreliable those are. Believe it or not, that wasn't a ghost gorilla you saw walking through the park, even if it makes for a great story. The fallibility of memory and perception are a one-two punch to any hubris we may harbor about the reliability of what we think we know.

4. Skeptics' Guide Entry: Pareidolia

Section: Neuropsychological Humility
See also: Apophenia

Pareidolia refers to the process of perceiving an image in random noise, such as seeing a face in the craters and maria of the moon.

> If you look at any walls spotted with various stains or with a mixture of different kinds of stones, if you are about to invent some scene you will be able to see in it a resemblance to various different landscapes adorned with mountains, rivers, rocks, trees, plains, wide valleys, and various groups of hills. You will also be able to see diverse combats and figures in quick movement, and strange expressions of faces, and outlandish costumes, and an infinite number of things which you can then reduce into separate and well-conceived forms.
>
> —Leonardo da Vinci

At some point in your life, probably when you were young and carefree and had more time than you knew what to do with, you lay on the ground and looked up at the clouds. Clouds are beautiful, their structures are fascinating, and they can give you a little perspective on how massive the world really is. But it's also fun to try to find images lurking in the white vaporous billows.

While animals and faces are common patterns to see floating overhead, no one actually thinks (or should think) the detailed shapes of clouds are anything but random. We intuitively understand that when we "see" a bunny rabbit in a cloud, we are just imposing that pattern onto the randomness. But this phenomenon goes much deeper than just children imagining a sky menagerie, and it reflects how our brains process and interpret information.

The term for this phenomenon is pareidolia, which refers to the perception of familiar yet meaningless patterns in random stimuli or noise. It usually applies to seeing visual patterns, but sometimes the term is used to refer to other sensations, such as sound (in which case it might be called, fittingly, audio pareidolia).

The technical term for the more general phenomenon of seeing patterns where they do not exist is apophenia, the tendency to see illusory patterns in noisy data. The information doesn't even have to be sensory; the pattern can be in numbers or in events. (In this way conspiracy theories can result from apophenia—seeing a nefarious pattern in random or disconnected incidents.)

There's nothing inherently wrong with seeing a face in a taco shell; it's just a by-product of our evolved perceptual systems, like many of the other illusions to which humans fall prey. Our skills in this regard are so nuanced and powerful that even multimillion-dollar petaflop supercomputers still struggle to match us.

Neurologically there are two important reasons for the human tendency to see patterns in noise. The first is that our brains (unlike computers) are organized for massive parallel processing. This is an ideal arrangement for finding patterns, making associations, and sifting through large amounts of data (see chapter 17 on data mining).

Second, as we learned in chapter 3, our perception is an active constructive process. Part of this process is taking an image and then quickly sifting through our catalogue of all possible matches, finding the best match, and then assigning it to the image. That blob looks like a horse, so your brain matches it to a horse and then backfills the details to make it look even more like a horse.

This works for speech as well. You hear sounds that your brain interprets as phonemes (parts of speech). It then searches through its database of phonemes and words until it finds the best match, and then that is what you hear.

Expectation plays a huge role in this process. That's why, once your friend says, "Hey, don't you see the dragon in that cloud? There's its head," the image pops into existence. Your brain found the pattern, and its construction of that image snaps into place. Or, if someone tells you that if you play "Stairway to Heaven" backward you can hear Robert Plant say, "Here's to my sweet Satan," then you'll hear the supposed devil worship.

Though pareidolia can manifest itself in many ways, involving any of our senses, it's the simple human face that is the poster child for this phenomenon. I remember watching a horror anthology series once in which a woman kept seeing ominous faces in the patterns on her ceiling. She asked if anyone had ever wondered why we tend to see faces more than anything else in these patterns. The answer, on that particular show, was that the faces were demons from another dimension. The real answer is far more interesting, if less spooky. Our pattern recognition skills in general are quite robust, but we have an especially sensitive knack for seeing faces.

There is a known neurological reason for this affinity for human faces: A dedicated part of the visual association cortex, the fusiform face area (FFA), specializes in recognizing and remembering them. Damage to the right FFA—from a stroke, for example—may cause a condition known as prosopagnosia, which is an inability to recognize faces. People with severe prosopagnosia cannot even identify their spouse or family members by sight alone. There is also developmental prosopagnosia, which is a relative deficit and can be mild.

It's no wonder that a facial pattern is preferred by the human brain. We can see this even in young babies. They will spend more time looking at a human face than another image of similar complexity.

It's easy to imagine why evolutionary selective pressures would favor this hyperability to see faces, given that we are such a social species. Our ancestors who were better able to quickly distinguish friend from foe, or determine the emotional states behind faces, probably had a survival advantage. Face and face-like recognition actually occurs subcortically (in the deep parts of the brain). This subconscious analysis appears to happen even before the image is passed along to other parts of the brain for more intricate processing. It's clear why this would be an advantage—quickly recognizing that someone is quite pissed at you and about to bash in your brains can do wonders for your survivability.

The most famous face seen as a result of pareidolia has got to be the Face on Mars. In 1976, NASA's *Viking* spacecraft was imaging Mars when it produced a picture of a mesa or butte in the Cydonia region that looked like a face. The scientists knew the face was pareidolia even if they didn't know that specific word. They were used to what the tricks of light and shadow could produce on the varied terrain of Mars. But popular culture enthusiastically absorbed the Face on Mars and gave it a life of its own. Books such as *The Mars Mystery* and *The Monuments of Mars* have been written about it, and countless "documentaries" have discussed the significance of that face and what it means for the history of Mars and life on that planet. (Um... nothing?)

The "face" is little more than a half-darkened visage with only one eye, a mouth, and a dot for a nose visible. The nose was actually a data dropout in the transmission that happened to be placed where the nostril would be. When NASA took a higher resolution image in 1998, it became obvious that the face was just an eroded pile of rocky detritus, no more an intentional face than the bumps on your ceiling.

Other worlds in our solar system and their surface features are also a great source of raw material for pareidolia. NASA has imaged Kermit the Frog, Bigfoot, and a giant smiley face on Mars. There is a nice picture of Homer Simpson on Mercury, and countless "alien artifacts" on the moon and elsewhere. UFO conspiracy theorist Richard Hoagland (you have to say "Hoaaaglaaand" as if you are Colonel Klink from *Hogan's Heroes*) has practically based his entire career on pareidolia of NASA images.

Even on Earth there are impressive examples of pareidolia, of which the app Google Earth has made an easy pastime. My favorite is Medicine Hat, Canada, which shows a profile of a woman apparently wearing earbuds (the wire of the earbuds is an access road).

Perry and I once investigated the face of the Virgin Mary on a tree in Hartford near where we live. It was just the usual swirly patterns in tree bark, but a little bit of pareidolia turned that into a face, and cultural belief did the rest. Thousands of faithful camped out around this tree, convinced they were witnessing a miracle. To Perry and me it was simply tree bark—and a rather bland example of a quirk of brain processing.

When looking at these popular examples of pareidolia, it seems they cannot just be random. But that's all part of the trick of how your brain constructs these patterns. Details that don't fit the pattern are deemphasized. Those that are important to the pattern are made more prominent. Missing details are filled in. Your brain connects the dots. It's amazing how few details are needed to suggest a face, and even an emotional expression, to our pattern-seeking brains. Even as little as a couple dots for eyes and some kind of line for a mouth is enough for our brains to see Elvis or the Pope.

Pareidolia can be fun, but if you aren't aware of our penchant for and love of patterns, an interesting and diverting illusion can feed into a delusion. As we will see, some illusory patterns are more nefarious than just seeing a bunny rabbit in a cloud.

5. Skeptics' Guide Entry: Hyperactive Agency Detection

Section: Neuropsychological Humility
See also: Coincidence

Hyperactive agency detection is the tendency to interpret events as if they were the deliberate intent of a conscious agent rather than the product of natural forces or unguided chaotic events.

> There is a very remarkable inclination in human nature to bestow on external objects the same emotions which it observes in itself, and to find everywhere those ideas which are most present to it.
>
> —David Hume

Something doesn't quite add up. The most powerful man in the world, John F. Kennedy, was taken out by a lone nutjob of no previous consequence? A jet flies into the Pentagon and yet the expected debris isn't visible? And why can't I see stars in the NASA *Apollo* moon landing photos?

Some hidden agent must be at work, conspiring to deceive and carry out a sinister plot.

At least that's how our brains are predisposed to think, and some of us more than others. The psychologist Justin Barrett coined the term "hyperactive (or hypersensitive) agency detection device"—HADD—to describe this tendency. Understanding that HADD is an intrinsic part of human nature is part of the core knowledge base of the skeptic.

Psychologists and neuroscientists in recent years have demonstrated that our brains are wired to distinguish things in our environment that are alive from those that are not alive. But being "alive" (from a psychological point of view) is not about biology, it's about agency—something that can act in the world, that has its own will and can cause things to happen. Sure, this is a property of living things, but that's not how our brains sort things out. We can perceive agency in nonliving things if they are acting as if they have a mind of their own. As previously discussed, even our visual system separates the world into things with agency and things without. So on a fundamental level, our brains treat agents differently than objects—from the very moment we see them.

Bruce Hood, author of *SuperSense*, details the psychological studies that have documented and described the human tendency to think of objects differently than agents. We imbue agents with an essence— a unique living force—even as infants. Objects are just generic things and are totally interchangeable, while agents have their own unique essence. Interestingly, children can come to view a favorite toy (a stuffed animal, for example) as having the properties of an agent and will treat it like a living thing. This reinforces the notion that the distinction we make is not between living and nonliving so much as agent versus object. This likely also explains why we can watch a cartoon and react emotionally to the characters as if they were real— they're not living, but we see them as agents.

According to Barrett, HADD works in part by detecting any movement that is non-inertial—something that seems to be moving of its own volition. We then assume the object is acting with agency and react accordingly. This likely provided an evolutionary advantage— it's better to assume the rustling in the bushes is not the wind but a hungry lion. So perhaps we are descended from hominids who were more paranoid than others and had hyperactive agency detection, because such primates were less likely to be eaten by predators.

HADD detects more than movement: It can perceive a pattern in otherwise unrelated events, details that defy easy explanation, or consequences that seem out of proportion to the alleged causes. When HADD is triggered we tend to see a hidden agent working behind the scenes, making events unfold the way they do and perhaps even deliberately hiding its own tracks. When HADD is triggered and we think we identify a hidden agent, it speaks to us in a very primal way. For some people the perception of hidden agency becomes overwhelming, dominating all other thought processes. We know these people as conspiracy theorists. But there is a little conspiracy theorist inside each of us.

Studies have demonstrated that HADD is more likely to be triggered when a stimulus is ambiguous. Therefore it tends to be our default assumption—an object is an agent until we are sure it's just an object. Also, our HADD becomes more active in situations where we have less control (which contributes to superstitious beliefs).

Barrett and others have speculated that HADD may be important to the development of religion—where God is the ultimate invisible agent. So far, this hypothesis has not been significantly researched, but it does seem plausible. Seeing natural or random events as the will of an agent is HADD.

Skepticism, in many ways, is a filter on HADD. Skeptics ask themselves, Is it really true? We see many patterns, but only some of those patterns represent underlying reality. We need a process to sort out which ones are real—that process is science and skepticism.

I think it's liberating to understand that in most situations, rather than there being an invisible supernatural or implausible agent at work, the simpler explanation is that (as was most elegantly stated on a bumper sticker) "shit happens."

6. Skeptics' Guide Entry: Hypnagogia

Section: Neuropsychological Humility
See also: Abductions

Hypnagogia is a neurological phenomenon in which the dreaming and waking states are fused, producing unusual experiences often mistaken for paranormal ones.

> Waking consciousness is dreaming—but dreaming constrained by external reality.
>
> —Oliver Sacks

In an interview, actress Jessica Alba recounted a terrifying experience:

> There was definitely something in my parents' old house... Something definitely took the covers off me and I definitely couldn't get off the bed...I felt this pressure and I couldn't get up, I couldn't scream, I couldn't talk, I couldn't do anything...I don't know what it was. I can't really explain it.

Jessica Alba's experience, while certainly unusual, is by no means unique. Perhaps you've experienced something similar. I've had similar episodes many times, usually when I'm sleep-deprived. One moment you're fast asleep, the next you're awoken by a sound or a feeling or even a smell that doesn't seem right. Now that you're awake, you find yourself incapable of moving. You can't lift your arms or turn your head. You feel there is something in the room, a nefarious presence, likely unfriendly and certainly uninvited. You gasp for a deep breath but it feels like something is pressing on your chest, inhibiting your ability to take that breath to call out for help. No matter how hard you try, you can't make a sound. You feel prone and vulnerable, at the mercy of whatever might be causing your powerlessness.

It's a terrifying experience and it happens to many millions of people. Most importantly, it is a relatively well-understood series of phenomena that requires no paranormal or pseudoscientific explanation.

Hypnagogia is a neurological phenomenon that can occur when a person is waking up (hypnopompic) or going to sleep (hypnagogic). It's an in-between state where the person is neither fully awake nor fully asleep. In this state, very realistic images and sounds can be experienced. Although visual and auditory hallucinations are most common, experiences can range from hearing your name whispered to ones involving all the senses, including touch. They are, in essence, dream experiences that are occurring while you're semi-awake. These waking dreams can be bizarre and terrifying.

Also associated with hypnagogia is temporary paralysis, often referred to as sleep paralysis. Regular sleep contains periods of REM (rapid eye movement) dream sleep, during which the brain stem inhibits the motor neurons in our spinal column, preventing any movement except for the eyes. Normally this only occurs during dream sleep. The apparent reason for sleep paralysis is our safety. Without this safeguard, we would act out anything we were dreaming and could possibly injure ourselves. Sleepwalkers have this problem. Paralysis persists during waking dreams because the affected neurons have not been reactivated immediately as they normally would be. Therefore waking dreams are, in a sense, a fusion of normal wakeful consciousness and the distinctive characteristics of dream sleep. We are indeed awake, but the paralysis and bizarre imagery typical of dreaming have not yet released their hold.

How common is sleep paralysis? In their 2011 paper *Lifetime Prevalence Rates of Sleep Paralysis*, sleep researchers Brian Sharpless and

Jacques Barber combined more than thirty studies on sleep paralysis prevalence from a variety of cultures and groups, giving them a large sample size of over 36,000. They concluded that approximately 8 percent of people experience sleep paralysis, rising to around 28 percent in high-risk groups (those who have a poor or disrupted sleep pattern, such as students) and up to 35 percent in those suffering from psychiatric disorders such as anxiety and depression. That's a significant portion of the population.

Internal vs. External Experience

For the critical thinker the important lesson here is that we use our brains to experience the world, and our brains sometimes glitch. A hypnagogic episode is like a brain glitch—it's stuck between two states. The biggest problems come from assuming that the phenomenon is a reflection of external reality rather than an internal experience happening in the brain.

Having had many hypnagogia episodes myself, I can tell you they can be dramatic and scary, even when you know what they are. It's understandable that someone who has been through such an event, outside their normal experience, would search for an extraordinary explanation.

We know this has been happening throughout history, as there are many historical accounts, and even paintings, depicting a typical hypnagogia event. People interpreted these strange experiences in line with their culture. In medieval Europe, people thought they were being visited by demons that sometimes sat on their chest and paralyzed them with their dark powers. When ghosts were more popular, episodes were explained as visitations from spirits who would steal your energy. In Newfoundland they called such nighttime visitors the "old hag." In fact the term "nightmare" derives from such beliefs—"maere" is the Old English word for a female demon who suffocates people in their sleep.

In more modern cultures, of course, people are too sophisticated to believe in hags or demons, so they blame such "visitations" on aliens. Perhaps the best example of this is author Whitley Strieber, who believes he was abducted by aliens and recounts his experience in his 1987 book, *Communion: A True Story*. He recounts that he awoke from sleep, sensed someone in his room, and was immediately paralyzed. He felt as if he were awake and not dreaming. These features fit a classic description of hypnagogia. He also documents many other accounts of people who claim they were abducted, the majority of which mirror this experience.

Clearly the cause of hypnagogic experiences is neurological and doesn't require any external explanation—no ghosts, demons, hags, or aliens. In fact, any time a person recounts a weird experience that happens around sleep, you should be skeptical.

But there are many other ways for the brain to glitch. We're all susceptible to some degree to illusions and hallucinations. Our brains are easily compromised by extreme emotion or even just a little sleep deprivation. How we direct our attention is subject to glitchiness and even deliberate manipulation (see, for example, all stage magic).

The real lesson from our knowledge of hypnagogia is that we can't assume our brains are always functioning flawlessly and therefore whatever we think we experienced must be real. And sometimes a ghost story is just that: a story.

7. Skeptics' Guide Entry: Ideomotor Effect

Section: Neuropsychological Humility
See also: Dowsing, Facilitated Communication

The ideomotor effect is an involuntary subconscious subtle muscle movement driven by expectation, which creates the illusion that the movement is due to an external force.

> If someone is able to show me that what I think or do is not right, I will happily change, for I seek the truth, by which no one was ever truly harmed. It is the person who continues in his self-deception and igno- rance who is harmed.
> —Marcus Aurelius

There are many ways to fool yourself, and sometimes your own body is an accomplice in this deception. Take dowsing, for example. According to the British Society of Dowsers: "To dowse is to search, with the aid of simple handheld tools or instruments, for that which is otherwise hidden from view or knowledge. It can be applied to searches for a great number of artefacts and entities."

While there are various handheld devices used by dowsers, many of them choose a method utilizing two thin metal "divining rods." Imagine, if you will, two pieces of wire hanger, each in the shape of an L. The dowser grips the wires by the small lines of the L and points them forward parallel to each other, like a sheriff in the Wild West holding a pair of revolvers waist high. The wires are held loosely, loose to the point that it feels like they can start swinging on their own. The dowser then concentrates on whatever it is they claim they can find. In most cases, they are "divining" for underground water; however, some dowsers claim they can locate precious metals, subterranean tunnels, or even missing people. They then proceed to walk around until the wires cross. When the crossing occurs, this denotes the location of the find, or if you prefer, "X marks the spot."

The "gift" of dowsing is as convincing to the dowser as any other thing they perceive as being real. By now you can see the theme developing here—the experience may seem real, but is it a manifestation of real forces out there in the world or just a glitch of brain wiring? Is there a mystical force that reaches up from the ground and influences the movement of the divining rods? Since such an explanation would require some new physics, perhaps there is a simpler process at work.

The ideomotor effect describes the subtle, unconscious muscular movements people make that result in the moving of objects. Noted parapsychology critic Dr. Ray Hyman has written extensively about the ideomotor effect, which he summarized in the following way:

> Under a variety of circumstances, our muscles will behave unconsciously in accordance with an implanted expectation. What makes this simple fact so important is that we are not aware that we ourselves are the source of the resulting action. This lack of any sense of volition is common in many everyday actions.

People first became aware of ideomotor actions back in the early 1800s. But it wasn't until 1852 that the "influence of suggestion in modifying and directing muscular movement, independently of volition" was termed "ideomotor action" by the psychologist/physiologist William B. Carpenter.

The ideomotor effect explains more than just dowsing. People who swing little pendulums over objects (to, for example, access divine guidance) or over an expecting mother's belly (to detect the sex of the fetus) never consider ideomotor forces. Children and adults alike who are fascinated with Ouija boards experience the ideomotor effect when they "lightly" place their fingertips on the planchette, see it "magically" slide around the board, and believe that spirits are communicating with them.

It may seem of trivial importance to investigate dowsing, pendulums, and Ouija boards. These pursuits could be excused as light entertainment at best or harmless misunderstanding at worst. But even seemingly silly claims, such as dowsing, can have serious real-world effects.

UK businessman James McCormick reportedly made over £50 million (about $75 million) selling a novelty golf-ball finder that was really just a fancy divining rod (a grip with a free-swinging rod attached). He took this device (the ADE651), initially sold to find lost golf balls, and marketed it as a bomb detector to locate explosive devices in Iraq and elsewhere. The device is worthless, but amazingly it is still in use by security services in some parts of the world. In Iraq alone, the device was likely responsible for hundreds of deaths. McCormick currently sits in jail on a ten-year sentence for fraud.

Ignorance of the ideomotor effect can have devastating effects in other arenas as well. Facilitated communication (FC) is perhaps the most obvious example—that is the practice of helping a language-impaired client to communicate by supporting their hand as they point to letters on a board or type on a keyboard. As I describe further in chapter 26 ("Witch Hunts"), FC is nothing more than self-deception through the ideomotor effect. Facilitators may think they're allowing their impaired clients to communicate, but the communication is coming entirely from their own minds. FC is nothing more than a Ouija board with a cognitively impaired person.

FC is a frightful waste of resources. Not only does it simply not work, it has also caused unmeasurable psychological damage, as parents were given the false hope that their severely impaired child was actually intellectually normal. But most significantly, there are many cases of criminal accusation—even courtroom testimony—occurring via FC.

Scientists who aren't aware of the ideomotor effect and other mechanisms of self-deception can be taken in. Dr. Steven Laureys is a successful researcher studying coma patients. He has done some great work, but in 2009 he became convinced that a patient he was studying, Rom Houben, who had been in a coma for twenty-three years, was able to communicate through FC. He claimed that Houben was mistakenly diagnosed as being in a persistent vegetative state when in fact he was "locked in"—conscious but paralyzed.

The problem lay in the fact that the facilitator was the only source of evidence that Houben was conscious. Videos circulated online showed the facilitation, during which Houben is often not looking at the keyboard. His eyes are often closed. Yet Houben is supposed to be controlling the typing, leading the facilitator to the correct letter, which is impossible if he isn't looking at the keyboard. Meanwhile, the facilitator moves his hand across the keyboard quickly and precisely. This would be impossible even if Houben were neurologically intact, let alone severely paralyzed.

I actually e-mailed Laureys to express my concerns about this case, but he was convinced it was genuine. However, to his credit, Laureys later did a blinded experiment and demonstrated that the facilitator, not Houben, was doing the communicating. But for a time, even an accomplished neurologist like Laureys was fooled by a basic neurological effect.

The ideomotor effect is a simple and well-understood phenomenon. Like waking dreams and illusions, it's just one example of how our own brains can fool us. It also speaks to the need to always be aware of the complex relationship between what is happening out there in the world and the internal models of reality that our brains create. Especially when we experience something unusual or extraordinary. Before hypothesizing about spirits, new forces, or supernatural abilities, consider that—just maybe—your brain (like the brain of every other human on the planet) has some quirks and foibles. That is the meaning of neuropsychological humility.

METACOGNITION

Most years we Rogues attend a large science and skeptical conference in Las Vegas, Nevada. The irony of hosting a conference dedicated to critical thinking in a city whose main industry is essentially based on sloppy thinking has not been lost on us. Sure, some people may be there for the entertainment, but billions of dollars flow into that city because many people rely on flawed thinking.

The hope to "beat the odds" is based on logical fallacies and cognitive biases. Superstition and belief in "luck" (which is highly encouraged by the casinos) are rampant. The casinos themselves don't trust in luck—they know that cold, hard math will grant them a steady flow of income. Our intuitions about risk, our gut feelings about probability and statistics, and our common sense about cause and effect leave us vulnerable to those with a little knowledge of psychology and a working calculator.

Even when our memory and perception are working reliably, we still have to reason with the information we have. That is a trip down a twisting road with blind alleys and many pitfalls. Fortunately, psychologists have spent decades working out many of the ways our thinking goes astray. In this next session we'll give you a road map for that snaking path, a guide to help you think about your own thinking, so you won't have to trust in chance, your lucky rabbit's foot, or that underwear you wore when you won the third-grade spelling bee.

8. Skeptics' Guide Entry: Dunning-Kruger Effect

Section: Metacognition
See also: Overconfidence Bias

The Dunning-Kruger effect describes the inability to evaluate one's own competency, leading to a general tendency to overestimate one's abilities.

> The greatest enemy of knowledge is not ignorance—it is the illusion of knowledge.
>
> —Daniel J. Boorstin

Remember the character David Brent, played by Ricky Gervais in the British version of *The Office* (an analogous character was played by Steve Carell in the American version)? David is a classic character because he is only a slight exaggeration of people we all know in real life—too incompetent and full of himself to have any idea how thoroughly incompetent he actually is.

Why does it seem that the world is full of people who think they are smart and effective when in fact they are ignorant and inept? Is this perception just confirmation bias (which is detailed in chapter 12) or is there some truth to it? In 1999 psychologist David Dunning and his graduate student Justin Kruger published a paper in which they describe what has come to be known (appropriately) as the Dunning-Kruger effect. In a 2014 article discussing his now famous paper, Dunning summarized the effect: "Incompetent people do not recognize—scratch that, cannot recognize—just how incompetent they are."

He further explains:

> What's curious is that, in many cases, incompetence does not leave people disoriented, perplexed, or cautious. Instead, the incompetent are often blessed with an inappropriate confidence, buoyed by *something* that feels to them like knowledge.

As you can see in the chart, the most competent individuals tend to underestimate their relative ability a little, but most people (the bottom 75 percent) increasingly overestimate their ability, and everyone thinks they're above average. I sometimes hear the effect incorrectly described as "the more incompetent you are, the more knowledgeable you think you are." As you can see, self-estimates do decrease with decreasing knowledge, but the gap between performance and self-assessment increases as your performance decreases.

The Dunning-Kruger effect has now been documented in many studies across many areas. There are several possible causes of the effect. One is simple ego—no one wants to think of themselves as below average, so they inflate their self-assessment. People also have

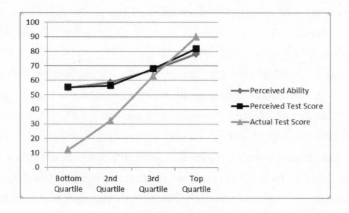

an easier time recognizing ignorance in others than in themselves, and this will create the illusion that they are above average, even when they're performing in the bottom 10 percent.

The core of the effect, however, seems to be what Dunning describes: Ignorance carries with it the inability to accurately assess one's own ignorance. Dunning also points out something that rings true to a veteran skeptic:

> An ignorant mind is precisely not a spotless, empty vessel, but one that's filled with the clutter of irrelevant or misleading life experiences, theories, facts, intuitions, strategies, algorithms, heuristics, metaphors, and hunches that regrettably have the look and feel of useful and accurate knowledge.

This accurately describes the people with unscientific or unsupported beliefs I confront daily. Just read the comments to any post on the SGU's Facebook page and you'll quickly be subjected to the full force of Dunning-Kruger. (And if you actually did go and do that, I'm so sorry. It was mean to even suggest it.)

What I think Dunning is describing above—a conclusion with which I completely agree—are the various components of confirmation bias (see chapter 12). As we try to make sense of the world, we work with our existing knowledge and paradigms. We formulate ideas and then systematically seek out information that confirms those ideas. We dismiss contrary information as exceptions. We interpret ambiguous experiences as in line with our theories. We make subjective judgments that further reinforce our beliefs. We remember these apparent confirmations, and then our memories are tweaked over time to make the appearance of confirmation even more dramatic.

In the end we are left with a powerful sense of knowledge—but it's false knowledge. Confirmation bias leads to a high level of confidence: We feel deep in our gut that we are right. And when confronted by someone saying we're wrong or promoting an alternate view, there is a tendency to become defensive, even hostile.

The Dunning-Kruger effect is not just a curiosity of psychology, it touches on a critical aspect of the default mode of human thought and a major flaw in our thinking. It also applies to everyone—we are all at various places on that curve with respect to different areas of knowledge. You may be an expert in some things and competent in others, but you will be toward the bottom of the curve in some areas.

Admit it: Up to this point in the chapter, you were probably imagining yourself in the upper half of that curve and inwardly smirking at the poor rubes in the bottom half. But we are all in the bottom half some of the time. The Dunning-Kruger effect does not just apply to other people—it applies to everyone. That's why the world is full of incompetent, deluded people—we all are these people.

This pattern, however, is just the default mode. It is not our destiny. Part of skeptical philosophy, metacognition, and critical thinking is the recognition that we all suffer from powerful and subtle cognitive biases. We have to both recognize them and make a conscious effort to work against them, realizing that this is an endless process. Part of the way to do this is to systematically doubt ourselves. We need to substitute a logical and scientific process for the one Dunning describes above.

To give an illuminating example, I am involved in medical-student and resident education. The Dunning-Kruger effect is in clear view in this context, but with some interesting differences. At one review, every new resident felt they were below average for their class. Being thrown into a profession where your knowledge is constantly being tested and evaluated, partly because knowledge is being directly translated into specific decisions, appears to have a humbling effect (which is good). It also helps that your mentors have years or decades more experience than you—this can produce a rather stark mirror.

Still, we see Dunning-Kruger in effect. Even among these young professionals the gap between self-assessment and actual ability grows toward the lower end of the ability scale.

Medical education is a special case because self-assessment is a skill we specifically teach and evaluate. It's critically important for physicians to have a fairly clear understanding of their own knowledge and skills. We specifically try to give students an appreciation for what they *don't* know, and the seemingly bottomless pit of medical information is on constant display.

I remember as a resident seeing a massive two-volume tome just on muscle disease, and thinking, "Holy crap! That's how much I don't know about muscle disease." There are similar volumes on every other tiny aspect of medicine. It can be overwhelming.

As students and residents go through their training, we keep moving that carrot forward. As they get more confident in their basic skills, we have to make sure they don't get cocky. I talk to my residents specifically about the difference between competence, expertise, and mastery.

One specific lesson I try to drive home as often as possible, both in the context of medical education and in general, is this: Think about some area in which you have a great deal of knowledge, at the expert to mastery level (or maybe a special interest in which your knowledge is above average). Now, think about how much the average person knows about your area of specialty. Not only do they know comparatively little, they likely have no idea how little they know and how much specialized knowledge even exists. Furthermore, most of what they think they know is likely wrong or misleading.

Here comes the critical part: Now realize that you are as ignorant as the average person is in every other area of knowledge in which you are not expert. The Dunning-Kruger effect is not just about dumb people not realizing how dumb they are. It is about basic human psychology and cognitive biases. Dunning-Kruger applies to everyone.

In addition to the various aspects of critical thinking, self-assessment is a skill we can strive to develop specifically. But a good rule of thumb is to err on the side of humility. If you assume that you know relatively less than you think you do and that there is more knowledge than you are aware of, you will usually be correct.

9. Skeptics' Guide Entry: Motivated Reasoning

Motivated reasoning is the biased process we use to defend a position, ideology, or belief that we hold with emotional investment.

> Some information, some ideas, feel like our allies. We want them to win. We want to defend them. And other information or ideas are the enemy, and we want to shoot them down. —Julia Galef

Have you ever been in a heated political discussion? Heck, have you ever interacted with other human beings? Then you are likely familiar with the frustration of someone else twisting logic, cherry-picking or distorting facts, and being generally biased in their defense of a position. Of course, here's the thing: You do it too.

Psychological studies have shown that people treat different beliefs differently. For most beliefs, people actually are (mostly) rational at baseline. We tend to follow a Bayesian approach, meaning that we update our beliefs as new information comes to our attention. If we are told that some historical fact is different than what we remember, we'll quickly change our beliefs about that historical fact. Further, the more information we have about something, and therefore the more solid our belief, the more slowly we will change that belief. We don't just change from one thing to the next, we incorporate the new information with our old information.

This is actually a very scientific approach. I would not easily change my belief that the sun is roughly at the center of our solar system. It would take a profound amount of very reliable information to counter all the solid scientific information on which my current belief is based. If, however, I was told by a reliable source something about George Washington I'd never heard before, I would accept it much more quickly. This is reasonable, and this is how most people function day-to-day.

I described this process as "mostly" rational, because even at our best, people have cognitive biases and follow heuristics (or mental shortcuts; see chapter 11) that are not strictly valid. These biases constrain our reason. The constraint on reason, though, is far greater when dealing with the special set of beliefs in which we have an emotional investment. These are alleged facts or beliefs about the world that support our sense of identity or our ideology. They may be part of our worldview. Skeptics commonly call such beliefs sacred cows.

We all have narratives by which we understand the world and our place in it. Some narratives are critical to our sense of identity. Preferred narratives support our worldview, our membership in a group, or our self-perception as a good and valuable person. We have narratives and beliefs that serve our basic psychological needs, such as the need for a sense of control. When those beliefs are challenged, we don't take a rational and detached approach. We dig in our heels and engage in what is called motivated reasoning. We defend the core beliefs at all costs, shredding logic, discarding inconvenient facts, making up facts as necessary, cherry-picking only the facts we like, engaging in magical thinking, and using subjective judgments as necessary without any consideration for internal consistency. Collectively, these processes constitute motivated reasoning, something at which humans generally excel.

Motivated reasoning is triggered by what psychologists call cognitive dissonance. Cognitive dissonance theory was first proposed by Leon Festinger in 1957. He suggested that psychological discomfort results when we are presented with two pieces of information that conflict with each other. We hold a belief, and now we have information that contradicts that belief. Ideally, we would resolve the conflict rationally and objectively, changing the belief as necessary, depending on the nature and validity of the new information. When the belief is strongly and emotionally held, however, it becomes too difficult to change. If the belief is at the core of our worldview, then changing it might cause a chain reaction, magnifying cognitive dissonance.

It's emotionally easier to simply dismiss the new information, challenge its source, rationalize its implications, even invent a conspiracy to explain it. In fact, for common beliefs, motivated reasoning may come prepackaged for you. There are organized movements with itemized lists of rationalizations and misinformation that conveniently dismiss the evidence for evolution, global warming, vaccine effectiveness, the gender pay gap, the safety of genetically modified organisms, even the sphericity of the Earth. They are like a salve for cognitive dissonance—just apply liberally.

Motivated reasoning is further facilitated by the fact that much of our information is subjective or requires some judgment. No scientific study is perfect, so you can always point to limitations if you want to deny the conclusions. No source is impeccable, and people make mistakes, so perhaps this is one. Different sources say different things, so you can choose to believe the one that reduces your cognitive dissonance.

There are also numerous ways to interpret the implications of facts, even if the facts themselves are not in dispute. If someone is rich, then you can conclude (if you like them) that they are successful and savvy, or (if you don't like them) that they are corrupt and greedy. Someone can be either courageous or foolish, steadfast or stubborn, a strong leader or authoritarian. It all depends on your perspective.

In general, political opinions tend to fall into the "sacred cow" category. People tend to identify with their political tribe and want to believe that their tribe is virtuous and smart, while the other tribe is mostly made of lying idiots. Of course, these dichotomies occur on a spectrum. You can have a little bit of an emotional attachment to a belief, or it can be fundamental to your worldview and identity. You can be a little tribal in your political views, or hyperpartisan.

The Neuroscience of Motivated Reasoning

Much of the psychological research on motivated reasoning involves political opinions. What multiple studies have shown is that people pursue two distinct goals when forming political opinions: a directional goal and an accuracy goal. The directional goal leads them to the opinion that is consistent with their partisan identity. The accuracy goal leads them to try to make their opinion as correct as possible.

What researchers have found is that the more partisan the individual and the issue they are evaluating, the more directional their reasoning will be. They will weigh evidence more heavily if it supports their partisan identity. They will dismiss evidence that is against the party line. Individuals are easily primed (affected subconsciously by a stimulus) as well. If they're told that a bill is sponsored by their party, they are far more likely to think it's a good idea and endorse it than if they think the identical bill was sponsored by the opposing party.

Psychologists speak of this phenomenon as if describing a computer program. People will try to adjust their beliefs to simultaneously maximize the alignment of those beliefs with facts and maximize their emotional comfort. It's almost like a mathematical equation. The greater the emotional discomfort (cognitive dissonance) a fact presents, the harder we will work to rationalize it away.

Perhaps the seminal study of this phenomenon is from 2010, in which Brendan Nyhan and Jason Reifler purport to show a "backfire" effect—when confronted with information that contradicts a political opinion, subjects increased their belief in the misinformation. They specifically studied the claim that there were weapons of mass destruction (WMDs) in Iraq. However, a 2016 follow-up study by Thomas Wood and Ethan Porter found that the backfire effect was rare, occurring in only one question out of thirty-six: the claim about

WMDs. Even then, the effect depended on how the information was presented.

The backfire effect clearly requires more research—when is it triggered and how long does the effect last? Partisans might accept the correction in the short term and then revert to their prior views over time. But these results support the notion that there is a conflict between the directional and accuracy goals.

Neuroscientists have started looking at what's happening in the brain when subjects are confronted with facts that present a problem for their political beliefs, and when they're offered justifications that help to resolve the problem. Drew Westen et al. used functional magnetic resonance imaging (fMRI), which images brain activity, while challenging subjects with neutral information and information problematic for their political affiliation. They found that subjects used different parts of their brains in these two situations.

When confronted with ideologically neutral information, subjects used the rational cognitive part of their brain. When confronted with partisan information, a completely different part lit up, one known to be associated with identity, sympathy, and emotion. Also interesting, after subjects arrived at their motivated conclusion, relieving the negative emotions of the conflict, the part of their brain involved in reward became activated, giving them a nice shot of dopamine.

Motivated reasoning is a double whammy: It reduces the negative emotion of facts conflicting with our identity or beliefs, and it induces positive emotions associated with reward. This process can have a powerful conditioning effect, reinforcing the behavior.

A 2016 study by Jonas Kaplan, Sarah Gimbel, and Sam Harris adds a bit of confirmation to this basic understanding. Forty left-leaning subjects had their brains scanned while being presented with statements that were designed to be either political or nonpolitical. They were then confronted with counterclaims to contradict those facts, some of which were exaggerated or untrue.

One example of a political claim used is that the US spends too much of its resources on the military. The counterclaim is that Russia's nuclear arsenal is twice the size of the US's (which isn't true—Russia has 7,300 warheads to our 7,100). An example of a nonpolitical claim is "Thomas Edison invented the light bulb." One counter to that claim is that "Nearly seventy years before Edison, Humphry Davy demonstrated an electric lamp to the Royal Society." This is true, but an exaggeration in that Davy's incandescent bulb was not practical. Edison's wasn't the first light bulb, but it was one of the first ones with the properties necessary to make its wide use feasible. (Actually, English inventor Joseph Swan beat Edison to the practical light bulb and won a patent infringement case against Edison.)

The researchers looked at the activity in the subjects' brains when they were confronted with a counterclaim to a political and a nonpolitical opinion. When a political belief was challenged, more of the brain lit up, including areas known as the "default mode network" and the amygdala. The former may be involved in identity, the latter in negative emotions.

It's always difficult to interpret such studies. First, forty subjects is a small number, and fMRIs involve a low signal-to-noise ratio. Assuming the results are valid, we also don't know exactly what they mean. Just because we can see one part of the brain light up, that doesn't mean we know what it was doing. The same structure in the brain will participate in different overlapping networks with different overlapping functions. It does seem clear, however, that the brain responds differently to political and nonpolitical challenges in a way that is suggestive of an emotional response.

The researchers also assessed the degree to which the subjects changed their minds on the facts that were challenged. They changed them more consistently for nonpolitical than for political views. Again, there are lots of variables here, and no one study is going to account for all of them, but it's an interesting start.

This new study is consistent with prior research on the topic of motivated reasoning, and it points the way to further research. Follow-up studies with conservatives, with other ideologies (religious, social, historical), addressing other variables more directly (like how truthful or plausible the counterclaims are), and with larger numbers of subjects would all be helpful.

These studies looking at the psychology and the neural correlates of motivated reasoning provide an essential insight into how the human brain behaves, and they reinforce how challenging it can be to be a consistent critical thinker. This suggests we should make a specific effort to be more detached when it comes to ideological beliefs. Factual beliefs about the world shouldn't be a source of identity, because those facts may be wrong, partly wrong, or incomplete.

This is easier said than done. My strategy has been to focus my emotional investment on being skeptical. I take pride in being detached when it comes to factual claims, in following a valid process of logic and empiricism, and in changing my mind when necessary. I focus on the validity of the process, not on any particular claim or set of beliefs. In psychological terms, I try to follow the process of accuracy reasoning, not directional reasoning. This takes vigilance and practice, but it becomes more automatic over time.

There is also a need to remind ourselves that people who disagree with us are just people. They are not demons. They have their reasons for believing what they do. They think they're right just as much as we think we are right. They don't disagree with us because we're virtuous and they are evil. They just have a different narrative than we do, one reinforced by a different set of facts and subjective judgments.

This doesn't mean that all views are equally valid. It does suggest we should strive to focus on logic and evidence, not self-serving assumptions of moral superiority.

10. Skeptics' Guide Entry: Arguments and Logical Fallacies

Section: Metacognition

A logical fallacy is an invalid connection between a premise and a conclusion, where the conclusion does not necessarily flow from the premise(s) but is argued as if it does.

> I am convinced that the act of thinking logically cannot possibly be natural to the human mind. If it were, then mathematics would be everybody's easiest course at school and our species would not have taken several millennia to figure out the scientific method.
>
> —Neil deGrasse Tyson

The following is a comment I received on one of my blog posts on alternative medicine in which I was criticizing the Mayo Clinic for promoting quackery. It is a fairly typical, even mild, example of the kinds of arguments I see on a daily basis:

> Clearly the Mayo clinic is run by a bunch of quacks and charlatans that have no interest in the health and well being of the public. How dare they even consider something other than surgery or prescription medication to help people heal. More and more I am getting the feeling that articles like this are by fronts for the pharmaceutical companies. This article is clearly written out of fear; fear that there are ideas and practices out there that you

don't understand and your only resort is to lash out against those ideas. In this case you are attacking one of the most respected medical institutions in the world. Do you, mr. author, perhaps think that maybe you are the one that doesn't get it?

How is it that people can disagree so vehemently, even about facts (not just subjective opinions)? Often disagreements result simply from bad arguments, of which the above comment offers many examples, and we will explore them in detail below. The comment is not persuasive, because it is so poorly constructed. If the commenter was hoping to change my mind, they failed out of the gate because their arguments were fatally flawed.

Among the core skills of critical thinking are the ability to formulate a coherent argument that uses valid logic, to police the validity of your own logic, and to recognize poor logic in the arguments of others. Arguments using logic that is not valid are considered logical fallacies.

What Is an Argument?

If you've ever spent any time engaging with other people, then you have likely been involved in an argument, perhaps even on a daily basis. Arguing is one of those things that everyone does but few understand. Yet arguing is an essential skill of critical thinking. How we argue reflects the way we think, the methods we use to evaluate our own conclusions, and how we challenge the beliefs of others.

Even the very purpose of arguing is often misunderstood. I have arguments almost every day. This doesn't mean I verbally fight with others on a daily basis, but rather that I have discussions that involve either attempting to convince another of a specific conclusion or resolving different conclusions on a factual matter. In most of the arguments, the other person has staked out a position and they defend it jealously, as if they were a high-paid lawyer defending a client. But this adversarial approach isn't constructive. Rather, the parties of an argument should be trying to find common ground and then proceed carefully from that common ground to resolve any differences.

The first thing we should understand about a logical argument is that it follows a certain format. There are one or more premises, which are underlying facts that the argument takes for granted or is built upon. There is then some logical connection showing how these premises necessarily lead to a specific conclusion.

Sometimes people will make an assertion as if it's an argument. An argument makes some connection between a premise and a conclusion. An assertion just states conclusions (or premises) without supporting them. So, for example, saying that my criticism of the Mayo Clinic was based on fear of new ideas was simply an assertion. The commenter offered nothing to support this conclusion. My original argument, however, was that the Mayo Clinic claims to be evidence-based (to support only those treatments with adequate evidence that they work) and yet they promote acupuncture, which is not evidence-based (those are my premises). Therefore, they are failing in their stated mission (my conclusion). See the difference?

Here is another important thing to keep in mind: If the premises of an argument are true and sufficiently complete, and the logic is valid (in which case the argument is said to be "sound"), then the conclusion must be true. In order for the conclusion to be false, one or more premises must be false, or there are hidden or incomplete premises, or the logic is invalid. If homeopathic products contain nothing but water, and water cannot have any specific medical effects (other than hydration), then homeopathy is not medicine.

So, if an argument is sound, the conclusion is true. However, the converse is not true. An unsound argument can still have a conclusion that happens to be true, even though the argument doesn't support it. Someone might argue that the sun is a sphere because spheres are pretty—that isn't a sound argument, but the conclusion is still true.

If two people have come to different conclusions about a factual claim, then one or both must be wrong. Both cannot be correct. One or both must therefore have made an error in the arguments they used to come to their conclusions. The two parties should work together to examine their arguments and resolve any errors.

Keep in mind, this only works if the arguments are about factual claims, not subjective feelings or value judgments. There is no objective way to resolve a difference of opinion regarding aesthetics, for example. If you prefer Mozart to Beethoven, there is no way to prove Mozart's superiority with facts or logic. It is very helpful, however, to identify when a conclusion contains an aesthetic opinion or a moral choice. It avoids arguing endlessly over an issue that is inherently irresolvable.

First, Turn Your Sights Inward

All too often I see people use knowledge of argument and logic to deconstruct the arguments of others. This can easily turn into an attempt to find some flaw in the arguments of those they perceive as being against them, and then declare victory.

But a knowledge of argument and logic must first be used to examine one's own position. If someone disagrees with someone else, it is best that they first deconstruct their own position to make sure their premises are true, there are no hidden premises, and their logic is valid. They must be open to the possibility that they have incomplete or wrong information or that they've made a mental error.

Outside of certain venues, like debate club and perhaps the courtroom, where it is one's job to defend a certain position, what is really desired is to have a position that is valid and true. This requires transparently examining one's own arguments.

In other words, logic and arguments should be used as a tool, not a weapon. When logic is used as a weapon, it's far too easy to twist it subtly to suit one's ends.

It is, of course, also good to deconstruct the arguments of others. However, it is important to strive to be as fair as possible. This is called the principle of charity. Give the other the benefit of the doubt—take the best interpretation of their position possible and deconstruct that.

Again, the goal should not be to "win." No one is keeping score. The goal of critical thinking and skepticism is to have the most valid position possible, and this means accounting for the best possible arguments that challenge our position.

Examine Your Premises

As stated above, in order for an argument to be sound, all of its premises must be true. Often, different people come to different conclusions because they're starting with different premises. So examining all the premises on both sides of an argument is a good place to start.

There are four types of potential problems with premises. The first, and most obvious, is that a premise can be wrong. If one argues, for example, that evolutionary theory is false because there are no transitional fossils, that argument is unsound because the premise—no transitional fossils—is false. In fact, there are copious transitional fossils.

Another type of premise error occurs when one or more premises are unwarranted assumptions. The premise may or may not be true, but it hasn't been established sufficiently to serve as a premise for an argument. Identifying all the assumptions upon which an argument is dependent is often the most critical step in analyzing the argument. Frequently, different conclusions are arrived at because of differing assumptions.

Often people will choose the assumptions that best fit the conclusion they prefer. In fact, psychological experiments show that most people start with conclusions they desire, then reverse engineer arguments to support them—a process called rationalization.

One way to resolve the problem of using assumptions as premises is to carefully identify and disclose those assumptions up front. Such arguments are often called hypothetical or are prefaced with the statement "Let's assume for the sake of argument…" Also, if two people examine their arguments and realize they are using different assumptions as premises, then at least they can "agree to disagree." They will realize that their disagreement cannot be resolved until more information is available to clarify which assumptions are more likely to be correct.

For example, two people may disagree about whether or not the minimum wage should be increased because they are basing their conclusions on different starting assumptions. One person may assume that raising the minimum wage costs jobs, while another may assume that it doesn't. Until they resolve that difference, further arguments may be futile.

The third type of premise difficulty is the most insidious: the hidden premise. Obviously, if a disagreement is based upon a hidden premise, then the disagreement will be irresolvable. So, when coming to an impasse in resolving differences, it's a good idea to go back and see if there are any implied premises that haven't been addressed.

Let's go back to the transitional fossil example. Why is it that scientists believe we have many transitional fossils and evolution deniers (creationists or intelligent design proponents) believe we do not? This would seem to be a straightforward factual claim, easily resolvable by checking the evidence. Sometimes evolution deniers are simply ignorant of the evidence or are being intellectually dishonest. However, the more sophisticated ones are fully aware of the fossil evidence and use a hidden premise to deny the existence of transitional fossils.

When paleontologists speak of "transitional" fossils, they are referring to species that occupy a space morphologically and temporally between two known species. This may be a common ancestor, in which case the transitional fossil will be more ancient than both descendant species; or it can be temporally between two species, the descendant of one and the ancestor of the other. But in reality, we often do not know if the transitional species is an actual ancestor or just closely related to the true ancestor. Because evolution is a bushy process, and not linear, most of the specimens we find will lie on an evolutionary side branch (an uncle rather than a parent). But if they fill a morphological gap in known species, they provide evidence of an evolutionary connection and therefore qualify as transitional. For example, archaeopteryx may not be on the direct path to modern birds, but clearly it occupies a space that is transitional between the-ropod dinosaurs and modern birds, and one of its close relatives is a direct ancestor of modern birds.

When evolution deniers say there are no transitional fossils, their unstated major premise is that they are employing a different definition of "transitional" than is generally accepted in the scientific community. They typically define transitional as some impossible monster with half-formed and useless structures. Or they may define transitional as only those fossils for which there is independent proof of their being a true ancestor rather than simply closely related to a direct ancestor—an impossible standard.

Another hidden premise in their argument is the notion of how many transitional fossils there *should* be in the fossil record. They, of course, can always assume an arbitrarily high number in order to claim that there aren't enough.

The fourth potential problem with premises, one that can be very subtle, is when a premise contains a subjective judgment. You might take as a premise that a particular news source is reliable. It is reliable according to what criteria and whose judgment? Subjective evaluations too easily allow for tweaking the premises to fit the desired conclusion.

Let's look again at the commenter's criticism—they "felt" as if articles criticizing the promotion of pseudoscience in medicine were just fronts for Big Pharma. This allows them to dismiss arguments they don't like. Their feeling is subjective; it is based on their preex-isting conclusion, which then becomes an assumed premise.

In this way subjective premises easily become circular reasoning—they are simply a filter through which we see the world in order to confirm what we already believe.

Logical Fallacies

Even when all the premises of an argument are reliably true, the argu-ment may still be unsound if the logic employed is not valid—if there's a so-called "logical fallacy." The human brain is a marvelous machine with capabilities that, in some ways, still outperform the most powerful of supercomputers. Our brains, however, do not appear to have evolved specifically for precise logic. There are many common logical pitfalls that our minds tend to fall into unless we are consciously aware of these pitfalls and make efforts to avoid them.

Also, because there is a tendency to start with desired conclusions and then construct arguments to support them, many people will happily draw upon logical fallacies to make their arguments. In fact, every time someone tries to form an argument to support a conclusion that isn't true, they must either employ a false premise or a logical fallacy. Remember, a sound argument (one with true premises and valid logic) cannot lead to a false conclusion. So they have no choice but to use false premises or bad logic if they want to defend a false conclusion. In order to avoid using logical fallacies to construct unsound arguments, we need to understand how to identify fallacious logic.

Below are the most common logical fallacies, with examples of each, to help us all avoid bad arguments. I should note that these are all informal logical fallacies, which means that the logic is questionable, but its validity often depends upon context. They're not strictly invalid all the time.

Formal logical fallacies are always invalid. For example, if I take the premises that A = B and B = C, and I conclude that A does not equal C, that is a formal logical fallacy. The conclusion is always wrong.

Informal logical fallacies, as you will see with the examples, are more...squishy.

Non Sequitur

From Latin, this term translates to "it doesn't follow," and it refers to an argument in which the conclusion does not necessarily follow from the premises. In other words, a logical connection is implied where none exists. This is the most basic type of logical fallacy, and in fact all logical fallacies are non sequiturs.

So, for example, if I said that Swiss cheese is healthy for you because it's yellow, that is a non sequitur because there is no logical connection between the premise and the conclusion.

Argument from Authority

The basic structure of such arguments is as follows: Professor X believes A, Professor X speaks from authority, therefore A is true. Often this argument is implied by emphasizing the many years of experience of the individual making a specific claim, or the number of formal degrees they hold. The converse of this argument is sometimes used too, that someone does not possess authority and therefore their claims must be false. (This may also be considered an ad hominem logical fallacy—see below.)

In practice this can be a complex logical fallacy to deal with. It is legitimate to consider the training and experience of an individual when examining their assessment of a particular claim. Also, a consensus of scientific opinion does carry some legitimate authority. But it is still possible for highly educated individuals and a broad consensus to be wrong—speaking from a position of authority does not make a claim necessarily true.

Remember my commenter above, who appealed to the authority of the Mayo Clinic—they are a respected institution, therefore anything they support must be legitimate. That argument does not counter my specific criticisms of the clinic's behavior, however, in promoting a treatment not supported by evidence.

This logical fallacy crops up in more subtle ways too. For example, UFO proponents have argued that UFO sightings by airline pilots should be given special weight because pilots are trained observers, reliable characters, trained not to panic in emergencies.

Oftentimes, the opinions of Nobel laureates are given great weight. This has historically led to frankly pseudoscientific positions taken up by such laureates later in their careers being taken seriously only because of the support of the laureate. The most notable example of this is Linus Pauling.

Linus Pauling won two Nobel Prizes: the Nobel Prize for chemistry in 1954 and the 1962 Nobel Peace Prize for his activism against nuclear weapons. Pauling was a brilliant researcher, of that there is no doubt. But later in life he descended into quackery, advocating for megadoses of vitamin C to fight off infections, including the common cold. This was part of his broader support for "orthomolecular" medicine—a term he coined. According to Pauling, substances that occur naturally in the body can be used to fight disease.

This is the perfect example of a brilliant scientist stepping outside his area of expertise and applying the wrong principles to another discipline. The concept of orthomolecular medicine may make sense to a chemist who is focused on the chemical activity of biological substances. But medical researchers are likely to find such ideas hopelessly naive (even accounting for the time period). Pauling failed to support his ideas with clinical research and failed to recognize the need to translate a basic scientific understanding of things like biochemistry to actual clinical applications. Our understanding is advancing, but the body is ridiculously complex and net clinical effects therefore need to be measured. We cannot simply extrapolate from our basic science knowledge to clinical claims, as Pauling did.

Saying that orthomolecular medicine must be valid because it was supported by a scientist of the stature of Pauling would be an argument-from-authority logical fallacy.

At the same time, saying that we should take very seriously the consensus of scientists who believe that life is the result of organic evolution would not be fallacious. That is a solid consensus (more than 98 percent of scientists) built upon a mountain of evidence, examined and argued for more than a century. Casually dismissing a solid scientific consensus as an "argument from authority" is a misuse of the logical fallacy. It is also an excellent example of how important context is in evaluating informal logical fallacies.

This is a good time to bring up another aspect of informal logical fallacies—the "lumper versus splitter" debate. Do we lump similar types of logical fallacies into one broad category, or do we split them up and give each slight variation its own name? I tend to be a lumper myself. I think it is important to grasp the essence of a fallacy and see how it applies broadly.

For example, there are many types of argument from authority. You can appeal to the authority of popular opinion (which may be called separately an argument *ad populi*). Or you could appeal to celebrity, or to the implied authority of the future, when your opinions will allegedly be validated. Any time you rely on the source of an argument or position rather than facts and logic, that may become an argument from authority.

Argument from Final Outcome

Such arguments (also called teleological) are based on a reversal of cause and effect, because they argue that something is caused by the ultimate effect that it has or the purpose that it serves. Christian creationists have argued, for example, that evolution must be wrong because if it were true it would lead to immorality (also a false premise).

Naive evolutionary thinking can also be teleological—for example, the argument that birds evolved feathers so that they could fly, when in fact, feathers likely had to evolve prior to flight, and bird ancestors did not know they would ultimately use feathers for flying. Evolution cannot look forward. Protofeathers must have evolved for a purpose they served at the time, like insulation.

Post Hoc Ergo Propter Hoc

This is perhaps the most common of logical fallacies. It follows the basic format of A preceded B, therefore A caused B, assuming cause and effect for two events just because they are temporally related (the Latin translates to "after this, therefore because of this"). This logical fallacy is frequently invoked when defending various forms of alternative medicine—I was sick, I took treatment A, I got better, therefore treatment A made me better. This is a logical fallacy because it is possible to have recovered from an illness without any treatment.

Confusing Correlation with Causation

This is similar to the post hoc fallacy in that it assumes cause and effect for two variables simply because they occur together. This fallacy is often used to give a statistical correlation a causal interpretation. For example, during the 1990s both religious attendance and illegal drug use were on the rise. It would be a fallacy to conclude that therefore religious attendance causes illegal drug use. It is also possible that drug use leads to an increase in religious attendance, or that both drug use and religious attendance are increased by a third variable, such as an increase in societal unrest. It is perfectly possible that both variables are independent of each other, and it is mere coincidence that they are both increasing at the same time.

In essence there are always four possible interpretations of any apparent correlation. The first is that the correlation is not causal at all. The second is that A causes B. The third is that B causes A. The fourth is that A and B are both caused by another variable, C. It's helpful to go through all such possibilities before concluding that any one causal pattern is true.

This fallacy is often combined with data mining: looking at large sets of data for apparent correlation. People usually underestimate the number of correlations that are likely to occur by chance alone when data sets are sufficiently large.

To illustrate this, there is a website made by Tyler Vigen called Spurious Correlations. You can enter a number of variables into databases linked to by the site, which will then display a graph showing how well they correlate.

For example, US spending on science, space, and technology between 1999 and 2009 tightly correlates with deaths by hanging, strangulation, and suffocation. There is also an excellent correlation between the number of people who drown by falling into a swimming pool and the number of movies in which Nicolas Cage appears. My favorite, however, is the tight correlation between organic food sales and autism diagnoses. Presumably, these are all chance correlations.

This fallacy—confusing causation and correlation—has a tendency to be abused, or applied inappropriately, to deny all statistical evidence. In fact, this constitutes a logical fallacy in itself: the denial of causation. This abuse is captured in the oft misquoted phrase, "Correlation does not imply causation." Well, in fact it does *imply* causation and may be due to a particular causation—sometimes A does cause B. A better quote would be that correlation does not *necessarily* imply causation, or that it does not, alone, *prove* causation. But it is one line of evidence for it.

To properly interpret the implication of an apparent correlation, we have to look at the evidence itself. If the correlation is part of an experiment in which all variables possible were controlled for, such as in a double-blind placebo-controlled medical trial, then any correlation found is likely to be a real effect of treatment. If the treatment correlates with a better outcome, and all placebo effects were controlled for, it is likely that this correlation does result from the causation of the treatment effecting a cure.

Second, even with purely epidemiological or statistical evidence, it's still possible to build a strong scientific case for a specific cause and effect. The weakness of such observational data is that confounding factors are not controlled for, so there may be many causes of an apparent correlation. However, we can look at multiple independent correlations to see if they all point to the same causal relationship. We can triangulate to the most likely cause.

For example, it was observed that cigarette smoking correlates with getting lung cancer. The tobacco industry, invoking the "correlation is not causation" logical fallacy, argued that this did not prove that smoking causes lung cancer. They offered as an alternate explanation "factor X," a third variable that causes both the urge to smoke and lung cancer. But we can make predictions based upon the "smoking causes cancer" hypothesis. If true, then duration of smoking should correlate with cancer risk, quitting smoking should decrease cancer risk, and smoking unfiltered cigarettes should have a higher cancer risk than filtered cigarettes. If all of these correlations turn out to be true, which they are, then we can triangulate to the "smoking causes cancer" hypothesis as the most likely possible causal relationship explaining the correlation between smoking and lung cancer.

Special Pleading, or Ad Hoc Reasoning

This is a subtle fallacy, often difficult to recognize. In essence, it's the arbitrary introduction of new elements into an argument in order to jerry-rig that argument so it appears valid. A good example is the ad hoc (literally, "for this") dismissal of negative test results. For instance, one might argue that ESP has never been demonstrated under adequate test conditions, therefore ESP is not a genuine phenomenon. Defenders of ESP have attempted to counter this argument by introducing the arbitrary premise that ESP does not work in the presence of skeptics. Or perhaps ESP starts and stops working as needed to explain random lucky streaks in the data. Or maybe ESP powers actually decline over time for some strange reason, which explains why ESP effects disappear when more rigorous scientific controls are put in place.

Bigfoot supporters have argued that Bigfeet are psychic, so they know when people are coming and hide. Or perhaps they can teleport. Or turn invisible at will. No bodies have been found because they burn their dead or bury them in well-hidden locations.

Carl Sagan gave perhaps the most famous example of this fallacy in his "invisible, floating, incorporeal, heatless dragon in his garage" argument. Essentially, he claims that there is a dragon in his garage, and then invents a special reason why each test for the presence of the dragon fails.

It's important to note that special-pleading arguments are not fallacious because they're necessarily wrong. It is quite possible that an acupuncture study failed to show effectiveness beyond placebo because the wrong needles or improper techniques were used. Such arguments are unconvincing because they are invented as needed, without independent evidence to support them. People are generally clever and can always think of some "My dog ate my Bigfoot evidence" type excuse.

Tu Quoque

The Latin phrase *tu quoque* translates to "you too." This is an attempt to justify wrong action because someone else does the same thing: "My evidence may be bad, but so is yours." This fallacy is frequently committed by proponents of various alternative medicine modalities, who argue that even though their therapies may lack evidence of effectiveness, more mainstream modalities also lack such evidence. That argument, of course, doesn't justify a treatment that lacks evidence. It is, furthermore, a false premise, as the level of evidence for mainstream therapies is often much higher than for those considered "alternative."

Ad Hominem

An ad hominem argument is one that attempts to counter another's claims or conclusions by attacking the person rather than by addressing the argument itself. True believers will often commit this fallacy by countering the arguments of skeptics with statements that skeptics are closed-minded (see sidebar, pages 72–73). Skeptics may fall into the same trap, dismissing the claims of UFO believers, for instance, by stating that people who believe in UFOs are crazy or stupid.

A common form of this fallacy is frequently present in the arguments of conspiracy theorists (who also rely heavily on ad hoc reasoning). For example, they may argue that the government must be lying because it's corrupt.

The term "poisoning the well" refers to a form of ad hominem fallacy. This is an attempt to discredit the argument of another by implying that they possess an unsavory trait, or that they're affiliated with beliefs or people that are wrong or unpopular. A well-known form of this, which has its own name—Godwin's Law or the reductio ad Hitlerum—refers to an attempt at poisoning the well by drawing an analogy between another's position and Hitler or the Nazis.

Again, remember the commenter at the beginning of this chapter. They felt I was a shill, that I was writing out of fear, and that I just didn't get it. They never countered any of my actual evidence or logic, they just attacked my motivations and character. That is a classic ad hominem fallacy.

It should be noted, however, that not all name-calling is a logical fallacy. If I impolitely state that someone with whom I disagree is a jackass, that's not an ad hominem logical fallacy. If I say their argument is wrong because they are a jackass, then that is a fallacy. But they may still be a jackass.

CLOSED-MINDED

Perhaps the most routine ad hominem fallacy directed at skeptics is the claim that we are closed-minded (which functions exactly like accusing someone of lacking faith or lacking vision). Using the charge of closed-mindedness to dismiss valid criticism is a fallacy, but it's also often an incorrect premise.

Skepticism isn't closed-minded, and the opposite of skepticism is not open-mindedness (it's gullibility). Scientists, critical thinkers, and skeptics can and should be completely open-minded, which means being open to the evidence and logic whatever it says. If the evidence supports a view, then we will accept that view in proportion to the evidence.

But being open-minded also means being open to the possibility that a claim is wrong. It doesn't mean assuming every claim is true or refusing to ever conclude that something is simply false. If the evidence leads to the conclusion that a claim is false or a phenomenon does not exist, then a truly open-minded person accepts that conclusion in proportion to the evidence. Open-mindedness works both ways.

Ironically, it's usually those accusing their critics of being closed-minded that tend to be the most closed. They are closed to the possibility that they are wrong.

Ad Ignorantiam

The argument from ignorance basically states that a specific belief is true because we don't know that it isn't true. Defenders of extrasensory perception, for example, will often overemphasize how much we don't know about the human brain. It's possible, they argue, that the brain may be capable of transmitting signals at a distance. UFO proponents are probably the most frequent committers of this fallacy. Almost all UFO eyewitness evidence is ultimately an argument from ignorance—lights or objects sighted in the sky are unidentified and are therefore alien spacecraft.

Intelligent design is almost entirely based upon this fallacy. The core argument for intelligent design is that there are biological structures that have not been fully explained by evolution, therefore a powerful intelligent designer must have created them. In this context, arguments from ignorance are often referred to as "god of the gaps" arguments, because God is offered as the explanation for any current gap in our knowledge.

Often the argument from ignorance is defended with the adage "Absence of evidence is not evidence of absence." While this sounds pithy, it's not strictly true. Absence of evidence is, in fact, evidence of absence. It's just not absolute proof of absence.

A more scientific way to look at this question is this: How predictive is the absence of evidence for the absence of the phenomenon in question? Well, that depends on how thoroughly we've looked and with what sensitivity. You can't ever prove a negative, but the more you look for something without finding it, the less likely it is to exist. We haven't found alien signals yet, but it's a big universe out there and we have only surveyed a tiny slice. On the other hand, we have scoured Loch Ness for decades without any signs of Nessie, so I am not holding my breath that a giant creature is lurking beneath the waves.

In any case, in order to make a positive claim, positive evidence for that specific claim must be presented. The absence of another explanation only means that we don't know—it doesn't mean we get to make up a specific explanation.

Confusing Currently Unexplained with Unexplainable

Because we don't currently have an adequate explanation for a phenomenon does not mean that it's forever unexplainable or that it therefore defies the laws of nature or requires a paranormal explanation. This is often tied to the argument from ignorance, assuming that our current ignorance will never be filled in by knowledge.

For example, it's often argued that we don't understand how the brain creates the phenomenon of consciousness, therefore the material causes of consciousness are unknowable and spiritual. Actually, neuroscience is progressing quite well, thank you, and we continue to deepen our knowledge of the brain.

False Continuum

A false continuum is the idea that because there is no definitive demarcation line between two extremes, the distinction between the extremes is therefore not real or meaningful. For example, there is a fuzzy line between cults and religion, therefore they are really the same thing.

This is like saying that no one is short or tall, that the very concepts of shortness and tallness are invalid because there is no sharp dividing line between the two. However, extremes can exist and be meaningful and clearly recognized, even when there is a fuzzy border in the middle.

This logical fallacy is often combined with a tu quoque logical fallacy. For instance, someone engaged in rank pseudoscience might argue that mainstream scientists sometimes break the rules too, by using small sample sizes, subjective outcomes, or similar bad methods. Therefore, there is no real difference between science and pseudoscience.

This type of argument is closely related to another logical fallacy, the false equivalency. Extreme fraud doesn't become okay because everyone cheats a little.

False Dichotomy

This fallacy involves arbitrarily reducing a set of many possibilities to only two. For example, evolution isn't possible, so we must have been created (which assumes these are the only two possibilities). This fallacy can also be used to oversimplify a continuum of variation to two black-and-white choices. Science and pseudoscience are not two discrete entities, for instance; the methods and claims of all those who attempt to explain reality fall along a continuum from one extreme to the other. Reducing all factual claims to either pure science or pure pseudoscience would be creating a false dichotomy.

False Analogy

A false or faulty analogy consists of assuming that because two or more things are similar in one way, then they are also similar in some other way, ignoring any important distinctions between the two.

One might say that the protein complex that moves the flagellum of a bacterium is like a motor. Creationists then argue that this is evidence against evolution, because motors are designed and therefore the flagellum must also have been designed. They are taking a metaphor that was intended to convey one meaning and extending it to another, turning a legitimate analogy into a false analogy. Flagella are like motors in that they have multiple moving parts that work together to create a specific motion. That is where the analogy ends, however. The flagellum is a biological motor, so it has a biological origin. Car motors are not biological; they are inanimate and therefore cannot evolve on their own.

False analogy can be a very subtle logical fallacy. You must always ask the question, Are the things being compared truly analogous, and even if they are analogous in some ways, are they analogous in the specific ways that are being claimed?

I think that false analogies are common because they represent a basic function of the human brain. Our brains operate largely by pattern recognition, and we frequently try to understand the novel by comparison to the familiar. Such comparisons are extremely useful, as is seeing commonality among disparate phenomena in order to perceive an underlying pattern. The trouble is that we're too good at pattern recognition and may see patterns where they don't exist. This carries over into the tendency to make analogies where they don't truly exist either.

There is a flip side to this fallacy—denying a legitimate analogy because the two things being compared are not the same in every way. A valid analogy means that two things are similar in an important way relevant to the argument at hand. It doesn't mean they are identical. In that case, they wouldn't be analogous, they would simply be the same thing.

However, some try to dismiss a valid analogy by finding a trivial or irrelevant difference between the things being compared. I might argue that vaccines produce immunity to a disease in the same way that an infection does. An anti-vaccinationist might then counter that this is a false analogy because infections occur naturally and vaccines are unnatural. This would be a "false false analogy" argument, since the manner of exposure doesn't necessarily affect the response of the immune system.

Genetic Fallacy

The term "genetic" here doesn't refer to DNA or genes but to origins. This fallacy consists of arguing against something because of where it came from, rather than considering whether or not it is valid in its current form.

For example, one might argue that the company Volkswagen is not a good company because it was created by Hitler. However, the company was created in 1937, and its origins eighty years ago likely have no influence on how it behaves today.

Similarly, the science of astronomy evolved out of astrology, but no one would seriously argue that astronomy is not a legitimate science because of its origins.

Inconsistency

It's not valid to apply criteria or rules to one belief, claim, argument, or position but not to others. Some consumer advocates argue that we need stronger regulation of prescription drugs to ensure their safety and effectiveness, but at the same time they argue that herbal drugs should be sold with no regulation for either safety or effectiveness.

This fallacy often takes the form of using inconsistent definitions of terms. In order to have a coherent discussion or formulate a valid scientific hypothesis, all terms must be unambiguously defined. Switching around a vague definition as needed causes confusion, perhaps the exact kind of confusion intended.

For example, some proponents of intelligent design use the term "information" inconsistently throughout their arguments. ID proponent William Dembski has tried to argue that evolution cannot create increased information in DNA without intelligent guidance. But he confuses mathematical information with philosophical meaning (semantic information). Mathematical information might describe the minimal bits of data necessary to fully describe something. A random and meaningless sequence actually takes more mathematical information to characterize because there are no patterns that can be summarized. Random mutation can therefore increase this type of information in DNA, even without any intelligent guidance. In his arguments Dembski bounces back and forth between these types of "information" without ever defining them. There are several types of information that have precise mathematical definitions. Any scientific discussion of information, therefore, must be careful to identify which type of information is being discussed at any time. By using vague and inconsistent definitions it can be made to seem as if there is a conflict where none exists.

Acupuncture proponents often use an inconsistent definition of the term "acupuncture." Generally, "acupuncture" refers to the sticking of thin needles into acupuncture points to elicit an alleged therapeutic effect. However, proponents will often use the term to refer to "electroacupuncture," which involves electrical stimulation and has its own effects, thereby muddying the waters.

When a 2009 study looking at the effectiveness of acupuncture for chronic back pain showed that needle location and even insertion do not affect the outcome (there was no difference between "real" acupuncture and placebo acupuncture), the authors claimed that "acupuncture" worked, it was just unclear by what mechanism. They essentially expanded the definition of acupuncture to include poking the skin randomly with toothpicks.

Naturalistic Fallacy

The naturalistic fallacy refers to the is/ought problem, confusing what is true with what ought to be true. This is not to be confused with the appeal-to-nature fallacy (discussed at length in chapter 14), which posits that being natural is in itself a virtue and anything natural is inherently superior to anything artificial.

The naturalistic fallacy usually involves moral judgments. For instance, one might refer to the behavior of animals to justify a particular human behavior as "natural." Animals kill each other, therefore it is moral for humans to kill each other. Moral judgments, though, are distinct from what happens to exist in nature. The whole point of ethics is to derive a fair system that allows humans to get along with one another in society. It doesn't matter what strategies specific animals, who do not have an ethical philosophy, evolved to survive.

This doesn't mean we can't use the natural world to inform our moral reasoning. It just doesn't determine what is moral.

Nirvana Fallacy

A familiar phrase that captures the essence of the Nirvana fallacy is that "the perfect is the enemy of the good." Essentially this style of argument starts with the premise that something is not perfect and concludes that it is therefore worthless. Vaccines don't work 100 percent, therefore they are not useful. Science is flawed, therefore we don't really know anything.

Another version of this fallacy is to argue that something doesn't work in a way or accomplish a goal that was never intended in the first place. For example, creationists have argued that natural selection doesn't increase variability, it only reduces it. That is true but irrelevant. Mutations and recombination increase variability, and then selection provides an adaptive force. These types of arguments are the equivalent of saying that the steering wheel of a car is useless because it cannot accelerate or slow down the vehicle.

No True Scotsman

The name of this fallacy comes from the traditional illustrative example: The argument is made that all Scotsmen are brave, and when a counterexample of a cowardly Scotsman is provided, the original claim is defended by saying, "Well then, he's no true Scotsman."

In essence this fallacy consists of making an argument true by arbitrarily altering the definition of a key term so that the argument is true by definition. It can therefore be considered a special case of the circular reasoning fallacy.

Another illustration might be if someone argued that all swans are white. When the example of black swans is raised, the person declares that black swans are not "real" swans, because swans are white.

Reductio ad Absurdum

In formal logic, the reductio ad absurdum (literally, "reduction to the absurd") is a legitimate argument. It follows the form that if the premises are assumed to be true, logic necessarily leads to an absurd (false) conclusion, and therefore one or more premises must be false. The term is now often used to refer to the abuse of this style of argument, when the logic is deliberately stretched in order to force an absurd conclusion. For example, a UFO enthusiast once argued that if I am skeptical about the existence of alien visitors, I must also be skeptical of the existence of the Great Wall of China, since I have not personally seen either. This is a false reductio ad absurdum because he is ignoring evidence other than personal eyewitness evidence, and also logical inference. In short, being skeptical of UFOs does not require rejecting the existence of the Great Wall.

Slippery Slope

This logical fallacy is the argument that a position is not consistent or tenable because accepting the position means that the extreme of the position must also be accepted. But moderate positions do not necessarily lead down the slippery slope to the extreme. This is common in politics. For example, some opponents of embryonic stem cell research have argued that allowing the use of human embryos in research (even those created for IVF that would otherwise be discarded) would inevitably lead to creating embryos specifically for research, a black market in human embryos, or even the forcible extraction of eggs from women for such research.

Straw Man

A straw man argument is one in which you construct a weak version of someone else's position so that it's an easier target for you to knock down. This is the opposite of the principle of charity, in which the strongest version of an opposing position is assumed. Often positions that no one actually holds are argued against. This can make it seem as if you have vanquished any objections to your own position, or it can be used to make the opposing side seem silly or easily refuted.

Some creationists argue that, because the eye must have evolved, the theory of evolution requires that humans must have had an ancestor with a nonfunctioning half eye. Of course, no scientist believes this. Rather, there is a plausible sequence of simpler but fully functioning eyes, with many examples from extant species.

The opening comment again provides a nice example. They characterized my position with this satire:

> Clearly the Mayo clinic is run by a bunch of quacks and charlatans that have no interest in the health and well being of the public. How dare they even consider something other than surgery or prescription medication to help people heal.

But of course I do not believe this, nor have I ever argued it. What I actually stated was that the Mayo Clinic, and in fact many academic medical centers, have a blind spot when it comes to unconventional treatments because they lack expertise in the specific treatments and in pseudoscience in general. There is also the common straw man that mainstream doctors only advocate drugs and surgery, when in fact we employ nutrition, physical therapy, lifestyle changes, and other interventions as part of standard care.

By attacking a straw man, one may feel better in the moment, but it accomplishes nothing.

Tautology

A tautology is an argument that utilizes circular reasoning, which means that the conclusion is also its own premise. The structure of such arguments is A = B therefore A = B, although the premise and conclusion might be formulated differently so the tautology is not immediately apparent. For example, saying that therapeutic touch works because it manipulates the life force is a tautology because the definition of therapeutic touch is the alleged manipulation (without touching) of the life force.

This fallacy is often called "begging the question," meaning that the premise assumes the conclusion, or that an argument assumes its initial point. Perhaps the most common example is to argue that we know the Bible is the literal word of God because the Bible says so.

WHAT'S THE WORD?

The term "begging the question" originated in the sixteenth century. It is a mistranslation of the Latin phrase *petitio principii*, which actually means to assume the initial point. Because of this mistranslation, the phrase "begs the question" is now often misused to mean "leads to the question." Basically, the phrase is a mess and almost no one uses it correctly, so it's perhaps best to avoid it.

Texas Sharpshooter Fallacy

The name of this fallacy comes from its most common illustrative example. A target shooter claims that he can always hit the bull's-eye. He aims at the side of a barn, shoots a hole in the wood, then goes up and draws a target around the hole he just made, giving himself a bull's-eye.

What this analogy refers to is when someone chooses the criteria for success or failure after they know the outcome. It becomes a form of post hoc reasoning, deciding that a certain piece of evidence is evidence for the conclusion you desire, but you decide that only after you know what the evidence is. In order for the evidence to be useful as a support for the conclusion, however, you would have to have determined the criteria for success ahead of time, based on valid principles.

Conspiracy theories make generous use of this fallacy. For example, some 9/11 conspiracy theories point to the size of the hole made by the plane that crashed into the side of the Pentagon as evidence for a conspiracy. They argue that if a jet truly hit the side of the Pentagon, the hole should be bigger, but there is no prior reason to conclude this. It's not as if we know what size hole a jet crashing into the Pentagon should make. Conspiracy theorists decided after they saw the hole that it was too small and therefore evidence for their conspiracy.

This can be a very subtle fallacy, one that essentially can occur anytime we reason backward from known outcomes.

The Moving Goalpost

The moving goalpost is a method of denial that involves arbitrarily moving the criteria for "proof" or acceptance out of range of whatever evidence currently exists. If new evidence comes to light meeting the prior criteria, the goalpost is pushed back further—keeping it out of range of this new evidence. Sometimes, impossible criteria are set up at the start—moving the goalpost impossibly out of range for the purpose of denying an undesirable conclusion.

Anti-vaxxers claimed that the MMR vaccine caused autism. When scientific studies shot down that claim, they moved on to thimerosal (a mercury-based preservative in some vaccines, but not in the MMR). They predicted that when thimerosal was removed from standard vaccines in the US in 2002, autism rates would plummet—they didn't. So they claimed that mercury from other sources, like coal factories, made up for the drop in mercury exposure from vaccines. When the evidence did not support that claim, they moved on to aluminum as the cause (nope). And now they just reference vague "toxins."

No amount of safety data on vaccines is ever enough. They just keep moving the goalpost.

Fallacy Fallacy

I've said it before, but just because an argument isn't sound doesn't mean the conclusion must be false. I might argue that global warming is real because the sky is blue. That's a non sequitur, making the argument unsound. But it may still happen to be true that global warming is a real phenomenon. When an argument isn't sound, it simply means the argument does not support the conclusion. The conclusion may be true or false. The fallacy fallacy comes from concluding that a claim must be wrong because one argument proposed for it is not valid.

The "fallacy fallacy" may also refer to a situation in which someone claims that another person committed a fallacy when they didn't. If I accuse you of using a straw man when in fact you fairly and accurately summarized my position, I would be committing a fallacy fallacy.

It's easy to fall into this trap because, as I stated, informal fallacies are context dependent and often require judgment to properly recognize them. When is a reference to the opinions of experts an appeal-to-authority fallacy? If I point out that someone was convicted of fraud, is that poisoning the well or legitimate information? It's possible to twist almost any argument to make it seem like a fallacy, if that is your goal, rather than simply using the most valid logic possible.

In the extreme, I regularly encounter social media thugs who think that if they can somehow portray another person's argument as fallacious, then they have "won." They commit both types of the fallacy fallacy, falsely portraying a sound argument as fallacious and then claiming the conclusion must therefore be wrong.

Applying Logic Every Day

Once I learned about common logical fallacies I started to notice them everywhere. This makes you a big hit at parties (ahem). This is partly the Baader-Meinhof phenomenon—once you stumble upon an obscure word or fact, you encounter it again and again, seemingly against the odds. This is part simple coincidence but also part perception bias. You've likely encountered the word or fact before, but you just didn't notice it.

When you become familiar with the common informal logical fallacies, you will likely encounter them everywhere as well—because they are common. You may be tempted to point fallacies out every time you think you hear one. I know my wife just loves it when we're having a heated "discussion" and I point out her logical fallacies.

Primarily I try to use my knowledge of logic to purge my own thinking of sloppy reasoning. It's difficult, but it's worth the effort. Understanding valid logic is also critical to dissecting all the news, opinions, and claims out there. Thinking without logic is like finding your way in the wilderness without a compass or map. You will probably go around in circles and not really get to where you want to go.

NAME THAT LOGICAL FALLACY

In a comment to a blog post I wrote detailing a negative study of ESP, one ESP proponent wrote:

@Steven Novella

Psi has been experienced by human beings throughout history and across all cultures. Our modern western culture is virtually unique in rejecting it. You're maintaining a position contrary to the collective experience of humankind.

You need therefore to advance some *reasons* for supposing it doesn't exist, not merely content yourself with criticizing the research.

How many logical fallacies can you point out in that brief statement? I see an unstated major premise, an appeal to popularity and antiquity, and an attempt at shifting the burden of proof.

11. Skeptics' Guide Entry: Cognitive Biases and Heuristics

Section: Metacognition
See also: Logical Fallacies

Cognitive biases are flaws in the way our brains process information. Heuristics are similar—they're rules of thumb or mental shortcuts that are not reliably true and therefore also lead to biased thinking.

> The confidence people have in their beliefs is not a measure of the quality of evidence but of the coherence of the story the mind has managed to construct.
>
> —Daniel Kahneman

The reality is that we're not Vulcans with mental constructs and processes of pure logic. Our thinking is too easily influenced by biases and superficial factors, such as how information is presented. Not just casinos, as mentioned in the opening of this section, but the entire marketing industry is based on the fact that people often make irrational decisions for emotional or quirky reasons they don't necessarily understand.

Cognitive flaws and biases can be mitigated by metacognition, which is, simply put, thinking about thinking. Skepticism is largely a systematic effort in metacognition, which means understanding how we think and avoiding common mental pitfalls. The human mind is like a sailboat on a sea with strong currents and a steady wind. We tend to just follow the currents of our biases and can easily be manipulated and blown about. Metacognition allows you to grab the rudder, to work the sails and head toward reliable conclusions, even against the current and into the wind.

In the previous chapter we discussed logical fallacies, arguments that involve invalid logic. In this one we'll look at two further types of flawed thinking: biases and heuristics. They're actually very similar—two sides of the same coin. A cognitive bias is a tendency to think in a certain way, to favor certain kinds of information, and to prefer some conclusions over others. Biases generally deviate your thinking away from reality.

Heuristics are mental shortcuts. They are rules of thumb that allow you to approximate a likely answer quickly, but they're not strictly true and often result in error. Heuristics may be helpful and may lead to the correct conclusion, but they are oversimplified. A flawed heuristic may result in a cognitive bias, and a cognitive bias may adversely affect a heuristic. Either way, you can think of these related phenomena as patterns of thinking that are biased or limited—they are currents in the sea of your thoughts. Being aware of them will help you take the helm of your own thinking.

Cognitive Biases

Have you ever wondered, Does it really matter if an item is priced at $19.99 instead of $20.00? The answer is yes, as ridiculous as that may seem when you examine it. One penny should not make a difference, but we have a leftmost digit bias—we're mainly influenced by the leftmost digit in any number. This likely comes from the tendency to deal with information overload by simplifying it. We tend to focus on the leftmost digit to give us a quick sense of how large the number is. Generally speaking, this isn't a bad rule, but it's easily exploited by using an unlikely but deliberate price ending in all 9s.

Then there's the handedness bias, which is a great example of how we often make decisions below the level of conscious awareness. When offered two equivalent choices (meaning there was no reason why the subject should prefer one choice over the other), right-handed subjects in a study were more likely to choose the option on the right, while left-handers were more likely to choose the option on the left. (If a right-handed person had an injured right hand, their choice shifted to the left option.)

This may reflect a deep aspect of our thinking called embodied cognition. We tend to understand the world in terms of physical relationships, which we then extrapolate to more abstract thinking. For example, we say that your boss is "above" you hierarchically. They're not physically above you, but we use this spatial analogy. An argument might be said to be "weak" or "strong," or if it's especially bad we might even call it "lame." You are "blind" to reality and "deaf" to the concerns of others. A theory can be "beautiful," and an idea can be "big." We use physical terms to describe, and therefore think about, abstract concepts that have no physicality.

Another important and pervasive bias is the framing bias. Your thinking about something can be dramatically affected simply by how it is presented to you. For example, what if a doctor told you that a procedure has a 10 percent death rate? And what if they told you it had a 90 percent survival rate? These are obviously identical, but people react very differently to the framing. Far more people will opt for a procedure that has a 90 percent survival rate than one presented with a 10 percent death rate.

Another example is a survey showing that "62 percent of people disagreed with allowing 'public condemnation of democracy,' but only 46 percent of people agreed that it was right to 'forbid public condemnation of democracy.'" These are, again, the same outcome.

Part of the reason for the difference is how we think about risk. Humans tend to be risk-averse when it comes to positive outcomes, and risk-seeking when it comes to negative outcomes. We don't like to miss out on a positive outcome, and so we'll take the sure thing—we will take a guarantee of $100 rather than a 50 percent chance of getting $200. However, we are willing to take a bigger risk in order to avoid a negative outcome. Patients with terminal illness, therefore, will take big risks with fanciful and unlikely treatments.

Related to our risk assessment biases is the gambler's fallacy. If you flip a fair coin heads five times in a row, most people will feel tails is now due. The bias is the thinking that past events influence future events, even when there is no causal connection. Since each coin flip is independent, it does not matter what happened before. The chance of another heads is still 50 percent. Casinos love this bias because it gives people the illusion that they can predict the next throw of the dice or the next outcome on a roulette table. Red is due, so let's bet it all.

We are also biased by our basic psychology, which caters to our emotional needs. We need to feel we are part of a group where we are liked and accepted. Thus there is an in-group bias—we are biased toward our group, making much more favorable judgments of our in-group than an out-group. Similarly, we are hugely biased toward ourselves. We will give ourselves every benefit of the doubt and inter-pret our own actions in the most favorable light.

We're also motivated to minimize cognitive dissonance, that uncomfortable feeling we discussed in chapter 9, which occurs when facts conflict with each other or with our desires. If we purchase an item, our assessment of the quality of that item goes up. We're try-ing to justify the purchase to ourselves. In fact, if we are given mul-tiple items to choose from, and then we discover our first choice is not available, we are likely to jump over our second choice and go for number three. This makes no sense, so why would anyone do this? It is likely because we already had to reject our second choice in favor of the first. Once we choose, we have to justify the choice by down-grading the quality of choice two. Therefore, number two is now worse than number three, which we then prefer. (Crazy, right? Silly humans.)

We also tend to assume that other people think like we do, a phe-nomenon called the projection bias. This is the way we think about and understand what other people are likely thinking. We tend to use our own mind as a template to make predictions about how other people will think and act. So if something bothers us, we assume it bothers other people as well. This is closely related to the consensus bias. We tend to assume that our own opinions are in the majority, that most other people share them.

At a friendly get-together, a distant relative of mine was shocked to find that my friend was a Republican. This relative laughed at themselves when they realized they just assumed everyone they met was a Democrat like them. In their mind, Republicans weren't real people they could run into at a party.

Hindsight bias can be a squirrely one. Once we know the outcome of a situation, that knowledge colors our interpretation of what happened and why. We tend to think that whatever happened was inevitable, destined to happen, even if it was a close call. Think about the news coverage of any recent presidential election. Once we know which candidate won, the pundits are happy to explain why this outcome was obvious. The losing candidate never had a chance.

Our thinking about evolution is also plagued by the hindsight bias. How often have you heard that dinosaurs were doomed to extinction because of some flaw or weakness? Mammals were clearly destined to dominate the world. Remember the movie *Ice Age*? The dodo bird was depicted as hopelessly defenseless and stupid, even suicidal. The fact of its extinction is apparently sufficient evidence for this. Hindsight bias is similar to post hoc reasoning—once we know the outcome, we are really good at inventing reasons to "explain" it.

Perhaps the biggest cognitive bias of them all is the confirmation bias. This bias is so significant we gave it its own chapter (which is how you know it's significant—right?).

Heuristics

Heuristics are a form of cognitive bias, but they serve a purpose and can be semi-useful. I tend to think of them as "90 percent rules." If someone calls me on the phone with a story that ends with them needing my credit card number, I assume it's a scam. I know it may not be, but it's a useful rule. Or if you always assume a situation is more complicated than you think, you will be right most of the time.

The key is to be aware of what rules you are using and recognize that they're not always applicable. Some unconscious shortcuts are biased and may be more misleading than useful.

The availability heuristic, for example, is the unstated assumption that if we can easily call an example of something to mind, it must therefore be common or important. If you can think of an instance when someone who worked at the DMV was rude, you will likely think that such behavior is typical. If you know someone who was mugged in New York City, you think muggings are common in the Big Apple.

Of course, neither of these conclusions is necessarily true—your own anecdotes are just that; they are quirky and not representative. We know intellectually that we need systematic statistical data to make judgments about how common something is, but we leap to conclusions anyway, based on the mental availability of an example.

A common heuristic with a statistical theme is the representativeness heuristic. We tend to think that someone or something likely belongs to a category if they have features typical of that category. While there is some truth to this, we overapply the rule and ignore two other critical pieces of information. First is the base rate—how common is that category? If the category is rare, then the probability of someone belonging to that category may still be low, even if they are typical for it. The second is predictive value—a feature may be typical of a category but not specific to it.

Sherlock Holmes famously used the example of the clopping of hooves to explain this principle to Watson. If you hear the clopping of hooves behind you on the streets of London, think horse, not zebra. This example has been borrowed by the medical culture—when medical students suggest a rare diagnosis, it's a "zebra."

Let's say a patient has palpitations, headaches, sweating, and high blood pressure. These are all symptoms that are very typical of a condition known as pheochromocytoma (a tumor that secretes adrenaline).

However, "pheos" are extremely rare. Therefore, even with typical symptoms, the patient is far more likely to have anxiety, or hyperthyroidism, or pretty much anything else that can cause these symptoms. Pheos are that rare. An inexperienced medical student, however, will follow the representativeness heuristic and suggest this diagnosis. Their experienced mentors will then chuckle, mutter something about zebras, and explain how damn rare pheos are. Oh sure, we'll still do the lab test to make certain, but no one will hold their breath.

In other words, having high blood pressure is not very predictive of having pheo, because it is not very specific to that one condition. Lots of people have high blood pressure and don't have this rare condition.

To take a more general example, and also the one Kahneman used when describing the representativeness heuristic, let's say that John is a computer nerd who likes to keep to himself and obsesses over tiny details. If John goes to Madeup University, what are the odds that he is an engineering major rather than a humanities major?

You might be tempted to say it is more likely he is an engineering major, but you can't know that and therefore may be wrong. What if 99 percent of the students at Madeup U. are humanities majors and only 1 percent are engineering majors? It is then more likely John is an atypical humanities major than a typical engineering major (the validity of such stereotypes aside).

Here is a simpler heuristic: the unit bias. It derives from a desire to simplify a complex world to make some reasonable first approximations that aren't always accurate. We tend to focus on one salient feature of an item or thing and use that feature as the one measure of value, quality, or quantity. In a 2009 study by Andrew Geier and Paul Rozin, they asked people to estimate the weight of others. They found that people relied solely on girth, while ignoring height, which led to significant inaccuracies.

The unit bias is also very common in the marketing of technology. Remember computer buying in the nineties and early aughts? Everything was about the speed, in megahertz, of the processor, as if that one number captured the overall power of the system. Other details like RAM, the motherboard, or the chipset, were too complex for the average computer buyer to worry about, so they focused on chip speed.

The same is basically true of digital cameras—it's all about the megapixels. Meanwhile, maximum resolution in megapixels is perhaps one of the least important details for most users. You're better off worrying about the quality of the lens. (I see you, techno-nerd, sweating the details of these examples. The point is, most people boil down the complex features of these technologies to a single number.)

Finally, we come to the anchoring heuristic. This one is also common in marketing. Let's say I show you a house and then ask you if you think the house is worth more or less than $100,000. I then ask you what you think the house is worth. I ask another person if they think the same house is worth more or less than $500,000, and then I ask them to guess the price. The person who was "anchored" to the $100,000 reference will guess significantly lower than the person who was anchored to the $500,000 price, even though they're assessing the same house.

Yes, we are that easy to manipulate. That's why commercials say, "Would you be willing to pay $100 for this fabulous product? How about $200? Well now you can get it for the ridiculously low price of $19.95." This is why psychologists suggest that in a negotiation you should make the opening bid, to anchor all further bidding. Anchoring goes beyond money as well. We also anchor things in time or place, and we anchor value judgments.

So can you really beat the odds at Las Vegas? No—the math is against you, and any belief that you can beat the math is based on cognitive biases, happily encouraged by the casinos (unless you use an actual analytical process like card counting, which is considered cheating). Next time you see a commercial, or watch the news, or simply chat with friends, try to get into the "metacognition zone." How are you processing and evaluating information? How are your judgments biased? How are you being psychologically manipulated?

Grab control of that rudder and go against the currents.

12. Skeptics' Guide Entry: Confirmation Bias

Section: Metacognition
See also: Selective thinking

Confirmation bias is the tendency of individuals to seek out or interpret new information as support for previously held notions or beliefs, even when such interpretations don't hold up to statistical scrutiny. If an athlete wears a particular pair of socks to each game because he believes it will improve his chances of winning, he may remember the times he won wearing those socks but forget the times he won wearing different socks or the times he lost wearing the so-called "special" socks.

> Ever since I first learned about confirmation bias, I've been seeing it everywhere.
> —Jon Ronson

We love to get feedback from our listeners, but a few years ago one SGU listener e-mailed to complain that a segment we were running at the time, *This Week in Skepticism*, was clearly biased toward women, likely as part of some feminist agenda. That sounded like a factual claim, so I quickly tallied up the hundred or so installments of that segment and found that the segment focused on a woman 15 percent of the time, on a man 45 percent of the time, and did not focus on an individual 40 percent of the time. I thanked the listener for pointing this out and vowed we would spend more time focusing on female scientists and skeptics.

How is it that this listener's perception was so at odds with the clear reality, and over something that mere counting could resolve? The answer is confirmation bias.

Confirmation bias is perhaps the most powerful and pervasive cognitive bias (which is why it gets its own chapter), and it's important to understand it thoroughly. Confirmation bias is the one bias to rule all biases, the mayor of Biastown, captain of the USS *Bias*, the Sith Lord of the bias side of the Force (okay, you get the idea).

Confirmation bias is a tendency to notice, accept, and remember information that appears to support an existing belief and to ignore, distort, explain away, or forget information that seems to contradict an existing belief. This process works undetected in the background of our minds to create the powerful illusion that the facts support our beliefs.

In an effort to make sense of all the information we encounter, we produce a specific worldview—a framework, narrative, or paradigm, as it were. Having an organizing and explanatory system for all the information we deal with is helpful, even necessary, but these processes can take on a life of their own. The narratives we use to understand the world don't just organize information, they curate and filter it to serve those very narratives.

Don't get me wrong, it's not as if we *intend* to manipulate the facts or to rewrite history. Confirmation bias is often subtle and nuanced, and because of this, it's especially pernicious. We talked previously about motivated reasoning (see chapter 9), which is more of a conscious process to protect our cherished beliefs. Confirmation bias is a more unconscious bias. This error in our thinking is constantly working in the background to undermine the rational process. It's lurking within us and we must always be on the lookout.

Consider this scenario: The leaves outside have begun to change color. You're swapping your T-shirts for sweaters. A chorus of coughing has begun to resonate throughout the halls of the office. It's flu season! Time for a quick trip to the pharmacy to get the annual flu vaccine. But years ago you stopped getting flu shots because you remember feeling really crummy the day after you got one. Your throat was a little raw, your nose got stuffy, and you had joint pain for half a week. Based on this anecdote, you concluded that the flu vaccine made you sick and wasn't worth it. This conclusion then became a filter through which you saw the world.

Later, your mom told you about her cousin's husband who broke out in a fever following his flu shot, forcing him to stay home from work. You remember this story (maybe even tweaking it in your memory—didn't he also have a rash?) because it confirms your belief. You see a report on the news that the flu vaccine is not a good match to that year's circulating virus and therefore not that effective—that one's a keeper. The next year, when the flu vaccine was a good match and particularly effective, you didn't notice, or you considered that to be an exception. You go out to dinner with some of your friends and the conversation turns to the season. Some of them recount how they always get vaccinated and rarely get the flu. Others tell of getting the flu despite the vaccine, or even getting side effects from the vaccine. Guess which stories you remember.

Relying on your subjective perception isn't only unreliable, it's biased by your existing beliefs. To find out if the flu vaccine really is worthwhile, you need to look at objective numbers. Each year, only about 0.0001 percent of people who get the vaccine have a severe allergic reaction (that's one person per million doses). And there's no way of knowing how many people contract the flu before the vaccine has reached its peak effectiveness. Influenza can incubate for up to four days, and the flu shot isn't fully effective until *two full weeks* after administration.

You probably have connections to a vast network of people, once you count second and third degrees of separation. It's likely that someone in that group (larger than you probably think) had a reaction to the vaccine or contracted the flu after getting the vaccine but before immunity kicked in. But significantly more people get vaccinated, and because of the protection this affords them, they don't get the flu. In the US alone, the Centers for Disease Control (CDC) reports that between 2005 and 2014, the flu vaccine prevented the deaths of more than 40,000 people. Flu vaccines work when you look at statistics and large numbers, but casually skimming your everyday experiences might not reflect that.

Here is another example, this one from psychologist Thomas Gilovich: A husband believes he always puts down the toilet seat. His wife, with whom he shares a bathroom, believes he never puts down the seat. They are working from the exact same set of data but arrive at opposite conclusions. Why? Well, the husband notices when he remembers to put down the seat. The wife notices when he doesn't. He doesn't notice when he forgets (because that's what it means to forget). She doesn't notice when he does, because it's a nonevent.

When referring to psychics we call this "remembering the hits and forgetting the misses." A psychic will throw out all sorts of statements, and the client is likely to remember only the ones that stand out for being accurate.

Confirmation bias works even more subtly in our everyday lives. We may have a narrative that blue-eyed people are rude. Whenever someone is rude to you, you may look to see what color eyes they have. If they have blue eyes, this confirms your narrative and you remember it as evidence to support it. If they do not have blue eyes, you dismiss that information as an exception and promptly forget the encounter. This is often how bigotry is formed and maintained.

The very concept of an exception implies a conclusion. It's not an exception—it's data, data that are as valid as the instances that appear to support your narrative. But we subconsciously treat confirming instances as data and disconfirming instances as exceptions. Or we may make a specific excuse, such as "The brown-eyed person wasn't that rude," or perhaps they were provoked in a way that would make anyone rude. The information is rationalized away and we eventually forget it. The blue-eyed person who was rude to you, however, is a story that lives forever, proving your narrative.

When I was working as an intern in the emergency room on a particularly busy night, a nurse noted how frantic the ER was. She then commented to the room, "It's a zoo here tonight. Is there a full moon out?" The answer was no, no full moon. She responded with a shrug and went about her business. To her that was an uninteresting nonevent, but if there had, by chance, been a full moon, that would have powerfully confirmed her prior belief that the ER is busier under maximum lunar influences.

This is why confirmation bias has such a strong effect. It gives us the confident illusion that we are following the evidence. In reality, our beliefs are manufacturing the evidence. In the end we may be extremely confident in a totally false belief.

Confirmation bias works with many other biases and processes in a synergistic way. There is so much information out there in the world. We encounter numerous events, people, and bits of data every day. Our brains are great at sifting this data for meaningful patterns, and when we see one, we think, "What are the odds? That cannot be a coincidence, so it confirms my belief." Rather, the odds are very good. The possibility that you would encounter something confirming your belief was almost certain, given the number of opportunities. In this way, confirmation bias dovetails with data mining and pattern recognition.

Another factor that plays into confirmation bias is the use of open-ended criteria, like ad hoc or post hoc analysis. We often decide after we encounter a bit of information that this information confirms our belief, and we retrofit the new data into our belief as confirmation. We may call this subjective validation—the tendency to use subjective criteria to validate our prior beliefs.

There's a closely related bias called desirability bias. We are biased toward information that supports what we want to be true, even if we don't already believe it. In a 2016 study conducted just prior to the presidential election, Ben Tappin and his fellow psychologists asked subjects which candidate they preferred and which candidate they thought would win. They were then presented with new polling data favoring one candidate or the other. The people surveyed were much more predisposed to accept polling data that favored the candidate they wanted to win rather than the candidate they believed would win. Confirming their desires was more powerful than confirming their beliefs.

What this study suggests is how complex human thinking is and therefore how difficult to study. I don't think we can make any general statements about how people treat "beliefs." Psychologically, there are different kinds of beliefs—people treat emotionally held beliefs differently from emotionally neutral beliefs. We happily update the latter when we receive new information, but we cling tightly to the former and may even tighten our grip in the face of disconfirming information (a backfire effect). Our emotions trigger motivated reasoning in some cases but not others.

So, in the polling study, some of the subjects may have held the belief that their candidate would win as an emotional belief. Fewer people probably had an emotional attachment to the belief that their candidate would lose. As new data came in, people would cling to the notion their candidate would win, but not to the idea that they would lose. But even this conclusion is too simple. If you recall, Donald Trump was claiming that the election was rigged. It's therefore possible that some Trump supporters believed Hillary Clinton would win because the election would be rigged. When new polling showed Trump might actually win, they could easily shift to the conclusion that Trump was popular enough to win even a rigged election.

People can change their opinion in the face of overwhelming evidence, however, if it's sufficient to overcome their motivated reasoning. I've engaged many times with people who deny global warming. When confronted with the solid evidence that the Earth is in fact warming, some deniers will still cling to denial that the warming is real, but others will retreat to the position that even though the Earth may be warming, we don't know if humans are causing it. Or even if we know humans are causing it, we don't know that the consequences will be bad. Or even if the consequences will be bad, there is nothing we can do about it.

These softer positions, however, are held reluctantly, and although they may now technically constitute prior beliefs, I doubt that confirmation bias would support them. Rather, the denier will still jump on any evidence that seems to support what they want to believe and will quickly revert to the belief that global warming is not even happening when given the chance.

I think that desirability bias and confirmation bias are two sides of the same coin and are not easily disentangled. In reality, there will often be a complex web of competing beliefs and desires. Furthermore, not all beliefs are the same, as there is a spectrum of emotional and identify implications for specific beliefs. Further still, in this study they were asking for people's predictions about what would happen, which may also introduce new biases, such as a potential optimist or pessimist bias.

The following comments come from one practitioner on a medical forum dedicated to body asymmetry disorders as an explanation for chronic conditions such as multiple sclerosis and chronic fatigue syndrome. See if you can spot confirmation bias at work:

> I personally have serious doubts if there is anything like "MS" or MS being an "autoimmune disease."
>
> Literally everyone I have seen and treated with "MS" was actually a TMJ [temporomandibular joint] dysfunction patient. I am anxiously looking for my first "Real MS Patient."
>
> I am currently treating at least 20 "MS" patients. Many of them have had short lived benefits from CCSVI [chronic cerebrospinal venous insufficiency]. They all appear to be substantially improving with TMJ correction and most symptoms previously labelled "MS related" have disappeared.
>
> I believe that most "MS lesions" are a consequence of CSF [cerebrospinal fluid] leaking into the brain stroma—they are not lesions in the neural tissue as such, otherwise the symptoms would persist...

Imagine taking a slice of pathological tissue around any infective lesion[.] One is bound to see an increased immune function. To go on to interpret it as an autoimmune disease would be pure folly. The immune system needs help—not an assault on its function [which the drugs do].

I remain to be corrected on my hypothesis.

Sounds pretty confident. It's also all utter nonsense. The commenter, a practitioner of dental therapy, is arguing that multiple sclerosis and some other diseases don't really exist and are just jaw and tooth problems. Meanwhile, we have over five decades of research telling us that multiple sclerosis (MS) is an autoimmune disease—the result of the immune system attacking one's own tissue, in the case of MS the brain and spinal cord. There are literally thousands of published studies supporting this conclusion, with multiple independent lines of evidence. The commenter, who clearly knows very little about immunology or autoimmune disease, tries to dismiss some of this evidence, such as that from biopsy or autopsy examination. He argues that seeing inflammation around MS lesions in the brain doesn't prove the lesion is autoimmune. It could just be the immune system doing its job.

Whenever you hear a claim like this about a mature science, you have to ask yourself, So all the researchers dedicated to this discipline over decades missed this simple idea? If something apparently obvious occurs to you as a nonexpert, it's a good bet that it has occurred to people who have spent their careers researching the issue. The difference between reactive immune activity and autoimmune activity is a basic concept—it's thought of every time a pathologist looks at immune activity on a slide and wonders, Is this reactive immunity or a primary inflammatory lesion? MS is an autoimmune disease. There is evidence of chronic central nervous system immune activity, lesions are inflammatory, and immunosuppressant treatment works.

The commenter claims they never met a real MS patient. What are their criteria? Are typical brain lesions and inflammatory markers not enough? He implies that if someone who is diagnosed with MS shows improvement in any of their symptoms with TMJ correction, then they don't have MS. He doesn't seem to be aware of the illusory power of placebo effects. Any treatment, if looked at in an open-ended way, will seem to work for almost any condition. That's the power of confirmation bias. Only by carefully studying a treatment in a blinded fashion, to eliminate the effect of all biases, can we get reliable information about the real effects of the treatment.

What published research is there to support the claims of this commenter? Very little—a couple of pilot studies (meaning preliminary-type studies that are not blinded, controlled, or otherwise rigorous). Preliminary studies themselves are little more than confirmation bias: They have a huge positive bias and are likely to support the researcher's hypothesis. They are not confirmatory, meaning they really aren't evidence at all but just an exploration to guide later research.

The commenter is further impressed by the fact that he can speculate wildly about an "explanation" for apparent MS lesions. This is just another form of confirmation bias—we believe that because we can think of an explanation, this in itself is a sort of confirmation (the entire field of astrology is based upon this process). We tend to underestimate our own creativity and ability to invent explanations for things post hoc. The aptitude to do so says absolutely nothing about the viability of our beliefs, because we can invent post hoc explanations for anything. The real question is: Is there any objective evidence for the hypothesis? In this case the answer is no.

Confirmation bias also works hand in hand with our tendency to look specifically for evidence that confirms, rather than disconfirms, our beliefs. Even when specifically testing a hypothesis or claim, we tend to frame our tests in a way designed to confirm the hypothesis. If we think classical music will make plants grow, we might test this idea by playing music for our plants and seeing how well they grow. You should be thinking, "But wait, we need a control not exposed to classical music." You need some basis of comparison to know what effect the music is having. The notion of a control group comes from scientific thinking and is not intuitive.

Less obvious is the "toupee fallacy." Some people believe they can always tell when someone is wearing a toupee (you can substitute many things for this: breast implants, contact lenses, if they're lying, etc.). Confirmation of this belief comes from all the times when someone was wearing a toupee and they noticed.

But here's the thing: They are, by definition, not aware of cases in which someone had a toupee and they didn't notice. Also, unless they went up to every person they identified as a toupee wearer and asked if they indeed had a toupee (or pulled their hair), they don't know if they had any false positives. Unless they systematically tested their ability to identify toupees, they really have no idea how good they are at such detection.

Let's say you make the observation that people who drive sports cars are more likely to have cats as pets. In order to test your hypothesis, you randomly ask people you encounter with sports cars if they own a pet and what type. You then tally up how many sports car owners have cats.

You might think you have tested your hypothesis, but you haven't. You have only looked to confirm your hypothesis. You also have to ask people who don't own sports cars to see what pets they own (if any).

The power of confirmation bias is that it works tirelessly in the background, filtering vast amounts of data, until you have a compelling illusion that the evidence supports your belief. In a way, your own brain is gaslighting you, convincing you that reality is different than it actually is and giving you false confidence. It can be incredibly difficult to shake a belief that is supported by a huge construct of confirmation bias. Doing so would involve admitting that you can be profoundly wrong about something that you are sure is correct because you've just seen too much evidence in its favor. It can be unsettling to realize how vulnerable we are to such biases—but that's a humbling and necessary step toward critical thinking.

Wason Selection Task

One of the pitfalls of learning about critical thinking and pseudoscience is that your confidence can easily turn into cockiness. It's important to remember that knowing about cognitive biases doesn't make you immune to them. After reading this book you won't be magically protected from error or bias. You won't be able to assert your opinions as if they are the authoritative Truth.

The metacognitive skills we discuss are tools. It takes vigilance and a lifetime of practice and reflection to put them to use. Doing this will help you minimize your flaws and biases, but you can never be rid of them.

Here's a fun task that, for some of you at least, will demonstrate that understanding metacognition is not necessarily enough to avoid making mental errors.

The task is this: There are four cards on the table before you. They each have a letter on one side and a number on the other. I have a hypothesis—every card that has a vowel on one side has an even number on the other. The four cards are showing A, 7, D, and 4.

The question for you is this: Which cards do I need to flip over in order to test my hypothesis? Go ahead and decide before reading ahead to the answer.

This was a psychological study performed by P. C. Wason in 1966. In his original study only 10 percent of subjects got the right answer (which means they turned over all the cards they needed to and none of the cards they did not need to). Later replications yielded similar results.

The correct answer is that you need to turn over A and 7 but not D or 4. The reason is, these are the two cards that can falsify your hypothesis. If A doesn't have an even number on the other side, or 7 has a vowel, the hypothesis is wrong. D and 4 are irrelevant, because they cannot test your rule. Most people will turn over 4 to see if there's a vowel on the other side, but this will still not prove the rule, and the 4 can't falsify the rule because it's okay to have an even number opposite a consonant. Instinctively, however, people look for evidence to confirm their hypothesis rather than evidence that can disprove it.

An interesting wrinkle to the Wason selection task is that the results differ depending on context. If, for example, one side of the cards has ages, and the other what someone is drinking, people have no problem figuring out if there is any underage drinking going on. Our instincts are better suited to social interactions than abstract problems, apparently.

But the core lesson remains—if you want to test your hypothesis, try to prove it wrong. Do not only look for evidence to prove it right.

13. Skeptics' Guide Entry: Appeal to Antiquity

Section: Metacognition
See also: Logical Fallacies

The appeal to antiquity is a special form of the appeal-to-authority fallacy. In this case the alleged authority is the assumption of ancient wisdom, or the notion that an idea that has stood the test of time must be valid.

> It is not the antiquity of a tale that is an evidence of its truth; on the contrary, it is a symptom of its being fabulous.　　　　　—Thomas Paine

---

We have an understandable fascination with the past. The pyramids of Egypt, the Great Wall of China, and the Colosseum of Rome have all stood the test of time and stand as monuments to ancient civilizations. Something reaches inside us, an eternal thread that helps us connect, in a rather tangible way, to civilizations long since passed away. Perhaps it makes us feel connected to something immortal, or at least far greater than a fleeting human lifespan.

It's not just monuments, walls, and buildings that have survived for thousands of years. Many ideas and notions from millennia ago survive and flourish in the twenty-first century as virtual monuments to ideas of the past. Some people still believe in astrology—prophecy based on the positions of planets relative to the constellations in the sky. They turn to astrology to help them make everyday decisions...

even politicians do it! Astrology represents a vision of the universe devoid of any real knowledge of cosmology or the forces of nature. It's a superstitious belief from a time before modern science existed, when the basics of logic, such as cause and effect, weren't routinely applied.

Other examples can be found in ancient health treatments such as acupuncture. The modern conception of acupuncture involves inserting very thin needles through the skin at alleged acupuncture points in order to have a beneficial physiological effect. This idea is only about a century old, however. Previous to this modern rebranding of acupuncture, it was more obviously superstitious—it was actually a form of bloodletting using lances or large needles, and acupuncture points were also connected to Chinese astrology, linking our bodies to the heavens. Despite this clearly documented reality, modern acupuncture is frequently marketed as an ancient treatment that is thousands of years old. The assumption is that if people have been doing something for so long, there must be something to it.

Typically, those employing this fallacious argument will try to rescue it by arguing that it is valid because antiquity implies that the method has stood the test of time. This argument, however, isn't valid, because it contains a major unstated premise that isn't true—namely, that time will test such modalities as a matter of course. History has shown this is a false assumption.

Astrology has been shown, thoroughly and scientifically, not to deliver on its promises of prophecy or accurately describing the personality of individuals. The effects of acupuncture have also been shown, quite convincingly, to be entirely placebo.

But culture has a lot of inertia—people believe things because that is what they are told, because those things are already believed. Further, there are many mechanisms of self-deception that perpetuate false beliefs. Millions of people can believe something for thousands of years, even if that belief has no basis in reality. An uncontroversial example is bloodletting, which is part of an entire system of treatment called Galenic medicine. This is based on the notion of the four humors: blood, phlegm, yellow bile, and black bile. In this system all illness is due to an imbalance of these four humors, and interventions are meant to restore balance. Bloodletting reduces excess blood, purgatives remove excess bile, etc.

This system of belief survived for about two thousand years, during which time it was extensively practiced. Medicine based on the four humors only faded away in the wake of scientific medicine. Not until the medical profession explicitly adopted scientific methods to test their ideas did the four humors lose favor—time itself was not enough.

The allure of the ancient survives even in cultures that are forward-looking. It's arguable that our rapidly advancing technology has caused people to put a lot of faith in new ideas. Those same people, however, can simultaneously respect the new and the ancient. This is manifest in the frequent marketing of products and services as having a combination of modern science and ancient wisdom. "Scientists are discovering what the ancients knew all along," we are told.

The bottom line is that time itself doesn't establish whether or not an idea is valid or a claim is true. We still need to rely on logic and evidence. Monuments are just that—icons of the past frozen in time. Our ideas, however, should occasionally be allowed to thaw.

14. Skeptics' Guide Entry: Appeal to Nature

Section: Metacognition
See also: Logical Fallacies

The appeal to nature is a logical fallacy based upon the unwarranted assumption that things that are natural are inherently superior to things that are manufactured. Additionally, it relies upon a vague definition of "natural."

> If it's natural, it has to be good for you. Well, bird shit and gravel are natural, but I won't eat them!
> —James "The Amazing" Randi

There seems to be something intrinsically positive about "nature." When we think of nature, we envision an environment that is pure, wholesome, and unmodified by the artificial forces and perversions of humans. We conjure images of lush green forests, clear blue rivers, and vast plains of tall grass.

We infuse our perception of nature with inherent virtues and redeeming qualities, while simultaneously ignoring any inconveniences equally inherent. In other words, if something is "natural," we feel it must be good for us and can do no harm.

This is the informal logical fallacy called the appeal to nature. It is distinct from the naturalistic fallacy, which refers mainly to the idea that whatever is in nature ought to be (the "is/ought" problem) as outlined by David Hume and discussed in chapter 10 on logical fallacies.

Actually, the term "naturalistic fallacy" originates with British philosopher G. E. Moore in his 1903 book *Principia Ethica*. His point that qualities associated with goodness are themselves good was a bit narrower than Hume's. In other words, if good things tend to be visually attractive, then visual attractiveness must also be good. (This is not strictly true, as attractiveness may be incidental to moral goodness.)

The appeal-to-nature fallacy is related to the naturalistic fallacy as a special case—some things that are good are natural, so all things natural must be good. The flip side of the appeal-to-nature coin is that anything that is "unnatural" is inherently tainted to some degree. In the words of surgical oncologist David Gorski:

> Basically, the idea that underlies the appeal to nature is a profane worship of nature as being, in essence, perfect, with anything humans do that is perceived as somehow being "unnatural" being viewed as, at the very least, inferior and at the very worst pure evil.

The appeal-to-nature fallacy falls apart quickly upon examination of its two main pillars. The first pillar is that things in nature tend to be good for humans. The second is that we can operationally define "natural." Nature has no special love for humanity. There is no scientific reason to think that nature is anything but indifferent to our happiness and fate. The universe simply does not care about us.

In evolutionary terms, every organism and every species is out for itself. Plants, for example, have evolved all sorts of chemicals that are designed to be poisonous to animals as a deterrent to eating them. In this sense, nature is actively trying to kill us. The single most toxic substance known (gram for gram), botulinum toxin, is all natural. Some natural poisons, in purified and carefully measured doses, can be exploited for their physiological effects. We call such poisons "drugs." But make no mistake: The plants in which we find these substances use them for chemical warfare against anything that would eat them.

Some plants did evolve edible parts, like fruit, so that animals will helpfully disperse and fertilize their seeds for them. Even then, the vast majority of what we consider edible fruits are the result of hundreds or even thousands of years of human tinkering to select against the pesticides and compounds that would make us sick. There is little "natural" about our food supply.

In order to get around the obvious conclusion that things in nature are either indifferent to us or actively out to kill us, some herbalists have taken a decidedly creationist approach. They argue that God or some benevolent being created natural things specifically to serve humanity. It takes that extreme level of hubris to conclude that nature is inherently healthful to humans.

Everyone knows that nature is not benign on some level. I don't know any adult who would walk into their backyard or the forest and eat a random plant they couldn't identify. Chances are very good you will get very sick, and you might even die.

What about the second pillar—how do we even define what is natural? There are few foods that people regularly eat that are as they occurred in nature, such as wild game, some flowers like dandelions, and wild varieties of raspberries and mushrooms. Almost everything else has been significantly transformed, often beyond recognition. Are apples natural? Well, it depends on your definition. There is no objectively right answer. Apples have been highly altered by human cultivation for millennia. What if I process those apples to make apple sauce? What if I add a little sugar to the apple sauce? How much processing makes something unnatural?

Further, what is the difference between a molecule of vitamin C that was purified from rose hips and a molecule of vitamin C that was manufactured in a lab? By definition, nothing. Atoms and molecules don't know where they came from. Their chemical and biological properties are not dependent on their source.

Despite the fact that there is no valid basis for the appeal-to-nature argument, it's often invoked by the proponents of nonscientific therapies and health treatments. Even more commonly, it rears its head when discussing food. The US Food and Drug Administration (FDA), the agency responsible for the safety and efficacy of our food, medicine, and other consumables, refers to "natural" as follows:

> From a food science perspective, it is difficult to define a food product that is "natural" because the food has probably been processed and is no longer the product of the earth. That said, FDA has not developed a definition for use of the term natural or its derivatives. However, the agency has not objected to the use of the term if the food does not contain added color, artificial flavors, or synthetic substances.

In other words, in the US anyway, there is no legal definition of the term "natural" and the FDA has decided to just look the other way if companies decide to market their products with the claim that they are natural.

Literally centuries of this type of marketing have created a health halo around the term "natural," and it is effective (if deceptive) marketing to slap the term on anything possible. Companies will often market a product as "natural" and jack up the price as a result. Without any difference in the product itself, you pay a premium for the comforting words "all natural" on the packaging.

Entire industries are based on this false reverence. The organic food industry, for example, doesn't allow the use of synthetic pesticides but does allow the use of so-called "natural" pesticides. In many cases the natural pesticides are less effective and more damaging to the environment than their synthetic alternatives. Some natural pesticides, like copper sulfate and rotenone, can be highly toxic. And yet they're not viewed with the same caution, simply because they are "natural."

When we realize that the very concept of "natural" is loosely defined, and there is no a priori reason to assume that apparently natural things are safe or helpful for humans, it becomes apparent that the "natural" label is nothing more than clever marketing. Ideology and the health halo of being "natural" substitute for actual science and evidence. And that's actually the point—to encourage you not to think or ask for evidence but to be reassured by a comforting, if ultimately meaningless, label.

15. Skeptics' Guide Entry: Fundamental Attribution Error

Section: Metacognition
See also: Correspondence Bias

The fundamental attribution error is a cognitive bias in which we ascribe other people's actions to internal factors such as personality while rationalizing our own actions as being the result of external factors beyond our control.

> We should be careful when interpreting the behavior of others. What might appear to be laziness, dishonesty, or stupidity might be better explained by situational factors of which we are ignorant.
>
> —Robert Todd Carroll

Have you ever believed in something you later came to realize was utter nonsense? I personally used to believe in extrasensory perception (ESP), alien visitation, the Bermuda Triangle, and a lot of other wacky pseudoscience. It's an almost unavoidable experience, as each of us starts our life with no information and must learn as we go. Sometimes silly or erroneous beliefs linger into adulthood. How do we explain this?

Many factors can result in someone maintaining a baseless belief. The belief may be common in your culture or social circles. Understanding why the belief is wrong may require specialized knowledge outside your experience. You may have been deceived by bad reporting or the slick presentation of misinformation. Or the belief may simply reflect a random gap in your life experience.

But what if someone else holds an obviously credulous belief? Then we may be more inclined to simply conclude they are gullible and ignorant or have a nefarious purpose for professing the belief.

See the difference? We're quick to conclude we're the victim of circumstance, and we tend to be acutely aware of external factors that influence our thoughts and behavior. For other people, however, we tend to assume they are driven predominantly by internal factors having to do with their disposition and qualities. This is the fundamental attribution error (also called the correspondence bias).

Part of this bias comes from a simple lack of information. Life is full of quirky details. You will be keenly aware of the recent events and factors that may influence your behavior, but you'll not be as aware of the details influencing someone else's. We have all seen parents lose it with their children in public. It's easy to be judgmental, to assume that they are bad parents or they lack patience. Then of course you may find yourself in public trying to manage a child in the middle of a meltdown while rushing to get three things done, and you're pissed at your spouse, who left these things for you to do. You may also be having a particularly stressful time at work. Perhaps someone close to you just died. The stress from one or more of these external factors may overwhelm what is otherwise an average or even above-average level of patience. Your normal parenting style gives way and you snap. Onlookers, unaware of all of this, simply assume you're a habitually abusive parent.

Screenwriters often take advantage of this fact. They will put you in the shoes of a character, so you can see all the external factors leading them into an unusual situation. The character will then be forced, or at least highly motivated, to behave in an abnormal way and must suffer the scornful or quizzical gazes of others. You sympathize with the character because you saw what they went through, and you condemn the onlookers for being judgmental and jumping to conclusions.

In everyday life, we are the onlookers. We are never aware of all the factors that influence someone else's behavior.

Even though I am very mindful of this bias, I still catch myself falling for it all the time. This one requires a lot of vigilance. For skeptics, perhaps the most common manifestation of the fundamental attribution error is to assume that others believe nonsensical things because they are gullible or ignorant.

We may assume someone is selling snake oil out of pure greed, but we need to resist the temptation to assume motivation on the part of others. We usually don't know and should reserve judgment. Perhaps they had a particularly powerful personal experience that happened to be very misleading, for example. Perhaps they're being influenced by someone close to them.

This asymmetry in how we view ourselves versus how we view others is important to keep in mind while you read this book or otherwise explore critical thinking. There is a tendency to apply these principles to others rather than, or at least more than, to ourselves. It's not just other people's memories that are flawed, it is also our own. Not only are other people guilty of overestimating their knowledge, we overestimate our own too.

We are all very charitable to ourselves. Imagine if we were as habitually charitable to others. With an open mind and before reaching any conclusions, we can ask other people why they did or said what they did. It's also okay to simply withhold judgment, to recognize that life is complex and we likely don't have enough information to judge a situation. We love to have opinions about celebrities or people in the news, often based on the flimsiest information. When a more complete story emerges, it's often different than what many people assumed.

Anyone who engages in social media witnesses the attribution error on a regular basis. Not only is there a tendency to assume other people's motivations; we hastily infer their arguments and positions, based upon the pigeonhole into which we think they fit. Without listening to what they are actually saying, charitably interpreting that, and giving them an opportunity to clarify their position, we risk attributing a position to them that they don't have, attacking a straw man, and then looking foolish. I've seen these exchanges rapidly degrade into mutual accusations of being a troll. There are real trolls out there, but sometimes trolling is in the eye of the beholder. Sometimes we can be the troll.

The fundamental attribution error often crops up with conspiracy theories. The apparently strange behavior of other people is offered as evidence of a conspiracy. Why did the fire marshal give the order to "pull it" in reference to World Trade Center Building 7 on 9/11? Was he giving the order to blow the building? Actually, he was giving the order to pull his people out of the building. Viewed from the outside, you can make anything strange look sinister and ascribe internal motivations to people. Knowing the circumstances is often enough to shed light on their behavior.

While it does take vigilance, this bias is surprisingly simple to correct. First, recognize that you never have all the information. Next, withhold judgment and give other people the benefit of the doubt. Imagine that the other person is a character in their own movie. Find out what the plot of that movie is before you make yourself into a cliché.

In the next chapter we will see how this principle applies to the universe, not just to other people.

16. Skeptics' Guide Entry: Anomaly Hunting

Section: Metacognition
See also: Conspiracy Thinking, Data Mining

An anomaly is something that sticks out because it doesn't seem to make sense or it appears to contradict established knowledge or scientific theory. The fallacy of anomaly hunting comes from looking for anything unusual, assuming any apparent anomaly is unexplainable, and then concluding that it is evidence for one's pet theory.

> Perhaps the problem is a psychological one of not recognizing when to stop searching for hidden causes. Nonetheless, I suggest that the study of conspiracy theories, even the crazy ones, is useful, if only because it forces us clearly to distinguish between our "good" explanations and their "bad" ones.
> —Brian Keeley

At the moment that bullets were fired into JFK's motorcade, a man can be seen standing on the side of the road near the car, holding an open black umbrella. It wasn't raining (although it had rained the night before) and no one else in Dallas was holding an umbrella. The odd behavior of this person who became known as the Umbrella Man is a true anomaly. There doesn't appear to be any obvious, or even rational, explanation for his behavior. It also seems extremely coincidental that such odd behavior would occur right where JFK was shot.

It seems superficially reasonable to conclude that the two events are connected, that perhaps the Umbrella Man was part of a conspiracy.

The real explanation is very interesting, but you could never guess it without specific knowledge. Conspiracy theorists, however, jumped on the Umbrella Man as an anomaly that could only be explained by invoking a conspiracy.

One of the commonest and most insidious bits of cognitive self-deception is the process of anomaly hunting. A true anomaly is something that can't be explained by our current model of nature; it doesn't fit into existing theories. Anomalies are therefore very useful to scientific inquiry because they point to new knowledge—the potential to deepen or extend existing theories.

For example, the orbit of Mercury couldn't be explained by Newtonian mechanics—it was a true anomaly. It and other anomalies hinted at the fact that Newton's laws of motion were incomplete in a fundamental way. Astronomers invented a number of hypotheses to explain the observed anomalies, including the existence of a hidden planet on the other side of the sun from Earth.

Eventually Einstein developed his general theory of relativity, which accounts for gravity. Einstein's equations showed that Newton's laws weren't wrong, but they were incomplete. They were a special case of a deeper reality, one in which relativistic effects from strong gravity or high relative speed are measurable. Mercury is close enough to the sun's gravitational field that relativistic effects have to be accounted for.

Pseudoscientists—those pretending to do science (or maybe even sincerely believing they are doing science) but who get the process profoundly wrong—use anomalies in a different way. They often engage in anomaly hunting, which is actively looking for apparent anomalies. They're not, however, looking for clues to a deeper understanding of reality. They're hunting in service to the pseudoscientific process of reverse engineering scientific conclusions.

The logic works like this: "If my pet theory is true, then when I look at the data I will find anomalies." The unstated major premise of this logic is that if their pet theory weren't true, then they wouldn't find them. This is naive, however, because apparent anomalies are nearly ubiquitous.

Another component of this line of argument is the broad definition of "anomaly." What, exactly, counts as an anomaly? Is it anything that seems weird to you, or only phenomena that cannot be explained even after exhaustive searching? Using a broad definition and looking at a large data set, it's almost guaranteed that one will find apparent anomalies. Finding them proves nothing. Even in a world in which the JFK assassination was carried out by a lone gunman without any deeper conspiracy, someone poring over all the events surrounding the assassination would find many things that seem strange or out of place.

The assumption that anomalies must be significant rather than random is an error in the understanding of statistics, a form of innumeracy. It may also incorporate the lottery fallacy, which involves asking the wrong question. The name "lottery fallacy" is based on the most common illustrative example. If John Smith wins the lottery, our natural tendency is to consider the odds against John Smith winning (usually hundreds of millions to one). However, if your concern is whether or not the lottery is truly random or whether some supernatural force needs to be invoked in order to explain the results, then the correct question is, "What are the odds that anyone would have won?" In this case they're close to one to one (at least over a few weeks' time).

The fallacy comes when we confuse a priori probability with posterior probability—once you know the outcome, asking for the odds of that particular outcome. This is perhaps more obvious when we consider the odds of someone winning the lottery twice. This occurs regularly, and when it does, the press often reports the odds as being astronomical. They're usually erroneously considering the odds of one particular person winning on two successive lottery tickets, and they calculate the odds of John Smith winning twice rather than the odds of anyone anywhere winning twice (the odds are actually quite good and match the observed rate).

I keep using the phrase "apparent anomaly" because something may look initially like an anomaly but upon closer inspection one can find a prosaic explanation. In real science an anomaly is only declared after exhaustive efforts to explain it within existing theories fail.

In 2011, physicists in the OPERA experiment thought they detected neutrinos (elementary particles that very weakly interact with matter) traveling faster than the speed of light. That would be a genuine and profound anomaly, as Einstein's theories clearly hold that nothing can move faster than light, and so far, that theory has held up. The scientists, however, didn't declare the death of relativity. They looked exhaustively for anything that could explain the anomaly without invalidating Einstein. When they failed, they appealed to their colleagues in the broader scientific community for help.

Others replicated the experimental setup with different equipment and found that neutrinos do not travel faster than light. Eventually the OPERA physicists discovered a faulty cable in the timing mechanism—the anomaly was a technical failure. Einstein's theories lived another day.

What pseudoscientists do is look for *apparent* anomalies—things that can't be immediately explained or (even worse) are just coincidences. There is no reason to assume that your instinct is a good guide for what should happen during a very unusual event, such as a passenger jet plowing into the side of the Pentagon. Often the conclusion that something is an anomaly derives from a lack of familiarity or expertise. We may, for example, be unfamiliar with conditions in exotic environments, and something may seem like an anomaly simply because you lack the specialized knowledge (scientific, technical, historical) to know the true explanation. There is often a lack of humility in the way pseudoscientists and conspiracy theorists declare a complex event or feature to be an anomaly based only on their limited knowledge. They also often look at the edges of detectability, where data become fuzzy and anomalies are easier to imagine. Think of the blurry photos of Bigfoot or flying saucers, with believers looking at details smaller than the resolution of the images and declaring them to be real. That blob you can't identify doesn't prove the existence of Bigfoot; it might just prove the existence of your thumb covering the lens.

Once an apparent anomaly (broadly defined) is found in the data, Jonny Pseudoscientist will then tend to commit a pair of logical fallacies. First, he confuses unexplained with unexplainable. This leads him to prematurely declare something a true anomaly without first exhaustively trying to explain it with conventional means. Second, he uses the argument from ignorance, saying that because the anomaly cannot be explained, his specific pet theory must be true. Many people don't recognize Venus in the night sky, and it may look unusual because of atmospheric conditions. "I don't know what that hazy object in the sky is," says the believer, looking right at Venus, "therefore it must be an alien spacecraft."

Examples of Anomaly Hunting

As a skeptic in the trenches, as it were, I have encountered countless, often humorous, examples of anomaly hunting. Perhaps my favorite is from a "psychic" lecture we attended in our early days as activist skeptics. The conversation turned to alien visitations, prompting one attendee to proclaim that crop circles were compelling evidence. "Some of them are perfect circles. How can you make a perfect circle?" To her this was an anomaly requiring an extraordinary explanation.

"Um, what about compasses?" I asked, trying (probably unsuccessfully) to be delicate. "Remember those things we used in grade school to draw perfect circles?" A standard technique of crop circle makers is to use a stake in the middle of the circle with a rope tied to it connected to a board that is then used to stomp down the wheat or other crop—a makeshift compass.

Ghost Anomalies

Almost the entire field of ghost hunting—looking for evidence that the earth is haunted by the spirits of the departed and other vaporous beings—is a giant exercise in anomaly hunting. Even before the days of reality TV, self-styled ghost hunters would go into allegedly haunted houses and essentially look for anomalies, sometimes with scientific equipment they didn't know how to use.

If they encountered a cold spot in the house, they overlooked any simple explanations for the drop in temperature and proclaimed the phenomenon "ghost cold." They may perhaps make a perfunctory effort to account for other explanations (there was no window open) to give the illusion that they're doing meaningful investigation.

Ghost hunters also like to measure electromagnetic fields (EMFs). What frequency should they try to detect? Whatever their equipment happens to measure will do just fine. They walk around the property with their EMF "Specter Detector" and declare any activity they find as anomalous and therefore evidence of ghosts. They seem to be uninterested in the fact that EMFs are ubiquitous. We have electrified our world, and flowing electricity generates EMFs. Even a piece of iron will set off an EMF detector. The real challenge would be to find a location that has no detectable EMF.

This ties in with our discussion of confirmation bias in chapter 12. I've never seen a ghost hunter explore a house that wasn't supposed to be haunted to get some control data on cold spots, EMF activity, and photographic artifacts like "ghost orbs" (i.e., lens flares). They specifically seek anomalies to confirm their prior belief in ghosts.

The JFK Assassination and the Umbrella Man

Let's get back to the Umbrella Man, a real anomaly and an incredible coincidence. The simple fact is that people do strange things for strange reasons. There's no way to account for all possible thought processes of every person involved in an event. Often the actions of others seem unfathomable to us. Our instinct is to try to explain the behavior of others as resulting from mostly internal forces, and we tend to underestimate the influence of external factors, as we discussed in the last chapter. We also tend to assume that the actions of others are deliberate and planned rather than random or accidental.

Conspiracy theorists have essentially formalized the tendency to assume agency, deliberateness, and sinister motivations in the unique details of events. When anomalies, like the Umbrella Man, are inevitably found, it's assumed that they are evidence of a conspiracy. Remember the lottery fallacy—conspiracy theorists tend to ask, "What are the odds of a man standing with an open umbrella right next to the president when he was shot?" Rather, they should be asking, "What are the odds of anything unusual occurring in any way associated with the JFK assassination?"

What, then, *is* the explanation for the seemingly bizarre actions of the Umbrella Man? The man (Louie Steven Witt) was asked to come forward and explain his actions before Congress, and he testified that the umbrella was a protest of Joseph Kennedy's appeasement policies when he was ambassador to the Court of St. James's in 1938–39. The umbrella was a reference to the umbrella often carried by British prime minister Neville Chamberlain. This is not as random as it may seem. An open umbrella was a common protest of appeasement policies. According to the Historical Society:

Umbrella protests first began in England after Chamberlain arrived home from the conference carrying his trademark accessory. Wherever Chamberlain traveled, the opposition party in Britain protested his appeasement at Munich by displaying umbrellas. Throughout the 1950s and 1960s, Americans on the far Right employed umbrellas to criticize leaders supposedly appeasing the enemies of the United States. Some politicians even refused to use them for that reason. Vice President Richard Nixon banned his own aides from carrying umbrellas when picking him up at the airport for fear of being photographed and charged as an appeaser.

In the early 1960s there were likely still people around who were angry at any attempts to appease Hitler and the Nazis prior to the start of WWII. The umbrella isn't such a random detail after all. Unless you knew this very specific fact of history, you would have no chance of figuring out why Louie Steven Witt was standing next to JFK's path on that day holding an open umbrella. It would be an anomaly to you.

Moon Landing Hoax

There are those who claim that we never sent astronauts to the moon—that the entire thing was an elaborate hoax by the United States, meant to intimidate our rivals with our spacefaring prowess. As is typical of most grand conspiracy theories, there is no actual evidence to support the claim. None of the many people who would by necessity have been involved have come forward to confess their involvement. No government documents have come to light, no secret studios have been revealed. There's no footage accidentally revealing stage equipment.

What the moon hoax theorists have is simply another example of anomaly hunting. They point to the lack of stars in the moon's sky, the visibility of astronauts with the sun behind them, and the non-parallel shadows of different objects lit by the same source. These all derive from the fact that the moon is an unfamiliar location for photography, and the apparent anomalies all have straightforward explanations. The stars are simply washed out by daylight (the sky looks black because there is no atmosphere). The landscape is uneven, hence the nonparallel shadows. And the moon's surface is highly reflective, providing the fill light necessary to make the front of astronauts visible even when the sun is behind them.

Some moon hoaxers claim they can see the American flag waving in the breeze, even though there is no atmosphere on the moon. The lack of an atmosphere, however, allows the flag to flap for a long time once moved by an astronaut. Without air to dampen the oscillations, they continue.

One more technical point often raised is the claim that the astronauts would have been killed by radiation from the Van Allen belts and cosmic rays. That was the claim made by Bill Kaysing, one of the early hoax proponents, in his self-published book *We Never Went to the Moon: America's Thirty Billion Dollar Swindle*. However, this is simply untrue. The *Apollo 11* astronauts received a total of only about 11 millisieverts of radiation (a lethal dose is around 8,000 millisieverts, or 8 sieverts). NASA's limit for lifetime exposure to radiation is 1 sievert, about what astronauts would get on a one-way trip to Mars. What saved the *Apollo* astronauts was the brief total overall time exposed to radiation. The longest *Apollo* missions lasted less than thirteen days.

Beyond the lack of evidence for a conspiracy, and the non-anomaly anomalies, there is a huge plausibility problem with the moon hoax conspiracy. Why haven't other countries, like Russia, ever come forward with evidence that their tracking of the mission doesn't support NASA's story? And where did all the moon rocks come from? (Don't say meteorites; those would look different due to their travel through the atmosphere, and we would never have found so many from the moon!) It's unlikely the US could have pulled off the hoax. Some have argued it would have been easier to just send astronauts to the moon than to execute such a deception.

There is also undeniable evidence of human artifacts on the moon. Anyone with the equipment and knowledge can fire up a laser and bounce the beam off a corner reflector left on the lunar surface.

Conspiracy theorists long have asked why, if we went to the moon, there are no pictures of the landing sites from telescopes. Well, telescopes don't have the resolving power, but moon probes do. The Lunar Reconnaissance Orbiter has taken pictures of *Apollo* landing sites, showing equipment left behind and the tracks made by the astronauts. Of course, these pictures are just dismissed as fabricated.

The evidence is overwhelming and undeniable that NASA sent multiple missions to the moon, leaving behind footprints and equipment and bringing back rocks and history. But there are those who deny it, and they largely use anomaly hunting to justify their outlandish conspiracy theories.

Flat-Earthers

This is perhaps the most surprising recent phenomenon—people in the twenty-first century, in developed and educated parts of the world, who actually think that the Earth is flat. And yes, there are people who really think this—they are not just punking the rest of us. Interviews suggest that the modern flat-earthers, at least those motivated enough to attend a flat-earth convention, are first and foremost conspiracy theorists. But they justify their belief with an endless series of apparent anomalies.

Flat-earthers point out, for example, that it is possible to see cities tens of miles away with a good pair of binoculars, although the alleged curvature of the surface of the Earth would make the horizon much closer. When you are standing on the ground with your eyes at a height of five feet seven inches (1.7 meters), the horizon is almost three miles (about 4.7 kilometers) away. Standing on a 300-foot tower expands the horizon to more than 21 miles (about 34 kilometers), assuming a level terrain. The distance you can see depends greatly on your height, and even being on a small hill can significantly extend your range.

But this is not enough to explain all examples. There is another phenomenon at work—refraction through the atmosphere. The air tends to bend light like a lens, so that it partially curves downward toward the surface of the Earth. This can significantly increase the range of one's vision. You are literally seeing over the horizon.

Of course, this was sorted out in the nineteenth century. But flat-earthers are more than a couple centuries behind on the science—they are a couple thousand years behind. That is when some basic observations showed that ships sink below the horizon, that the shadow of the Earth on the moon is always a curve, and that vertical sticks will leave shadows of different length in different locations because of Earth's curvature. But I guess those ancient observations are all a NASA conspiracy.

My favorite example of evidence that destroys the flat-earth nonsense is simple and personal. On a trip to Australia I took several pictures of the moon, which is upside down compared to my usual perspective in the Northern Hemisphere. This is because I was literally upside down compared to someone in the opposite hemisphere—because Earth is a sphere. There is no explanation for this simple observation in the flat-earth model. How many millions or even billions of people have seen this evidence firsthand? How deep would this conspiracy need to go?

What these few examples demonstrate is that by using anomaly hunting and other mental tricks, humans have an almost unlimited capacity to deny the obvious and convince themselves of almost any claim, even one that can be debunked just by looking up at the sky.

17. Skeptics' Guide Entry: Data Mining

Section: Metacognition
See also: Anomaly Hunting

Data mining is the process of sifting through large sets of data looking for any possible correlation, many of which will occur by chance. While this is a legitimate method for generating hypotheses, such data are not confirmatory and the method is easily abused.

> Torture the data, and it will confess to anything.　　　—Ronald Coase

---

In February 2002, an Australian financial services group, Suncorp-Metway Ltd., released the results of a study looking at 160,000 driving accident claims. They correlated the claims to astrological sun signs and found that Gemini, Taurus, and Pisces were the most accident prone, while Capricorn, Scorpio, and Sagittarius were the least. But don't worry, insurance companies aren't using astrological sun signs to determine insurance rates (yet).

The results of this study, and studies like it, were promptly praised by astrologers as evidence their arcane art is real, and entirely ignored by everyone else as obvious nonsense. What this study *actually* shows is the pitfall of data mining to look for potential correlations that may ultimately mean nothing.

The problem of data mining is both common and often subtle and missed. It's rooted in both the nature of human brain function and a common logical fallacy. The former is that of pattern recognition and the latter the fallacy of confusing correlation with causation.

It's critical to recognize that the world contains many appearances of patterns that don't reflect underlying reality. We are confronted with a tremendous amount of information about the world around us. There exist true patterns in this information, such as the recurrence of the seasons, but much of it is random. Random information, furthermore, is likely to contain patterns by chance alone. As Carl Sagan eloquently pointed out, randomness is clumpy. We tend to notice clumps. They stick out as potential patterns.

We inherently use a two-step process in evaluating the data that surround us. First we have hyperactive pattern recognition—a tendency to see any possible patterns, erring on the side of false positives. This would serve the purpose of minimizing the chance of missing real patterns that may be important. Then we evaluate those potential patterns to see if they are in fact real—do they make sense and do they comport with what we already know (psychologists call this process "reality testing")? Our brains evolved to capture everything and then weed out the fake patterns, but we are better at step one than step two.

This is why we need science. Science is partly the task of separating those patterns that are real from those that are accidents of random clumpiness. Science is formalized reality testing.

Data mining refers to the process of actively looking at large sets of data for any patterns (correlations). But since random data are clumpy, we should expect to see accidental correlations even when there is no real underlying phenomenon causing the correlation. This often occurs in the context of statistical analysis of compiled data, whether gathered in a study or from databases of biographical or other information. Mining large sets of data dramatically increases the probability of seeing apparent patterns.

One methodological pitfall of data mining is not determining ahead of time what potential correlations are being searched for—so any correlation counts as a hit. Functionally this is the same as just noticing a pattern even though you weren't actively looking. For example, a doctor may notice that they've seen a cluster of a particular rare disease recently. Or someone may notice that they typically have a bad day at work on Tuesdays. We are all, in effect, mining the data of the world around us every day, subconsciously looking for patterns.

Such patterns may be real—may reflect an actual underlying cause—but they're more likely to be random clumpiness. The reason for this is the vast number of potential correlations that could happen by chance. There are so many, we should expect to be confronted with patterns, whether we are actively or passively mining the data, many times a day.

From a statistical point of view, you cannot simply calculate the odds of a particular correlation occurring by chance alone. That particular correlation may be incredibly unlikely—with odds of thousands or even millions to one against its occurrence by chance. This may seem compelling, but it's misleading because the wrong question is being asked (this is the lottery fallacy we discussed previously). That question assumes you specified the pattern you were looking for ahead of time. If you didn't, then the real question is: What are the odds of *any* correlation occurring in this data set?

Therefore, any pattern or correlation that is the result of searching (again, whether actively or passively) a large data set for any potential correlation should be viewed as only a *possible* correlation requiring further validation. Such correlations can be used as the beginning of meaningful research (not the conclusion). Legitimate scientific procedure is to then ask the question, Is this correlation real? The doctor in the example above should ask, "Is this a real outbreak of this rare disease or a random cluster?" To confirm a correlation, you must ask ahead of time, What is the probability of this specific correlation occurring? and then test a new or fresh set of data to see if the correlation holds up. Since you're looking for a specific predetermined correlation, now you can apply statistics and ask what is the probability of that correlation occurring by chance.

There is still another statistical pitfall to avoid, however. When looking at the new data set, you can't include the original data that made you notice the correlation in the first place. The new data have to be entirely independent. This is to avoid simply carrying forward a random correlation that was mined out of a large set of data.

Remember that astrological study about driving accidents? That was never truly replicated, but others have looked at similar data. In 2006, Lee Romanov, president of InsuranceHotline.com, looked at 100,000 insurance claims and found that Libra, Aquarius, and Aries were the worst drivers, while Leo, Gemini, and Cancer were the best—completely different results from the Suncorp-Metway study—as if the correlations were all entirely random.

Data mining errors occur in science all the time. They're most prominent in epidemiological studies, which are basically studies that sift through large data sets to look for correlations. To be fair, most of the time they're presented properly as preliminary data that needs verification. Such correlations are then run through the mill of science and either hold up or they don't. But this process may take years. Meanwhile, the media often reports such preliminary correlations as if they were conclusions, without putting them into proper scientific context. Scientists and the institutions that support them are also often to blame, for example sending out press releases to announce an interesting new health correlation before it has been verified. Because of this, the public is treated to an endless parade of possible correlations coming from scientists and is largely unaware of what role such data play in the overall process of science.

While data mining is a nuisance in mainstream science (competent scientists and statisticians should be well aware of it and know how to avoid it), it's endemic in pseudoscience. Astrology is an excellent example. Studies alleged to support astrology are built almost entirely on data mining, and they always evaporate when valid statistics or independent tests are applied.

It's also a huge factor in our everyday interaction with the world around us. We are compelled by the patterns we see. They speak to us. Our "common sense" often fails to properly guide us, as we tend to err hugely on the side of accepting whatever patterns we see. The only way to navigate through the sea of patterns is with the systematic methods of logic and testing that collectively are known as science.

Otherwise we might be fooled into thinking that subjective patterns of stars in the sky might make us more or less likely to get into a car accident. I think I'll stick to wearing a seat belt.

18. Skeptics' Guide Entry: Coincidence

Section: Metacognition
See also: Innumeracy

The term "coincidence" refers to the chance alignment of two variables or events that seem to be independent, especially if it seems as if the occurrence defies the odds.

> Coincidence is the science of the True Believer.　　　　　　—Chet Raymo

Have you ever had a dream that eerily seemed to "predict" the events of the next day? Let's say you dream of a friend you haven't seen in ten years, and the next day, out of the blue, they call you. Is that too much of a coincidence to be just a random occurrence? Is this evidence that you are psychic? (Spoiler: no.)

Coincidences (the chance alignment of two or more variables), even remarkable coincidences, are not all that surprising. In fact, most are inevitable occurrences with no special significance at all. Despite this, we tend to feel as if there must be some explanation for apparently amazing coincidences. There is, but the explanation, like many things we've been discussing, is internal not external.

There are many simple reasons for the common misinterpretation of coincidences. We've already talked about the human tendency to see patterns, how we subconsciously (or even consciously) mine large sets of data looking for correlations, about our tendency to see deliberate causes where none exist, and our bias toward data that confirm our existing beliefs. We also have a poor innate grasp of probability and the laws regarding truly large numbers, and this causes us to be more amazed with coincidences than we should be. Hence the easy jump to a special or even metaphysical explanation. Understanding simple probability can demystify many apparent coincidences.

What are the odds of two people sharing the same birthday in a room containing twenty-three people? Often people will guess one in thirty or more. Surprisingly, it's only about one in two. If you have seventy-five people in a room, the probability of two of them sharing a birthday is 99.9 percent. Not knowing this has caused many people to conclude that they share some special link or that a supernatural force brought them together. Rather, there's just a mismatch between our intuition about probability and the actual math.

This mismatch is part of a more general phenomenon of innumeracy, or our general lack of intuition for probability and randomness. Here's another fun example: If I am flipping a fair coin, which of the following sequences is more likely: HHTHTTTHT or HHHHH-HHHH? Most people would choose the former, but the real answer is that they are equally likely. Each flip is independent, with a 50 percent chance of an H or T on every one. Matching either sequence exactly is therefore equally likely, but our intuition says that the more random-looking sequence must be more probable.

You can also think of coincidences this way: There are over eight million people living in New York City. That means a 1-in-8,000,000 occurrence should happen to someone in the city every day.

Not only is our intuition about probability terrible—and it's even worse when we deal with large numbers—our memory kicks in to further distort things. Dramatic experiences leave a much more indelible mark on our minds than any other. Therefore it's only natural to remember unusual experiences and forget routine ones.

Returning to our friend who calls soon after you think about him, this event becomes much less striking if we consider how many times we've thought of friends who did not then call. Further, how many details are there in all of your dreams in a typical month? How many details do you encounter in a typical day? That any two of those details would seem to align every now and then isn't remarkable. It would be truly remarkable if such things never happened.

Over the course of a lifetime you should, by chance alone, accumulate several extremely unlikely coincidences—unlikely if considered in isolation but not as part of the countless experiences that make up your life. Consider again the eight million people living in New York City, or the billions living on the planet. There must be some really amazing stories out there, and those are the ones that will be told and remembered.

Our memories also tweak such occurrences so that they seem even more outstanding. Once we have a story, our memories will change to make the story more dramatic or to reinforce the central theme. Perhaps you dreamt of your old friend James, but when Fred called the next day, your foggy memory of the dream changed to fit this new information, and suddenly you remember dreaming of Fred. In the retelling, the story of your prophetic dream may become even more dramatic, with the addition of extra details that seem uncanny.

Alleged psychics exploit this phenomenon. A common ploy (often called the Jeane Dixon effect) is to make dozens of predictions knowing that the more that are made, the better the odds that one will hit. When one comes true, the psychic counts on us to conveniently forget the 99 percent that were way off, making the correct predictions seem much more compelling than they really are. This is a conscious or deliberate form of subjective validation, or, put more simply, fraud.

Superstitions

The common belief in superstitions is reinforced by these phenomena of pattern seeking, faulty memory, and a poor intuitive grasp of probability. Just add a dollop of magical thinking and you have a superstition. Unfortunately, it may be no more than a coincidence that the Patriots won the last two games you watched while wearing your blue sweater. Belief in superstitions is further driven by both a need for control and feeling a lack of control. Research shows that the more people feel at the mercy of forces beyond their control, the more they reach for superstitious beliefs to provide the illusion of control. This can be extremely counterproductive, as they may put their efforts into magical rituals rather than, say, studying for an exam.

It's not my contention that all coincidences have no meaning and should be ignored—indeed, truly unlikely events may have some underlying significance, and the search for their causes could be a useful endeavor. However, the vast majority of what we experience turns out to be much more probable than it appears, if analyzed critically. It's important, therefore, not to trust your intuitions about probability. Do the math... or look to see if someone has already done it for you.

It really is just a coincidence that your friend from the dream called the next day.

The Monty Hall Problem

The Monty Hall problem is a logic/probability puzzle that's famous for being simple yet so counterintuitive that most people get it wrong, even well-known mathematicians.

Here's the typical setup of the puzzle:

A contestant in a game show is presented with three doors. Behind one is a prize, like a new car. The other two conceal something less desirable, like goats.

The contestant is asked to pick one of the three doors, and they will win the prize behind that door. Once they pick, the omniscient host (well, at least he knows what is behind each door) will open one of the two doors the contestant did not pick, a door that invariably reveals a goat.

Here's the question—would changing their choice to the other remaining door increase the contestant's chances of winning the prize (assuming they desire the new car more than a goat)?

Think about what you would do in this situation while I offer some background.

If you have more decades of TV watching under your belt than you'd care to admit, you probably remember Monty Hall and his hosting gig on the game show *Let's Make a Deal*, which was popular in the 1970s. The show would often offer real prizes or gag prizes (like goats) behind doors and curtains that contestants had to select. The origin of the puzzle was not this game show however. It was originally posed by Steve Selvin in a letter to the *American Statistician* in 1975.

So why did this even become a thing? Probability puzzles are not exactly in high demand in our culture. I don't suspect that Selvin's letter to the editor went viral in the seventies; nor did the game show itself popularize the eponymous puzzle. The immense popularity of the Monty Hall problem can be directly traced to a genius writing for *Parade* magazine.

Marilyn vos Savant was listed in the *Guinness Book of World Records* for four years as having the highest IQ—228. Her column, "Ask Marilyn," answered countless questions over the years, but none of her answers produced the outrage seen when she described the proper solution to the Monty Hall problem. The real surprise was not the thousands of angry letters themselves but the high percentage of them that were from mathematicians and PhDs.

For example:

You blew it, and you blew it big! Since you seem to have difficulty grasping the basic principle at work here, I'll explain. After the host reveals a goat, you now have a one-in-two chance of being correct. Whether you change your selection or not, the odds are the same. There is enough mathematical illiteracy in this country, and we don't need the world's highest IQ propagating more. Shame!—Scott Smith, Ph.D. University of Florida

("Ask Marilyn," *Parade*, September 9, 1990)

The correct answer, which vos Savant gave and which caused such a hubbub, was that the best strategy was to change your choice of door once a goat has been revealed.

If you're new to all this, that probably doesn't sound at all correct. You are probably thinking, "There are two closed doors. One has a prize and the other a goat. That means it's a 50/50 chance. Switching can't improve the odds?"

This kind of thinking is called the "equal probability" assumption, and it's a common intuition that's often incorrect (see Ruma Falk, 1992). We have a strong desire to spread out probabilities equally among all the unknowns, and that simply doesn't apply to the Monty Hall problem. That's because when the goat was revealed, it gave us new information that we didn't have when we made our initial choice, and that insight makes all the difference.

Say you pick door #1 and the goat is revealed behind door #2.

1. The door you picked clearly has a 1/3 chance of hiding the prize. That means the other two must have a 2/3 chance. If you decide to switch your door to #3, you are in essence picking both #2 and #3 at the same time, since you have the new information that a goat was behind #2. This switches your chance of winning from 1/3 to 2/3.
2. Imagine if the game had 100 doors hiding one prize and 99 goats. If you select door #1 and the host opens 98 doors showing 98 goats, would you switch your choice to the one remaining door? The answer seems obvious. The three-door version is the same thing, just with fewer doors.

3. If you're still not convinced and can't wrap your head around it, that's okay. Probability doesn't come naturally to us humans. Keep in mind, though, that computer simulations running this scenario thousands or millions of times clearly show that switching doors increases the odds exactly as I've described. If you still refute this answer in the face of evidence like that, maybe you're better off with the goat.

SCIENCE AND PSEUDOSCIENCE

One classic pseudoscience I have tangled with frequently is extrasensory perception (ESP), the notion that our brains contain the ability to read the minds of others, to see the future, or to view remote locations. Researchers who take this notion seriously call these abilities "anomalous cognition." Dealing with the claims of ESP proponents can be challenging for skeptics because ESP comes right up against the demarcation between science and pseudoscience. There are serious researchers conducting rigorous experiments and claiming to have significant results. I have often been confronted by believers who were well versed in this research and who challenged me with numerous citations.

The problem is, none of this research rises to the point of convincing evidence that ESP is real. But arriving at this conclusion requires a thorough understanding of what compelling scientific evidence is, and how an entire field of scientific research can fall short. Mainstream science often has the same type of shortcomings found in ESP research, although sometimes not as bad, and separating good science with some quality issues from fatally flawed pseudoscience can be tricky. ESP research sits smack in the middle.

To illustrate this, let's consider the 2011 research of psychologist Daryl Bem, who published a series of ten experiments that, he claimed, showed evidence that people can "feel the future" through extrasensory perception (ESP). On close inspection, his research is clearly full of pseudoscience, but it was accepted and published in a mainstream journal, prompting discussions about the state of modern science and the nature of pseudoscience.

Bem took standard psychological experiments and simply reversed their order. For example, he tested subjects' ability to recall words and then let them practice with a set of words. He claims that words were easier to recall if they were studied afterward—so the effect of studying would have had to travel back in time to influence performance.

ESP proponents were thrilled. Bem's studies followed rigorous scientific protocols, were published in a peer-reviewed journal, and showed impressive statistical significance. Skeptics were not impressed—reversing the arrow of time is as close to impossible as you get in science, and at the very least it is an extraordinary claim. It's much more likely that Bem simply made an error than that he overturned a fundamental scientific principle.

The psychological and wider scientific communities, however, were worried. Bem followed all the rules, and yet his results could not be true. That implies there might be something wrong with the rules—with science itself.

Everyone hoped that the conflict would be resolved with replications, which are the ultimate test of a scientific finding. If the effect is real, then anyone can find it. Most attempts at replicating Bem's studies (which, to his credit, he made easy by providing documentation for all of his methods) were negative. However, a few ESP proponents were able to replicate his results. In the end, nothing was resolved— each side had its replications to point to: ESP proponents argued that skeptics inhibited ESP through their negative vibes; ESP skeptics claimed that true believers were conducting biased research.

In an attempt to resolve the dispute, both sides agreed to a rigorous protocol for replicating Bem's research, to be carried out by believers and skeptics alike. If Bem's original studies were valid, both sides should find the same result with these carefully spelled-out procedures. Bem and his colleagues, however, speculated that skeptics might still block the ESP effect with their own negative ESP.

The results of these rigorous studies were presented at a parapsychological meeting in the summer of 2016. All the replications, both believer and skeptic, were completely negative. It's clear from Bem's own description of his research methods that he was doing it wrong. When those errors were fixed, the ESP effect vanished. (Pro tip: If a phenomenon vanishes under rigorous scientific protocols, it is probably not real.)

That should have been the end of the story, but in response to the negative data, Bem and his colleagues did a reanalysis and were able to coax some apparently positive correlations from the data. Whether or not intentional, when Bem reintroduced some of his original errors, an illusion of positive results was created.

This entire saga is a microcosm of science and pseudoscience, the pitfalls and shortcomings of modern science, the role of scientific rigor, and the need to consider plausibility. It reflects why we set our thresholds for compelling evidence where we do, and the many ways in which believers go wrong.

In the following chapters we'll explore the philosophical underpinnings of science, the nature of knowledge, and how we know what we know. With that as background, we will then delve into the methods of science and how easily they can be affected by bias and error. We will eventually see exactly what mistakes Bem made, why the final outcome was entirely predictable (at least to experienced skeptics), and how we can avoid such error in the future. Along the way we'll explore many different kinds of pseudoscience and types of scientific, critical thinking, and we'll look into philosophical errors.

Essentially this is the information Bem desperately needed in order to avoid becoming the poster child for pseudoscience that he is.

19. Skeptics' Guide Entry: Methodological Naturalism and Its Critics

Section: Science and Pseudoscience
See also: Materialism, Postmodernism

Methodological naturalism is the philosophical basis for scientific methodology that proceeds as if the universe follows natural laws in which all effects have a natural cause.

> Being a scientist and staring immensity and eternity in the face every day is as grand and inspiring as it gets.
> —Carolyn Porco

I think the most parsimonious explanation for all scientific knowledge is that there is an actual real physical universe out there. Reality is really real. This isn't all just a dream or a construct of our minds, either individually or collectively. The universe exists and it has rules that it always follows. Every effect has a cause.

I don't think there is anything else. There is nothing beyond the natural. Well, at least, if there is anything beyond the natural, we can't really know about it, by definition. Further, I don't think there is any compelling reason to hypothesize anything beyond the natural. Science must operate within the framework that all of this is true.

This position is called methodological naturalism, and it's a cornerstone of science and skepticism. It's an important concept to understand for several reasons. First, it is vital to be aware of implicit philosophical positions that you take: If you advocate for science, you need to understand the philosophical underpinnings of science. Second, there are those (as we'll see below) who oppose or deny science by attacking those philosophical underpinnings. You can't recognize this strategy or defend against it if you don't know the philosophy.

So let's dig deeper.

Materialism

The starting point for the philosophy of science is materialism. Put simply, it is the philosophical position that all physical effects have physical causes. There are no nonphysical or nonmaterial causes of physical effects. Historically, materialism has been defined as the philosophical belief that matter is all there is—the only type of substance that can be said to exist in nature. Materialism stands specifically in opposition to dualism and other philosophies that posit a spiritual or nonmaterial aspect of existence. For example, some ESP proponents believe that our minds exist in a spiritual realm in addition to the physical one, and it is in this spiritual realm that our "anomalous cognition" occurs.

Any viable modern definition of materialism must include energy, forces, space-time, dark matter, and possibly dark energy—and anything else discovered by science to exist in nature. In this way materialism is really just a manifestation of naturalism, the philosophy that says that nature (in all of its aspects) is all there is—there is nothing supernatural. In fact, the term "materialism" as broadly defined doesn't have much applicability today and is mostly used in the narrow context of how it applies specifically to consciousness (the notion that consciousness is what the brain does), where it stands in opposition to dualism (the belief that consciousness is a nonphysical thing unto itself).

Another way to look at it is this: If scientists discover that something previously believed to be supernatural exists, then it will become natural, as it will have been demonstrated to be part of nature. Some argue that materialism is therefore a useless tautology (a definition that references itself), but this misses the point. The definition of materialism is really more about method; it's about testable causes that we can investigate scientifically. "Supernatural," therefore, is untestable magic.

All of this is important because there is a broad "anti-materialist" movement that includes intelligent design (ID), dualism, and various healing pseudosciences. They generally don't like the fact that modern science has not validated their spiritual beliefs, so they attack the very basis of science itself. While they openly attack "materialism," they are more accurately defined as "anti-naturalist"—they don't believe that nature is all there is, and they want science to validate their faith in the supernatural. But I guess for propaganda purposes it is better to be against materialism than against nature.

Methodological Naturalism

The real debate is whether or not science is required to follow methodological naturalism (which it clearly does). Philosophical naturalism is the metaphysical belief that nature is all there is, while methodological naturalism is proceeding *as if* nature is all there is while remaining agnostic about the deeper metaphysical question. Methodological naturalism posits that nature is all that we can know, regardless of whether or not it's all there is (which by definition we cannot know). For example, it's possible that the entire universe was created five seconds ago by an omnipotent being who made it appear exactly as if it had evolved naturally over billions of years. There is no possible method that could discern such a reality from a universe that actually evolved over billions of years, so such a notion is outside the realm of methodological naturalism and science.

Does science require methodological naturalism? Yes. That philosophical fight was fought in centuries past—and the naturalists won. The fight is over. But the anti-naturalists want to resurrect it, and since they can't win in the arena of science, they want to fight in the arena of public opinion and then in the legal and academic realms.

The key feature of methodological naturalism that makes it essential to science is that its ideas are testable. Nonnatural causes are by definition non-falsifiable, and therefore scientific methods cannot act upon them. It's like the now famous cartoon of the mathematician writing out a very complex equation but in one part simply writes, "Then a miracle occurs." His colleague points to the comment and says, "I think you should be more explicit here in step two."

Science cannot say, "And then a miracle occurs." There's no way to do an experiment or make an observation that can test a miracle. Miracles, by definition, defy natural forces or explanations. They cannot be *constrained*, which is a necessary feature of any hypothesis that can be falsified.

Science doesn't say—and cannot say—that all life on earth was not created in an instant by an all-powerful designer. It is agnostic toward such a belief. It can only say that such a hypothesis is outside the realm of science, because it cannot be tested scientifically. That's methodological naturalism.

As evidence that the testability feature of methodological naturalism is the real issue at stake here is the *Kitzmiller v. Dover* trial involving the teaching of ID in public school science classes. Here is a long excerpt from the federal district court's decision, which makes it clear ID proponents want to get out of the testability criterion for science.

NAS [the National Academy of Sciences] is in agreement that science is limited to empirical, observable and ultimately testable data: "Science is a particular way of knowing about the world. In science, explanations are restricted to those that can be inferred from the confirmable data—the results obtained through observations and experiments that can be substantiated by other scientists. Anything that can be observed or measured is amenable to scientific investigation. Explanations that cannot be based upon empirical evidence are not part of science." (P-649 at 27).

This rigorous attachment to "natural" explanations is an essential attribute to science by definition and by convention. (1:63 (Miller); 5:29-31 (Pennock)). We are in agreement with Plaintiffs' lead expert Dr. Miller, that from a practical perspective, attributing unsolved problems about nature to causes and forces that lie outside the natural world is a "science stopper." (3:14-15 (Miller)). As Dr. Miller explained, once you attribute a cause to an untestable supernatural force, a proposition that cannot be disproven, there is no reason to continue seeking natural explanations as we have our answer. Idem.

ID is predicated on supernatural causation, as we previously explained and as various expert testimony revealed. (17:96 (Padian); 2:35-36 (Miller); 14:62 (Alters)). ID takes a natural phenomenon and, instead of accepting or seeking a natural explanation, argues that the explanation is supernatural. (5:107 (Pennock)). Further support for the conclusion that ID is predicated on supernatural causation is found in the ID reference book to which ninth grade biology students are directed, Pandas. Pandas states, in pertinent part, as follows:

"Darwinists object to the view of intelligent design *because it does not give a natural cause explanation* of how the various forms of life started in the first place. Intelligent design means that various forms of life began abruptly, through an intelligent agency, with their distinctive features already intact—fish with fins and scales, birds with feathers, beaks, and wings, etc."

...Stated another way, ID posits that animals did not evolve naturally through evolutionary means but were created abruptly by a non-natural, or supernatural, designer. Defendants' own expert witnesses acknowledged this point. (21:96-100 (Behe); P-718 at 696, 700 ("implausible that the designer is a natural entity"); 28:21-22 (Fuller) ("...ID's rejection of naturalism and commitment to supernaturalism..."); 38:95-96 (Minnich) (ID does not exclude the possibility of a supernatural designer, including deities).

It is notable that defense experts' own mission, which mirrors that of the IDM itself, is to change the ground rules of science to allow supernatural causation of the natural world, which the Supreme Court in Edwards and the court in McLean correctly recognized as an inherently religious concept. Edwards, 482 U.S. at 591-92; McLean, 529 F. Supp. At 1267.

The last paragraph is key—the anti-materialist/anti-naturalist movement is really about changing the ground rules of science (refighting the fight they lost in the past) to include supernatural explanations, but this is impossible within the necessary framework of science.

The Wedge Strategy

In a now infamous Wedge document (published in 1998 by the Discovery Institute), proponents of so-called "intelligent design" lay out their overarching strategy:

The proposition that human beings are created in the image of God is one of the bedrock principles on which Western civilization was built. Its influence can be detected in most, if not all, of the West's greatest achievements, including representative democracy, human rights, free enterprise, and progress in the arts and sciences.

Yet a little over a century ago, this cardinal idea came under wholesale attack by intellectuals drawing on the discoveries of modern science. Debunking the traditional conceptions of both God and man, thinkers such as Charles Darwin, Karl Marx, and Sigmund Freud portrayed humans not as moral and spiritual beings, but as animals or machines who inhabited a universe ruled by purely impersonal forces and whose behavior and very thoughts were dictated by the unbending forces of biology, chemistry, and environment. This materialistic conception of reality

eventually infected virtually every area of our culture, from politics and economics to literature and art.

It is crystal clear from this and other writings, as well as the history of the ID/creationist movement, that this is about ideology, not
science. ID proponents feel that their spiritual ideological worldview is
threatened by the findings of modern science, and so they have decided
to undermine it. They want this to be an ideological and cultural war,
because in the arena of science they lose. They therefore claim that
science (at least those sciences with which they feel uncomfortable)
is nothing more than the ideology of materialism. They attempt to
frame the conflict as one between traditional, moral, and God-fearing
spiritualism on one side and cold, amoral, mechanistic materialism on
the other. This is an emotional fight they feel they can win.

Their dilemma (as made clear by the recurrent failures of the
old-school creationist movement) has been that the institution of science appears to have a lock on public education, research funding,
mainstream publications, and even to a large degree public respect.
Therefore they decided—and this is clearly laid out in the Wedge
document—to fight fire with fire, to create their own "scientific"
institutions, their own scholars and publications and funding sources.
They set out to pretend to do science and to make scientific arguments (the thin edge of the wedge) so as to break into the established
scientific infrastructure, but their farce had a predetermined goal—to
undermine the materialist basis of modern science.

ID proponents began their efforts with evolution, but that was
only ever a means to an end—the thin edge of the wedge. Their
recent efforts have moved toward attacking modern neuroscience as
an explanation for consciousness and the big bang as an explanation
of the origin of the universe. Their beef is not just with evolution but
with the methods of science.

But they've already lost the underlying battle. Changing the battle-
field won't matter. Just as there is no crying in baseball, there is no
magic in science.

But What If the Supernatural Exists?

It is an interesting thought experiment to ask, What if we lived in a universe in which there were supernatural phenomena? We again are faced with the problem of how to define "supernatural," as you could always argue that if the supernatural existed it would be part of nature and therefore "natural." That's why we have to define the supernatural by how it behaves, not by what it is.

I think the best way to conceptualize the supernatural is as phenomena that transcend the knowable laws of the universe. This means they're not bound by those natural laws or they can suspend them in unpredictable ways: There are effects without natural causes.

If such a thing existed, how could we know about it? By definition, I think you can't, at least not scientifically. At best, science could identify anomalies that it could never resolve.

Science encounters anomalies all the time. Scientists love anomalies because they point the way to new discoveries, to new knowledge. But if the methods of science can't identify a cause, because there is no natural cause, it would be stuck in an endless loop of failed hypotheses. The anomaly would never be explained.

In fact, we can use this idea as a meta-experiment about science and methodological naturalism itself. If we live in a universe where every effect has a natural cause, then science should work well. We will encounter anomalies but they will eventually be explained. Science will progress. If we lived in a universe with supernatural phenomena, science would encounter persistent anomalies with which it could make no real progress.

If we look back at the last few centuries, I think the meta-experiment significantly favors science and naturalism. True believers will always have anomalies to point to, but they are always changing. As we explain phenomena, they move on to new mysteries. So far, science has not encountered any true anomaly that defies scientific exploration. This whole science thing is working out pretty well.

20. Skeptics' Guide Entry: Postmodernism

Section: Science and Pseudoscience
See also: Epistemology

Postmodernism, as it applies to science, is the philosophical position that science is nothing more than a cultural narrative and therefore has no special or privileged relationship with the truth.

> The history of science, like the history of all human ideas, is a history of irresponsible dreams, of obstinacy, and of error. But science is one of the very few human activities—perhaps the only one—in which errors are systematically criticized and fairly often, in time, corrected.
>
> —Karl Popper

The abstract to a 2006 paper criticizing evidence-based medicine begins:

> Drawing on the work of the late French philosophers Deleuze and Guattari, the objective of this paper is to demonstrate that the evidence-based movement in the health sciences is outrageously exclusionary and dangerously normative with regards to scientific knowledge. As such, we assert that the evidence-based movement in health sciences constitutes a good example of microfascism at play in the contemporary scientific arena.

The authors explain that using science as a basis for deciding which treatments are safe and effective is nothing more than fascism, oppressing "other ways of knowing" and denying non-Western cultures. Science, in their view, is nothing but a social construct.

This is an excellent example of the application of postmodernist philosophy to science. In this context, postmodernism is the notion that all ideas and beliefs can be best understood as subjective human storytelling—a narrative dominated by culture and bias, with no special relationship to the truth. When applied to science, it negates the implication of methodology and reduces all scientific research to a cultural narrative.

Philosophers of science have already rooted out the flaws in such reasoning (in philosophical parlance, postmodernism confuses the context of discovery with the context of later justification), but the humanities subculture within academia didn't get the memo. The notion that science is socially constructed is a convenient way to dismiss the findings of science that you don't like for ideological or any other reasons. Postmodernism, in practice, is the ultimate sour grapes of science deniers— "Well, all science is socially constructed anyway." Add in a little talk about fascism and oppression and you can make it all seem socially conscious.

Thomas Kuhn

One of the originators of postmodernist philosophy is Thomas Kuhn, who wrote *The Structure of Scientific Revolutions* in 1962. It was in that book that Kuhn introduced the word "paradigm" to the world.

Kuhn's basic point was that new ideas in science do not arise from any rigorous methodology but as a chaotic by-product of sociology and scientific enthusiasm. During "normal" periods of scientific progress, scientists work incrementally to tweak existing ideas. They're working within a theoretical framework—that framework is what Kuhn calls a paradigm.

One of Kuhn's main examples is the change from the Earth-centered universe of Ptolemy to the sun-centered (heliocentric) universe of Copernicus. He argues that Ptolemy's paradigm was just as good as Copernicus's paradigm at the time it was introduced. The "paradigm shift" occurred mainly for cultural reasons and because Copernicus promised that a more elegant system would eventually evolve out of his approach.

A paradigm in science is relatively stable over a period of time, but it may suddenly shift to a new paradigm for quirky reasons. Here's a critical bit: Postmodernism insists that evidence and ideas can only be evaluated *within* a paradigm, not *between* paradigms. The key question is this: Is one paradigm objectively better than another, and is there therefore real progress in science, or is it all subjective? The postmodernists claim it's all subjective.

Interestingly, even Kuhn does not believe that science is entirely subjective. He has famously said he is not, apparently, a Kuhnian. His point was more limited, that paradigms shift for quirky social reasons.

There are two main criticisms of the Kuhnian postmodernist position. The first is that it is a false dichotomy. You can't separate scientific progress cleanly into normal incremental change within a paradigm and dramatic paradigm shifts. Rather, scientific change occurs along a continuum. Some discoveries are bigger and more disruptive than others, but you can't lump them into two categories.

The second and more devastating criticism of the postmodernist position is that this interpretation of the history of science confuses the context of discovery with the context of later justification. It actually doesn't matter how scientists come up with their ideas. They can get them from science fiction, popular culture, or a drug trip. It simply doesn't matter. What does matter is how those ideas are tested—later justification.

Copernicus survived, and Ptolemy didn't, because Copernicus was more correct. Ptolemy believed the Earth was at the center of the universe and used a complex system of epicycles (the planets moving in their own small circles) to explain observations of the planetary positions. Copernicus preferred a different model with the sun at the center, but observations at the time were crude, and while the Copernican system was more elegant, it didn't fit the data any better than the Ptolemaic system at first. What matters, however, are the predictions that each system made and how well they matched later, more accurate, or more thorough observations. We accept the sun-centered model today because it fits precisely with highly accurate observations. If it didn't, we would have had to revise or replace it.

Keeping with this example, we see that as science progresses, theories are refined, typically in subtle ways. The history of science isn't one of wholesale replacement of one theory with another. That kind of complete change did occur when we were shifting from an essentially prescientific view to a scientific one. Once a scientific theory is well established, however, it's not replaced, only refined.

The Copernican system, which treated planetary orbits as circles, didn't fit better and better data. Then along came Kepler, who figured out the three laws of planetary motion, including the fact that planetary orbits are ovals with the sun at one focus. That model fit very well with observations of the time, but it didn't fit perfectly with later precise observations of the orbit of Mercury. Explaining those anomalies required Einstein's general theory of relativity.

One of my favorite examples of this type of objective progress in science (which I am borrowing from Isaac Asimov) relates to our knowledge of the shape of the Earth. The ancient Greeks figured out that our planet is a sphere. Better observations refined that model to include the fact that it bulges a bit at the equator, and therefore it's really an oblate spheroid. Still better satellite data showed that the Earth is a bit larger in the southern hemisphere. And of course we know the Earth has mountains and valleys. Over time, you can define the Earth's true shape in finer and finer detail.

These details, however, don't change the basic understanding of Earth as roughly a sphere. That is objectively its shape. This paradigm will never shift into a model of the Earth as a cube.

Each scientific discipline is not a world unto itself, but part of the overall scientific endeavor to explain reality. Since all science is exploring the same reality, it all has to agree. This is what E. O. Wilson called "consilience" in his book of the same name. Part of the reason the Copernican/Keplerian system survived is that it fits with other discoveries about the universe, such as gravity. The force of gravity can explain well how the planets move. There's no theoretical force that would keep the planets moving in Ptolemy's epicycles.

These aren't just culturally determined stories we tell each other. Science is a method, and ideas have to work in order to survive. But we occasionally encounter postmodernist arguments that essentially try to dismiss the hard-won conclusions of science. I guess if you're losing a fight over evidence and logic, it's easy to just sweep the board off the table and say none of it matters.

It is true that science is a human, and therefore cultural, endeavor. In this respect there is a kernel of truth to some of the more reasonable postmodernist claims. The institutions of science may be biased by prevailing cultural assumptions and norms. For example, racism was in the past justified by racially biased science.

This doesn't mean, however, that science cannot or does not objectively advance. Because the process of science is inherently self-critical and the methods of science are all about testing ideas against objective reality, cultural bias is eventually beaten out of scientific ideas.

Some medical treatments do in fact work, while others objectively do not, and science can eventually tell the difference.

21. Skeptics' Guide Entry: Occam's Razor

Section: Science and Pseudoscience
See also: Principle of Parsimony

The principle of Occam's razor, attributed to William of Ockham (1287–1347), states that when two or more hypotheses are consistent with the available data, then the hypothesis that introduces the fewest new assumptions should be preferred. In the original Latin, *"Non sunt multiplicanda entia sine necessitate,"* which translates to "Entities must not be multiplied without necessity."

> We are to admit no more causes of natural things, than such as are both true and sufficient to explain their appearances. —Isaac Newton

---

A UFO proponent once argued to me that Occam's razor, which he defined as preferring the simplest explanation for any phenomenon, actually favors the theory that aliens are visiting the Earth. The presence of aliens could potentially explain cattle mutilations, crop circles, reports of sightings and abductions, and a host of other strange phenomena. Skeptics come up with a separate explanation for each of these things. One explanation—aliens—was far simpler, and therefore a good skeptic should accept claims of alien visitation.

Occam's razor is a principle of logic that is often invoked but rarely properly understood, as in the case of this UFO enthusiast. It's a useful rule of thumb to help clarify one's thinking, not a strict logical necessity. Failure to understand Occam's razor, however, can lead to very sloppy thinking.

Occam's razor is often paraphrased as, "When there are multiple possible answers, the simplest should be preferred." However, the direct quote from William of Ockham is this: *"Numquam ponenda est pluralitas sine necessitate* [Plurality must never be posited without necessity]."

The problem with the paraphrase is that it misses the point of the rule, and often the simplest answer is not the best one. In medicine, for example, "simplest" is often translated into "fewest number of diagnoses." This is contradicted by Hickam's dictum, which states that "patients can have as many diseases as they damn well please." The fix, however, isn't in concluding, as some have, that Occam's razor has limited utility, but rather in understanding what the principle actually is.

Stated another (and more accurate) way, Occam's razor is the principle that the introduction of new assumptions should be minimized. "New assumptions" shouldn't be conflated with "additional explanations or diagnoses." That is the error.

As we get older we tend to accumulate diseases and disorders. Common things will occur commonly. Further, one disease might predispose to another, or multiple diseases might all stem from one underlying risk factor. For example, a patient may have insomnia, which predisposes to obesity, which further worsens sleep through sleep apnea. The obesity then results in type 2 diabetes, which has its own complications. In the end, the patient's complaints may stem from a long list of identifiable diagnoses, all of which are common and comorbid.

Occam's razor would favor giving the patient many different diagnoses to explain their symptoms rather than invoking one diagnosis to explain all their symptoms, if that one diagnosis were very rare, not well established, or even pseudoscientific. A rare or unknown disease is a big new assumption, while known common diseases are not, especially if they derive from the patient's other known conditions.

What a clinician shouldn't do (and this does violate Occam's razor) is introduce an entirely new disease or condition just to explain each individual sign or symptom of a patient.

Occam's razor is ultimately all about probability. Every time you introduce a new element to an explanation or make a new assumption, you reduce the probability that your explanation is correct. Clinicians try to think of every plausible way to explain the patient's total presentation, then they rank the possibilities from most likely to least likely. In ranking how likely each possibility is, you can imagine adding up all the new assumptions that each diagnosis would represent, in both number and magnitude, resulting in a total assumption burden. Therefore, one entirely new and extremely rare diagnosis might be far less likely than using three very probable new diagnoses.

When viewed in this way—minimizing the total burden of new assumptions (accounting for both number and magnitude)—Occam's razor is a very useful logical tool for any investigatory endeavor.

Getting back to aliens, Occam's razor provides a practical guiding principle whenever someone sees a light or object in the sky that they can't identify—by definition an unidentified flying object or UFO. We know there are a host of natural and man-made phenomena that could result in a UFO. Invoking such explanations doesn't introduce any new assumptions into our view of reality. On the other hand, since it hasn't been established that an alien technological race exists and is currently visiting the Earth, invoking alien spacecraft as an explanation requires the introduction of a massive new assumption.

What the UFO community has done is pull together many common and known phenomena (hoaxes, misinterpretations, hypnagogia) and interpret them as alien visitations. Once belief in aliens became popular, that drove a host of other phenomena, like crop circle hoaxes and people believing they've been abducted. The UFOlogists then argued that, well, there's too much going on for each to have a separate explanation, so there must be aliens.

We don't need to invoke space-traveling aliens to explain all this, however. Culture and human psychology are enough.

Part of what makes Occam's razor so useful is the fact that smart creative people can come up with an almost unlimited number of hypotheses that are consistent with available evidence. Often proponents of a particular pseudoscience will have to create a special explanation for every aspect of their theory that seems to contradict reality.

If you think aliens are abducting people, why don't they remember the abductions? Well, because their memories were carefully erased (but not completely, so they can be recovered through hypnosis). We have no unambiguous pictures or videos of spacecraft because the aliens are hiding, but not completely, so occasionally we get a suggestive glimpse. If the aliens are trying to communicate with us, why not just communicate? What's with the funky symbols in the wheat? Maybe they are just trying to ease us into the knowledge of their existence.

For each question, they introduce a new assumption to explain the apparent problem away. They don't just violate Occam's razor, they dig up the corpse of Ockham and pound his bones into dust.

Sometimes Occam's razor is all we have between us and belief in nonsense. It's often true that the elaborate and highly rigged theories of pseudoscientists aren't impossible. We may not even be able to prove they are not true. But they're often simply unnecessary. Once sliced away by an appropriate use of Occam's razor, there is no reason to reject the far simpler conclusion that aliens are simply not visiting the Earth.

22. Skeptics' Guide Entry: Pseudoscience and the Demarcation Problem

Section: Science and Pseudoscience
See also: Denialism

Pseudoscience refers to claims and procedures that superficially resemble science but lack the true essence of the scientific method. In practice there is a continuum from rank pseudoscience at one end to rigorous science at the other, with no sharp demarcation line in between.

> Science is simply common sense at its best; that is, rigidly accurate in observation, and merciless to fallacy in logic. —Thomas Huxley

---

Science vs. Pseudoscience

In the 1995 book *Ablaze! The Mysterious Fires of Spontaneous Human Combustion*, Larry Arnold attempts to make the case that human beings sometimes spontaneously burst into flames and are reduced to ash. He writes:

> Spontaneous Human Combustion: detested, disdained, denied, debunked by nearly all academicians and forensic professionals as simply "Impossible! The human body does not burn this way!

These are hoaxes! There is no way; no way, period." But what if the overwhelming consensus against SHC is wrong?

In order to explain this unlikely phenomenon, Arnold speculates that there is an unknown new particle he calls the "pyrotron." In the last two decades, however, SHC has not become an accepted medical condition, as Arnold had hoped, and physicists have not been using the Large Hadron Collider to find the pyrotron. Is this due to an unfair bias, or even a conspiracy, against SHC science? Or perhaps SHC hasn't earned serious scientific consideration? But why not? Why spend millions of dollars looking for the Higgs boson particle and not the pyrotron?

There are many similar claims relegated to the dustbin of pseudoscience: the aquatic ape theory, astrology, the Electric Universe, and extrasensory perception (including Daryl Dem's "Feeling the Future" research from the intro to this section), to name a few. One of the core skills of skeptical thinking is to tell the difference between legitimate science and these pretenders.

This is a critical problem. Science is perhaps the most important tool for separating out valid claims to knowledge from the ocean of fraud, nonsense, bias, and error. Science also enjoys a great deal of public respect and interest. A 2009 Pew survey found that 70 percent of Americans thought that science contributes "a lot" to society:

> Overwhelming majorities say that science has had a positive effect on society and that science has made life easier for most people. Most also say that government investments in science, as well as engineering and technology, pay off in the long run. And scientists are very highly rated compared with members of other professions: Only members of the military and teachers are more likely to be viewed as contributing a lot to society's well-being.

It's therefore not surprising that many people try to support their beliefs and claims with science. Companies will claim that their products are "scientifically formulated" and have actors in white coats make sciencey claims. Science only works, however, when you rigorously follow its methods. When someone follows what looks superficially like a scientific process but the quality is hopelessly low and the process is fundamentally flawed, or distorted to achieve a predetermined end, we call that pseudoscience.

Telling the difference between science and pseudoscience isn't always easy, as good clean science blends imperceptibly into blatantly absurd pseudoscience. The challenge of drawing an objective line between the two is what philosophers of science call "the demarcation problem." So far, there is no clear and concise solution to the demarcation problem that has been generally accepted by philosophers and scientists. There probably never will be, as the quality of science represents a continuous spectrum, not a simple dichotomy with an objective line of separation between two extremes.

Despite the insoluble nature of the demarcation problem, philosophers, scientists, and skeptics can establish a set of characteristics that define good science and another set of characteristics that define pseudoscience, and then use these characteristics to establish where along the spectrum from science to pseudoscience a particular theory or practice lies. In this way, theories and disciplines can be identified that clearly lie toward the legitimate science end of the spectrum, and others that can be safely relegated to the opposite extreme. While some theories will remain somewhere in the middle gray zone, others will simply combust spontaneously of their own nonsense and flawed methods.

Good Science

Sometimes it's easiest to understand what something is by defining what it is not. Pseudoscience is not good science, so what is?

Good science uses observations about the world that are as objective, quantitative, precise, and unambiguous as possible. It uses these observations to test hypotheses and specifically to try to disprove those hypotheses. Those that survive repeated genuine attempts to disprove them are used to build theories that provide an explanatory framework for how the world works and to make predictions about future observations.

Good science is therefore skeptical of its own ideas, humble and conservative in its conclusions, and always open to new data and new interpretations.

Science is rigorous when it carefully isolates variables so as not to confuse cause and effect. It considers all the evidence, not just the evidence that supports a favored idea. Its logic is internally consistent and its judgments unbiased. Science is about minimizing bias, and good experiments blind subjects and experimenters to avoid bias. It checks ideas against other experts. And all of this needs to be done in a transparent way—there is no secret knowledge or hidden methods.

By adhering to all these principles, science can slowly grind forward. But there are also countless ways for the gears to turn in the wrong direction.

Features of Pseudoscience

1) Working Backward from Conclusions

The fundamental feature that separates the process of science from pseudoscience is that science is a genuine search for what is true, regardless of what that might be, whereas pseudoscience begins with a desired conclusion and then works backward to verify only that conclusion.

Scientists, of course, have their preferred hypotheses, but they need to act as if they were dispassionate toward the findings of their investigations, with no vested interest in the outcome one way or another. They also need to design their experiments so that any bias they might have will not influence the outcome. Scientists should be their own most dedicated skeptics: They should work hard to disprove their hypotheses, seeking out disconfirming evidence, thinking of alternative hypotheses, and criticizing their own work.

Pseudoscientists, by contrast, are overwhelmed by their own bias. They seek to confirm their hypotheses and often design experiments to guarantee that confirmation. They make a lawyer's case for their beliefs, dismissing skepticism and alternatives, explaining away faults, and generously interpreting their results.

2) Hostility Toward Scientific Criticism, Claims of Persecution

Criticism is a necessary and healthy part of the scientific process. Anyone who has ever submitted a paper for publication in a scientific journal has been the target of such criticism. It's the primary mechanism by which standards are maintained and an important source of the self-corrective nature of science. Scientists, therefore, learn to be thick-skinned. They also learn how to focus their criticism on the logic and evidence of an issue, rather than making personal attacks against those proposing a view different from their own.

Pseudoscientists, by contrast, display clear hostility toward any such criticism. They view criticism as a personal attack, even when it's not. They tend to characterize criticism from mainstream scientists as supporting the status quo, hostility toward new or innovative ideas, or even a full-fledged conspiracy to suppress their ideas. They'll often dismiss the content of criticism by denouncing the philosophical basis of science altogether, or they'll deny the ability of science to pierce their arcane knowledge.

In short, they view the criticism of their ideas as a problem with science and scientists, not with the evidentiary or logical basis of their claims.

Jacques Benveniste, discussed in chapter 37 on N-rays, who published studies claiming to show evidence for water memory, didn't respond well when *Nature* magazine reviewed his laboratory procedures and concluded his controls were inadequate. He rejected the criticism as a "witch hunt."

Rather than accepting constructive criticism as a necessary part of the process, the pseudoscientist feels persecuted. In fact, it's so common for pseudoscientists to compare themselves to Galileo (who was persecuted and turned out to be correct) that this phenomenon has its own name, the Galileo syndrome.

Some pseudoscientists will complain about a scientific "orthodoxy" that unfairly rejects any ideas that are too radical. They may also claim that the world is just not ready for their genius, or that their claims are too disruptive. Some begin to believe in an elaborate conspiracy to suppress their ideas: It's the only explanation for why the scientific community isn't fawning at their feet, granting them accolades and awards for their genius. In the end, these are all excuses for the fact that they haven't done the rigorous science necessary to convince the scientific community of their ideas.

3) Making a Virtue out of Ignorance

Some pseudoscientists lack formal training in science. In centuries past, this wasn't a serious obstacle to performing cutting-edge science. Many scientists were independently wealthy gentlemen who made major scientific discoveries in basements or cottages that had been converted into laboratories, or by making basic field observations. Darwin, Galileo, and Newton all fit this mold. Today, the gentleman scientist is a rarity, although his image persists in the lay consciousness.

Cutting-edge science is too advanced for anyone to have a reasonable chance of making a significant contribution unless they first have sufficient education in science. The pace of change in any active field of research is so rapid that a researcher must keep in contact with the community of scientists through journals, meetings, and seminars, just to keep up.

This modern status of the practice of science is a double-edged sword. It's a testimony to the success of institutionalized scientific research and the progress that has been made so far. However, it also tends to alienate the amateur scientist and the public. The lay, or amateur, scientist must be content to sit on the sidelines and learn about exciting scientific discoveries in those books, journals, and lectures that are designed to distill this information for the public. No matter how great the interest, the private citizen can't just fire up a particle accelerator in their backyard and make important discoveries in particle physics. Professional scientists understand that even for them it's difficult to contribute meaningfully to a field even marginally outside their areas of expertise.

There are, however, citizen scientist projects in which interested people can participate. You can classify galaxies, find Kuiper Belt objects, and fold proteins. Trained scientists are responsible for maintaining a rigorous procedure, however.

Some amateur scientists aren't content to sit on the sidelines or be citizen scientists. They weave hypotheses and sometimes even carry out experiments in highly specialized fields. Often their conclusions are somewhat naive and betray their lack of formal training. Perhaps keenly aware of this potential shortcoming, the pseudoscientist is quick to turn a lack of training into a putative virtue by arguing that this is actually an advantage. The argument frequently put forward is that formally trained scientists are "brainwashed" into a narrow view of reality. They are, in effect, pawns of the status quo, unable to think outside the box. The untrained scientist, by contrast, is free to conceive of unique and truly innovative ideas. The truth becomes obvious to them, while trained scientists can't see the forest for the trees.

On my blog, *NeuroLogica*, I had a detailed debate with comic book artist Neal Adams. Adams believes that the Earth is hollow and growing, gaining matter over time. The full debate is amazing to behold, and the excerpted tidbits below are phenomenal interactions and classic examples of pseudoscience.

First, Adams is unfazed by the fact that his wild speculations aren't generally accepted by scientists. He wrote:

Of course, I'm going to be dismissed by most educated scientists. For me, I'm not quite as impressed by formal education. I can read and there are no books forbidden to me in the end. I can think, and I use many aspects of science in my work.

The flaw in this argument is that knowledge isn't necessarily constricting. It can also be liberating. The more one knows, the easier it is to learn more information. Knowledge provides intellectual tools that can be applied to the process of discovery. Also, knowledge of what is already known helps in evaluating the plausibility of new ideas.

I admit it's easy to understand how this view can be interpreted as elitist, and pseudoscientists often take great rhetorical advantage of this fact. Keep in mind, however, that I'm not making the argument from authority that scientists who hold advanced degrees are always right simply because of their training. I'm not even saying that someone without a degree can't be correct in a scientific claim—their claims should be judged solely on the basis of their logic and evidence. What I'm saying is that ignorance isn't an advantage, nor is it a virtue to be touted. It's a hindrance.

4) Reliance upon Weak Forms of Evidence While Dismissing More Rigorous Evidence

There is no way to remove the need for judgment from the process of science. You can't apply a mathematical formula to a claim to determine if it's valid (although people try). You have to weigh evidence, decide which kind of evidence is most significant, and determine which explanations are most likely.

It is with these kinds of judgments that scientists need to be most careful and pseudoscientists run into deep trouble. They can be so biased that they will casually dismiss conclusions based upon mountains of rigorous evidence, while citing the flimsiest of evidence to support their alternative claims.

In order to read and appreciate the technical literature, scientists must know how to evaluate the quality of an experiment, look for flaws in the design, and determine if the study has the power and sensitivity to detect what the experimenters are trying to detect. Reading original scientific research is highly technical and requires detailed knowledge of methods and statistics. This skill is critical, as many—if not most—published studies are in fact wrong in their conclusions. Most studies are preliminary and not very rigorous.

Pseudoscientists, by contrast, will tend to accept any testimony or anecdote that supports their desired belief. They will sometimes present large volumes of low-quality evidence, implying that a large amount of poor-quality evidence equates to high-quality evidence. Alternative medicine guru Andrew Weil, for example, supports the use of "uncontrolled clinical observations" in determining whether or not a treatment works. Such observations have a history, however, of being contradicted by later well-controlled and more reliable experiments. This is a lesson that good scientists have learned and that pseudoscientists deride.

A classic example of this can be seen in the field of UFO research. Believers often tout the great number of UFO sightings as compelling evidence that we're being visited by aliens. Skeptical scientists, however, are more compelled by the fact that there is not one single piece of high-quality evidence to support the alien hypothesis.

5) Cherry-Picking Data

Closely related to the use of poor data is the selective use of data. Scientific experiments are designed to take a complete look at a sequential set of data. Anecdotal evidence is by definition selective, because it is limited to self-reporting and is not a thorough analysis of all outcomes. Patients who die receiving a specific treatment, for example, aren't around to give anecdotes about their experience with that treatment. (This is called the survivor bias.)

There are many ways to cherry-pick data. Early ESP research is infamous for finding ways to select the data the researchers want. They introduced the idea of "optional starting and stopping," because, they claimed, psychic individuals need time to warm up before their powers work, and their powers will then work only until they become too fatigued, then they will stop working again. Therefore, researchers could look at a sequence of data (like guessing what cards a target was looking at) and decide when to start and stop counting the data. They could essentially mine the data for sequences that were statistically significant. They were simply cherry-picking the data they wanted and discarding the rest. There's a more conventional term for this procedure—it's called cheating.

6) Fundamental Principles Are Often Based upon a Single Case

This is really just an extension of the overreliance upon testimony and anecdote as a feature of pseudoscience. Some entire belief systems that pretend to be scientific base their fundamental underlying principles on a single uncontrolled observation. Any good scientist, before launching into a research career investigating some principle, will first make sure that the principle in question is correct, lest he risk wasting an entire career on a false idea. Certainly, before applying a basic principle to the real world—by, for example, using it to treat patients—it should be verified with repeatable experiments.

Some pseudosciences, however, have extrapolated an entire elaborate belief system around a single observation that was never verified. My favorite two examples are chiropractic and iridology. D. D. Palmer, the originator of chiropractic, reported that he "discovered" the primary underlying principle of chiropractic when he cured a janitor of his deafness by manipulating his neck, thereby relieving (he believed) the pressure on the auditory nerve. Palmer conducted no experiments of any kind to verify his assumptions, but rather extrapolated all of classic chiropractic theory and practice from this single case. D. D. Palmer was apparently unaware of the fact that the auditory nerve, and in fact the entire neurological pathway responsible for hearing, at no point passes through the neck.

Iridology has a similar history and is based upon the observation of a single owl. Apparently, Ignatz von Peczely, a Hungarian physician, noticed that an owl that had injured its wing had a particular fleck of color in the iris of its eye. He set the owl's wing, which later healed well. Dr. Peczely then noticed that the fleck of color in the owl's iris had disappeared. From this single observation, Dr. Peczely developed a system of diagnosing all human disease by the pattern of colors in the iris. (In truth, the story of the owl may be apocryphal and Dr. Peczely made up iridology out of whole cloth.)

7) Failure to Engage with the Scientific Community

Science is hard, and getting harder. We've picked most of the low-hanging fruit, answered all the big and easy questions, and we are now engaged in complex research to address more and more sophisticated and subtle scientific questions.

For this reason, it's difficult for any one person alone to make significant advances in our understanding. It's generally too challenging to address all possible flaws and errors, consider all possible alternatives, and look at a problem from multiple perspectives by oneself. The scientific community has a better chance of doing this as a whole. That's why scientists publish their data in peer-reviewed journals and discuss their ideas at meetings. They're testing their ideas, addressing criticisms, and accounting for new ideas.

Pseudoscientists tend to be disconnected from this process (or perhaps exist in a protected echo chamber of similar-believing pseudoscientists). They ferment their ideas by themselves, which often allows them to drift further and further from reality.

8) Claims Often Promise Easy and Simplistic Solutions to Complex Problems or Questions

One of the primary reasons for the psychological appeal of pseudoscience is that it provides a putative easy answer to a complex problem. Classic chiropractic, for example, states that all human disease is caused by spinal subluxations, and therefore spinal manipulation can be used to cure all human disease. Nutrition guru Gary Null, by contrast, claims that all human disease is caused by nutritional deficiencies, and therefore all disease can be prevented or cured by taking nutritional supplements.

In a broader sense, this feature relates to the psychology of belief. Pseudosciences all tend to have a particular psychological appeal, of which providing easy answers is just one example. Others include alleged evidence of a supernatural or spiritual world, confirmation of deeply held religious beliefs, illusion of personal empowerment or control, or simply the appeal of the fantastical or unusual.

9) Utilizing Scientific-Sounding but Ultimately Meaningless Language

All sciences have their technical jargon. The purpose of jargon is simply to express complex technical concepts in precise terminology. Subtle distinctions can often be very important, and colloquial language may not have the precision necessary to unambiguously convey the necessary meaning. Moreover, as new concepts and entities are discovered, new words must be invented to refer to them. There's also a tendency to render the resulting cumbersome terminology into a more convenient shorthand, which can add an additional barrier to understanding for the public. The most challenging aspect of popularizing science is often translating technical jargon into everyday language while minimizing the loss of precision and accuracy.

For example, as a neurologist I may describe a patient as having cerebellar ataxia instead of just saying they are clumsy. This is because clumsiness is a general phenomenon that may have many causes, while cerebellar ataxia is a specific neurological phenomenon that correlates with specific structures in the nervous system.

Pseudosciences often imitate real science by cloaking themselves in pseudojargon. The result is the frequent use of scientific-sounding terminology that lacks a precise definition (much like the technobabble in a typical episode of *Star Trek*).

For example, Gwyneth Paltrow's lifestyle brand Goop sells "Body Vibes stickers" they claim will promote healing:

> The concept: Human bodies operate at an ideal energetic frequency, but everyday stresses and anxiety can throw off our internal balance, depleting our energy reserves and weakening our immune systems. Body Vibes stickers come pre-programmed to an ideal frequency, allowing them to target imbalances. While you're wearing them—close to your heart, on your left shoulder or arm—they'll fill in the deficiencies in your reserves, creating a calming effect, smoothing out both physical tension and anxiety. The founders, both aestheticians, also say they help clear skin by reducing inflammation and boosting cell turnover.

Translation—"magic stickers."

10) Lack of Humility—Making Bold Claims on Flimsy Evidence

Successful science must intelligently combine almost giddy speculation with harsh conservatism. Scientists must invent hypotheses, which extend the limits of human knowledge, introduce entirely new concepts, or invent new aspects of nature. At the same time, they must not accept any conclusion unless it's rigorously supported by solid evidence and all other reasonable alternatives have been eliminated. In this way science attempts not only to move forward but also to build on a solid foundation.

For this reason, technical literature tends to utilize very conservative language and is careful not to endorse or promote any conclusions that cannot be rigorously supported. Those scientists who make premature claims are typically harshly criticized by their colleagues for doing so. The premature announcement to the press of the achievement of cold fusion by physicists Stanley Pons and Martin Fleischmann, for example, did great damage to their professional careers.

Part of this is also being humble. If you find a result that contradicts well-established scientific conclusions, your first thought should be that you made a mistake, not that you just overturned an entire field of science.

Pseudoscientists, by contrast, make use of bold claims, superlative descriptions, and unrestrained self-serving accolades. Their discoveries are often touted, for example, as being world-altering in their scope and their implications for humanity.

Getting back to our friend Neal Adams, he wrote:

> Imagine being me and seeing these unexplained ridge complexes
> EVERYWHERE and being the (almost) only person on Earth
> who knows what they all are, pull-aparts. The ONLY THING
> they can be.

He thinks he has a unique insight that allows him to look at geological formations and the map of the Earth and see the Truth. When it's pointed out to him that his ideas conflict with entire areas of science, he simply overturns them. The Earth grows by gaining matter, so gravity doesn't work like scientists thought, and neither does particle physics. He also wrote:

> Know how fast the moon goes over the face of the Earth?
> 1,000 miles per hour. I think the water compresses. Got some
> textbooks wanna be rewritten!

and

> There is no standard model. It is in fact all a little math game
> of where the densities are and it's all theorising. This is a very
> big topic and without exposing you to several papers and long
> involved discussion—let me simply say, at the end of the day...
> densities evolve to when and where iron is.

The standard model of particle physics is in my way? Gone!

11) Claiming to Be Years or Decades Ahead of the Curve

One dramatic red flag that someone is making claims that go way beyond the evidence is that the breakthroughs they are claiming would require years or even decades of supportive research, and yet there is little or no scientific paper trail to support them.

For example, starting in 2013 when writing his book *Immortal: Why CONSCIOUSNESS is NOT in the BRAIN*, and through 2017, Italian surgeon Sergio Canavero claimed he was close to performing the world's first "head transplant." (We can argue about whether or not this should be called a "body transplant.") In an interview with *Newsweek*, he said:

> I can only disclose that there has been massive progress in medical experiments, which would have seemed impossible even as recently as a few months ago…The milestones we have reached will undoubtedly revolutionize medicine.

This is a bold claim. In order for such an operation to be considered successful, Canavero must have found a way to regenerate the spinal cord. Otherwise the head would be attached to a completely paralyzed body.

Canavero is claiming not just that he has perfected the surgical technique of removing a head from one body and attaching it to another, but that he has solved the problem of spinal cord regeneration. This is more than impressive, given that labs around the world have been working on this problem for decades with only incremental advances, nothing really clinically useful.

Even more impressive is that Canavero and his colleagues apparently did this in secret, without publishing a single paper or earning a single grant to conduct the needed research. They did all the basic science, animal studies, and clinical trials without leaving a single word of evidence in the scientific literature.

The fact is, modern science is complex and requires significant infrastructure and collaboration. Before we figure out how to actually regenerate spinal cords, scientists will need to put a hundred pieces of that puzzle together. There will be hundreds, maybe even thousands, of published studies leading up to the clinical trials that eventually show success.

It's simply not feasible to have leapfrogged all this necessary science, making decades of progress in a few years, all from nothing.

12) Attempts to Shift the Burden of Proof Away from Themselves

It is generally accepted within the scientific community that anyone making a claim to any truth bears the burden of proving their claim. The more out of sync such a claim is with accepted reality, the greater this burden of proof becomes.

Pseudoscientists, often because they cannot prove their claims, frequently attempt to shift the burden of proof to those who are skeptical of those claims. They maintain that their claim must be accepted as true because it hasn't been proven false.

This shift in the burden of proof often takes the form of our logical fallacy friend known as the argument from ignorance, the one that says that if we don't yet know the cause of a phenomenon, it must be paranormal. For example, ghost hunters will often present photographs containing blobs or wisps of light and claim that because these photographic artifacts can't be precisely explained, they must be ghosts.

Neal Adams again, after I asked him for a bit of evidence to support his hollow-growing-Earth claims:

> Are you suggesting that I provide a new invention, or show you a discovery of some magic thing that no one else has ever made before, like a flying mechanical donkey?
>
> I don't have one, I have all the same things that are known by many people, like Darwin. Darwin didn't have to go to the Galapagos to observe what he did. He simply went THERE. The

evidence was and is everywhere. It was his eyes and his brain that drew the conclusions. The question is, will you take the chip off your shoulder long enough to see what I see and think with me?

13) Rendering Claims Non-Falsifiable

Sometimes the pseudoscientists not only shift the burden of proof away from themselves, they also attempt to make their claims immune to refutation. Often this is done through inventing reasons why the expected evidence doesn't exist or experiments are negative.

For example, homeopaths sometimes say that their products can't be tested against placebos because they only work as part of an entire treatment program. This makes it impossible to isolate homeopathy as a specific variable. Of course, when you do isolate homeopathic products, they don't work.

At times the claim is formulated in such a way that it cannot be falsified even in theory. Creationists, for example, are fond of saying that we cannot know the mind of God, therefore we cannot say why nature looks the way it does. God just happened to make a creature that looks halfway between a dinosaur and a bird for his own unimaginable reasons. Ironically, doing so doesn't rescue the claim from being pseudoscientific but makes it more so. A scientific hypothesis must be falsifiable. If it isn't, then it's "not even wrong." Being wrong in science is useful; it still helps us move toward the answer. Being not even wrong is worthless and is by definition not scientific.

14) Violating Occam's Razor and Failing to Fairly Consider All Competing Hypotheses

One critical step in the process of scientific discovery is to consider all possible explanations for any observation or phenomenon. Often a scientific hypothesis is considered to be probably true because it's the best and simplest current fit for all available data. If all reasonable alternatives haven't been considered, however, this conclusion is likely premature.

Pseudoscientists, because they are invested in a desired conclusion, will give only perfunctory consideration to competing hypotheses. Often one or two token alternatives will be put forward and summarily shot down, leaving the desired belief as the only possibility. UFOlogists are most notorious for this behavior: The classic "unidentified light in the sky" cannot be a plane or a star, therefore it must be an alien spacecraft.

As part of this, pseudoscientists routinely violate Occam's razor by dismissing simpler explanations for more complex or implausible ones that fit their beliefs. Remember Arnold and SHC—he casually dismissed simpler explanations for alleged cases, even when there were obvious external sources of flame. Perhaps the most dramatic example is a 1980 case from Chorley, England. The charred remains of an elderly woman were found in her apartment. Here's the kicker—her head was in the fireplace. She'd clearly fallen and hit her head on the grate.

The corollary to this shortcoming is the failure to reject a disproved but desired hypothesis. This is often the ultimate test of a scientist's objectivity—can they discard a cherished hypothesis when it's confronted with incontrovertible disconfirming evidence? For the pseudoscientist, no amount of disconfirming evidence will result in such rejection.

15) Failure to Challenge Core Assumptions

What often passes for research within pseudosciences addresses questions about the alleged phenomenon but isn't designed to test whether or not the phenomenon exists. This is what fellow science-based medicine writer Harriet Hall called "Tooth Fairy science." Imagine you perform a "scientific" study in which you carefully collect data on the amount of money the Tooth Fairy leaves behind and run sophisticated statistical analyses correlating the amount given with the type and size of the tooth and the age and gender of the child. This type of study would have all the trappings of serious science but would never address the key underlying questions: Is the Tooth Fairy real, and who is actually leaving the money?

We often find these types of studies within alternative medicine—how is a particular alternative treatment being used, by whom, and what are their attitudes toward it? Such studies don't address the questions of whether or not the treatment works and the underlying philosophy is valid. Once efficacy studies start to come back negative, we find that proponents stop doing them in favor of more "Tooth Fairy" type studies.

In a famous speech at Caltech in 1974, Richard Feynman referred to such practices as "Cargo Cult Science." He made an analogy to the preindustrial Melanesian tribes who, after World War II, would build grass huts and fake runways hoping to attract the supply planes that delivered goods during the war. Their rituals had only a superficial resemblance to actual landing strips and lacked the necessary technology and infrastructure.

Similarly, pseudoscience may superficially resemble the ritual of doing research, but it lacks the true essence of real science—rigorously testing hypotheses with evidence capable of proving them wrong. In short, pseudoscientists engage in motivated reasoning (see chapter 9). They'll commit endless special pleading to explain away experimental failures (like the presence of skeptics causing psychic ability to stop working).

By keeping in mind the characteristics outlined above, one can make a reasonable judgment about a claim or theory and determine where along the spectrum it lies from solid science to absurd pseudoscience.

Evolutionary science, for example, pulls together multiple independent lines of evidence, has been debated transparently for decades, and has stood the test of multiple observations that could have potentially disproved it. It remains the only viable scientific explanation for the life we see in the world.

At the other end of the spectrum is poor Neal Adams. He desperately wants to be taken seriously as a scientist, but he has done none of the actual work. He has nothing except wacky ideas based on his own superficial observations, and he blithely suggests a cascading transformation of virtually all of modern science to accommodate his wild notions.

Somewhere in the middle is Daryl Bem and his ESP research. He is following standard scientific protocol, to an extent, but is overwhelmed by his own biases. In the end he's created a great example of pseudoscience, with all the trapping of science but failing to rigorously prove his main hypothesis.

23. Skeptics' Guide Entry: Denialism

Section: Science and Pseudoscience
See also: Pseudoscience

Denialism or science denial refers to the motivated denial of accepted science using a series of invalid strategies.

> I deny nothing, but doubt everything.　　　　　　—Lord Byron

Denialism is a real thing, by which I mean that denialism is a definable intellectual strategy with consistent features that tend to cluster together. Denialism begins with the desire to deny an accepted scientific or historical fact, and therefore, like all pseudosciences, works backward from the desired conclusion.

In fact denialism is a subset of pseudoscience, one that tries to cloak itself in the language of skepticism while eschewing the actual process of scientific skepticism. Denialism exists on a spectrum with skepticism, without a clear demarcation between the two (similar to science and pseudoscience). People tend to use themselves for calibration—anyone more skeptical than you is a denier, and anyone less skeptical than you is a true believer.

At this point, science denial has become common in our society and may be more impactful than the promotion of pseudoscientific beliefs that are not true. There are ideological movements that deny the scientific consensus of anthropogenic global climate change, the modern synthesis of evolutionary theory, the germ theory of disease, that the brain causes consciousness, the existence of mental illness, that HIV causes AIDS, and the safety and effectiveness of vaccines.

There are even those who deny the existence of denialism itself, arguing that this is just a rhetorical device to shut down criticism of mainstream ideas. They miss the fact that denialism relies upon a set of invalid logical strategies—it's defined by these features, not the belief itself.

As you will see, many of the strategies used by deniers are insidious because they are extreme versions of reasonable positions. Some of their underlying principles are sound. It's their specific application that's the problem.

What follows is a list of some of the most prevalent denialist strategies.

Manufacture and Exaggerate Doubt

Doubt is key to skepticism and science. The absence of doubt is gullibility. This feature, most of all, is what makes denialism pseudoskepticism. The problem with the denialist approach is that doubt is not used as a tool of honest questioning but rather for undermining a belief one doesn't like.

This strategy can also be called "just asking questions" or "JAQing off." You can often tell the difference between science and denialism because, when true scientists ask a question, they want an answer and will give due consideration to any possibilities. Deniers, on the other hand, will ask the same undermining questions over and over, long after they have been definitively answered. The questions—used to cast doubt—are all they are interested in, not the process of discovery they're meant to inspire.

Of course, there is always doubt in science. Science is never 100 percent certain about anything, because science is not about certainty. It's not even really about proving things, but rather *dis*proving them. Science is also not directly about truth, but rather about building testable models that predict how the universe behaves.

Scientific theories become progressively more accepted as they survive serious attempts at proving them wrong. Such acceptance is provisional, however, as the next experiment or observation could potentially falsify any theory. Theories are favored when they have useful explanatory power and are consistent with other accepted theories as we slowly build one coherent model about how the universe works.

Scientific literacy means not only understanding what the best current scientific explanation of a phenomenon is, but also how sure we are that the current answer is correct, and how complete an answer it is (what does it leave unexplained?). Some theories are controversial, others simply unknown, while still others are fairly robust. At the extreme end of the spectrum are those theories that are, as Stephen Jay Gould noted, "confirmed to such a degree that it would be perverse to withhold provisional assent."

Deniers exaggerate our current level of doubt about a scientific theory and minimize what is known. They perversely withhold even their provisional assent. Often they take this strategy to the extreme of denying that we can know anything—they deny scientific knowledge itself.

As part of this strategy, they will likely appeal to the fact that scientific knowledge has changed over time. If scientists were wrong in the past, they can be wrong now. Of course, no one denies that current scientific knowledge is provisional—that misses the point entirely. They are essentially making a false analogy between a prior belief that wasn't well established and one that is currently very robust, even rock-solid.

Always Ask for More Evidence than Exists or Can Exist

Perhaps the core logical fallacy of the denier is moving the goalposts, a technique described in chapter 10: They ask for evidence, and when that evidence is provided, they demand still more evidence. Nothing will ever satisfy them.

This is very different from how science normally operates. After preliminary research clarifies the questions being asked and how the evidence relates to current theories, scientists supporting competing theories will often put their nickel down—they'll state exactly what evidence will refute their theory, refute competing theories, or will change their mind about which theory is superior. When the evidence comes in, scientists will actually change their mind. Of course, individuals don't always change their mind with the evidence, but enough do to shift the consensus to the theory supported by the evidence.

Deniers rarely do this. When their questions are answered or their demands for evidence met, they simply slide over to another question. Nothing will convince them because they've already decided on the answer.

For example, evolution deniers are fond of pointing out gaps in the fossil record or our knowledge of which species evolved from which ancestors. They take a snapshot of our current scientific knowledge and argue that the existence of gaps in that knowledge calls into question the more fundamental conclusions about what is happening. Because we cannot prove, for example, what group birds evolved from, evolution itself is questioned (the previously mentioned "god of the gaps" strategy).

Scientific theories, however, are better judged by how useful they are than what they can currently explain. What evolution deniers should be asking is not what evolution can explain right now, but how has it fared over the years? Specifically, if evolutionary theory is useful and correct, those gaps should be shrinking over time. How they change over time is more telling than how big they are at any one moment in time.

Anyone even casually familiar with evolutionary biology knows the answer to this question: Those gaps have been steadily decreasing. The gaps between humans and our closest ape relative, birds and dinosaurs, whales and terrestrial animals, fish and tetrapods, and many others have been filled in nicely in the last century.

Evolution deniers pointed to all these gaps in the past, and when they were filled in, they never acknowledged that fact. They just focused on another gap. One of their favorite gaps now is the one between bats and other mammals. When that one gets filled they'll move on to something else.

All this goalpost moving can be tiring, however, so some deniers use a simpler strategy—they simply ask for more evidence than can possibly exist.

Sticking with evolution as an example, they will ask for one fossil that proves evolution. This, of course, is impossible, because evolution is a complex process. There are similar calls from HIV deniers to point to one paper that proves the human immunodeficiency virus is the sole cause of the acquired immune deficiency syndrome. These claims require dozens, even hundreds, of individual studies added together.

Anti-vaccinationists who deny the safety and effectiveness of vaccines commonly ask for a randomized vaccinated-versus-unvaccinated study. While this may seem reasonable at first, they know such a study will never be done. They are essentially requiring a study that randomizes children to receive no vaccines at all. It's unethical: Vaccines are already part of the standard of care because they have proven benefits. In a clinical trial you cannot randomize subjects to not receive standard treatment.

They set the bar this high so they can deny the value of all other evidence that shows vaccines are safe. There are vaccinated-versus-unvaccinated studies, just not randomized, and there are many other types of studies showing the safety of vaccines.

This strategy also uses special pleading. If a study shows that there is no association between the timing of vaccines and the incidence of autism or any other outcome, they'll say that the effect is delayed or is from the vaccines the mother received while pregnant. If there's no dose response, that's because maximal negative effects are reached at tiny doses. If an ingredient is removed, then trace amounts are sufficient to have the negative effect, or some other ingredient is to blame. No evidence will ever be enough.

Use Semantics to Deny Categories of Evidence

As part of the strategy to deny evidence for the scientific theory they wish to dismiss, deniers will often play word games to exclude entire categories of evidence. For example, evolution deniers have argued that anything that happened in the past is beyond the reach of science. Science is about experiments, and you cannot run an experiment on evolution, therefore evolution is not even science.

This is an artificially narrow definition of science, however. Science is about using empirical evidence to test hypotheses. The past leaves traces of itself behind—evidence. We can ask falsifiable questions about what happened in the past, including evolution. Obviously, fossils are left behind, but so are the genes and other features of living things.

Those who deny the existence of mental illness play similar semantic games. They narrowly define illness as pathological disease, meaning that there has to be something objectively abnormal about cells, tissues, or organs. This description does apply to some diseases, but not all: There are disorders that are defined by the way some organ or system is functioning, but in the absence of clear pathology. Migraine headaches, for example, are a clear disorder without any diagnosable pathology.

There are many brain disorders, because brain function depends upon more than just the health of brain cells. Healthy brain cells may still be organized and networked in such a way that their function is disordered. The brain is the organ of mood, thoughts, and behavior. Disordered brain function may therefore lead to a mood disorder or thought disorder. We call such entities mental illness.

Mental illness deniers (such as Scientologists, who are against the psychiatric profession) don't speak meaningfully about the relationship between brain function and symptoms of mental disorders, but rather use semantics to deny that such things even exist.

Interpret Disagreements About Details As If They Call the Deeper Consensus into Question

To understand this feature of denialism you have to first understand something about science itself. As science progresses it tends to dig deeper into finer and finer details about how nature works, and hopefully achieves a more fundamental understanding. It's important to think of scientific knowledge as acting on different levels, with some levels being deeper than others.

For example, it was known since antiquity that certain traits can be passed down from parents to children. It's obvious that children look like their parents. It was thought, however, that the biological mechanism for inheritance was a pattern, a little homunculus curled up inside each sperm that was the template for the child.

Mendel and others discovered that some traits seem to be inherited discretely. When you cross a yellow pea with a green pea, you get some pattern of yellow and green peas, not mixed yellow-green peas. This trait didn't blend.

Not all traits work that way, but it showed that some are transferred as discrete units, or genes. At that point, we hadn't identified the molecule that carried this information. Some scientists thought it would be proteins, but it was eventually demonstrated convincingly that the molecule of inheritance is DNA.

That DNA carries genes and heritable information is now so well established that it can be considered a scientific fact. Yet there remained a great deal we didn't know about DNA and genes. We later discovered the genetic code, how that code is translated into proteins, and how DNA function is regulated. We still have many more details to learn about DNA.

But—and here is the critical part—nothing we currently don't know about DNA or will learn in the future can possibly change the basic fact that DNA is the primary molecule of inheritance. Scientists arguing about the details of gene regulation or stating how much we don't know about the evolution of the genetic code doesn't call into question this fundamental fact.

That, however, is exactly what many denialists argue. That is, in fact, the favorite strategy of neuroscience denialism, specifically the denial that consciousness is essentially the functioning of the brain.

Dualists—or those who think that the brain doesn't explain consciousness—point to what we currently don't know about how the brain produces consciousness, which shouldn't reduce our confidence in the conclusion that the brain does indeed cause consciousness.

The same is true for evolution. That life is the result of common descent through an evolutionary process has been demonstrated beyond any reasonable scientific doubt. There is no competing theory that can come even close to accounting for the evidence we have, or that has been as successful in predicting what we will find when we explore biology. Evolution is a slam dunk. But, of course, there is always complexity and argument in the details: What, exactly, evolved from what? What is the pace and tempo of evolution? Have we accounted for all the forces at work?

These are all interesting questions, but none of them call into question the fact that evolution occurred.

Deny and Distort the Consensus

Magnifying disagreements among scientists is easy, because such disagreements are always present. This strategy can take two basic forms. The first is to magnify the implication of the disagreement, as I described above. The second is to present a small minority dissent as if it's a mainstream controversy.

You can almost always find some scientists somewhere to disagree with even the most solid scientific consensus. I've argued that this is a good thing. Complacency can lead to stagnation in science, and it's always good to have someone shaking the tree. But such dissent needs to be put into context. Sometimes it's a genuine controversy and the science can go either way. Other times the science is solid and the dissent is insignificant.

Recent political controversies over anthropogenic global warming (AGW) have brought the consensus argument to center stage. Defenders of AGW argue that there is a strong scientific consensus that the globe is warming due to human activity. A 2013 survey of the published scientific literature, for example, described its method and findings as follows:

We analyze the evolution of the scientific consensus on anthropogenic global warming (AGW) in the peer-reviewed scientific literature, examining 11 944 climate abstracts from 1991–2011 matching the topics 'global climate change' or 'global warming.' We find that 66.4% of abstracts expressed no position on AGW, 32.6% endorsed AGW, 0.7% rejected AGW and 0.3% were uncertain about the cause of global warming. Among abstracts expressing a position on AGW, 97.1% endorsed the consensus position that humans are causing global warming.

This was the source for the famous "97 percent" consensus figure that is widely quoted (although it doesn't refer to 97 percent of climate scientists but rather 97 percent of published studies that either explicitly or implicitly expressed an opinion about AGW). This is in line with other research on the question. Another way to establish a scientific consensus is for scientific organizations to review the evidence and then make a determination. The 2013 Intergovernmental Panel on Climate Change report, for example, concluded with 95 percent confidence that AGW is real.

There is also a very strong consensus that currently available genetically modified organisms are safe for human consumption. More than twenty international scientific organizations have reviewed the research and independently come to this conclusion. For example, in 2013 the American Association for the Advancement of Science concluded that "the science is quite clear: crop improvement by the modern molecular techniques of biotechnology is safe."

Turning to evolution, 98 percent of the world's scientists in biological fields agree with the consensus that life is the result of organic evolution.

Deniers will not only deny that a consensus exists, they'll often deny that a scientific consensus is meaningful. They try to portray referring to a robust scientific consensus as an "argument from authority" logical fallacy. This, of course, is not the case. That fallacy is the result of inappropriately relying upon an individual as if they were a sufficient authority or relying upon the opinion of those without appropriate expertise (such as celebrities). It can also be a fallacy if it's used to deflect valid criticism or evidence.

For a nonexpert to cite a legitimate consensus of experts, however, isn't a fallacy. In fact, substituting your own nonexpert opinion in place of a robust consensus of recognized experts is the dubious approach.

Appeal to Conspiracy, Question the Motives of Scientists

It is easy to portray someone's motives as sinister. There's always some way to weave a tale about how scientists are biased or corrupt. Such claims can be manufactured out of whole cloth as needed.

While bias and corruption certainly exist, that doesn't mean it is reasonable to assume that any science with which you disagree can be casually dismissed as entirely the result of such corruption. But that's exactly what deniers do.

Deniers of global warming would, for instance, have you believe that climate scientists throughout the world decided to manufacture an elaborate hoax in order to increase their funding. To support their allegations, they engineered "Climategate" by trolling through thousands of hacked e-mails until they found some statements that could be taken out of context.

Anti-GMO (genetically modified organism) activists have employed the same strategy. Just read the comments to any article on GMO and see how long it takes for anyone defending the science of GMO to be labeled a Monsanto shill. US Right to Know (an anti-GMO group funded by the organic lobby) has used Freedom of Information Act (FOIA) requests to demand the e-mails of public scientists. They then scour these e-mails looking for anything that can be made to seem sinister. Any connection to industry, no matter how innocent or appropriate, is spun into a narrative of nefarious corporate shilling.

And, of course, evolutionary biologists just hate God.

At the reasonable end of the spectrum is the sensible requirement for full disclosure of potential conflicts of interest, so readers can judge for themselves the integrity of the source. This can easily slide into a witch hunt, however, with even the thinnest and most tenuous connection used to argue that a scientist is actually a paid insider and should be completely discounted.

Appeal to Academic/Intellectual Freedom

Personal freedom is highly valued in US culture (and elsewhere), so the appeal to personal freedom is especially effective, which explains its popularity as an argument. Laws meant to shield charlatans from being held responsible for the proper standard of care in medicine are sold as "health care freedom" laws. Creationist attempts to undermine the teaching of evolution are framed as "academic freedom." Anti-vaccinationists, of course, are constantly advocating for the parent's right to choose.

What all such movements miss, however, is that science, academia, and professions have standards. Maintaining standards isn't anti-freedom, but it is easy to misrepresent it as such and to portray all attempts at promoting high standards as "elitism."

Universities, for example, are under no obligation to allow any crank to teach nonsense in their name. They do, however, have a responsibility to their students to teach only academically valid material.

Argument from Consequences

Accepting a particular science may be inconvenient for a specific political or religious ideology, and pointing out this inconvenience in order to deny the science is also extremely common. Creationists argue that accepting evolution will undermine belief in God and even result in moral decay. Global warming deniers argue that accepting the "alarmist" claims about climate change will result in a government takeover of private industry. I characterize this strategy as an argument-from-final-consequences logical fallacy—evolution is wrong because if it were true society would suffer. This feature also often provides a clue as to the true motivation of the denial. The science is secondary: It's the moral hazard they're truly concerned about.

This is an inherently flawed strategy. If you truly wish to advocate for a particular moral or ethical position, the worst thing you can do is tie that position to a false scientific conclusion. Doing so allows opponents to attack your moral position by attacking the pseudoscience to which you have anchored it. You are far better off acknowledging legitimate science and advocating for your moral position on moral grounds. If you ideologically favor free markets, don't deny global warming, rather offer free-market solutions.

Some researchers have labeled this phenomenon "solution aversion" in the case of global warming: Reject the science because you don't like the proposed solutions. Again, it's a better strategy to focus on the solutions rather than deny the science.

It should be clear by now that I and my SGU colleagues are all strong advocates of science. It is clear from history that science is the most powerful tool we have for understanding the world and improving our position in it. But science requires courage—the courage to face reality and accept its findings, even if they upset us or are disruptive to our comfortable ideology.

Denialism seeks to deprive us of the power of science by attacking those findings the deniers lack the intellectual courage to evaluate honestly. But denialism does not refer to specific people as much as it does to a behavior, one we need to be vigilant against—in ourselves most of all.

24. Skeptics' Guide Entry: *P*-Hacking and Other Research Foibles

Section: Science and Pseudoscience
See also: Pseudoscience

There are multiple ways to bias the outcome of seemingly rigorous scientific studies. It is critical to be able to assess the reliability of research to know if its claims should be taken seriously.

> The nature of science is not that of a steady, linear progression toward the Truth, but rather a tortuous road, often characterized by dead ends and U-turns, and yet ultimately inching toward a better, if tentative, understanding of the natural world.
> —Massimo Pigliucci

I was watching my daughter Julia play Rock'em Sock'em Robots with my four-year-old nephew (Jay's son, Dylan). On one round, Dylan's robot knocked Julia's bot's head off, and he declared that he had won. On the next round, Julia struck first, and Dylan's robot dutifully popped his head. Again, Dylan declared that he had won. In response to Julia's questioning he explained that on that round the robot that lost its head was the winner.

We all remember this from our first ventures into group play—young children will sometimes try to change the rules after they see the results. We laugh now because their feeble attempts at rigging the game are so transparent and contrived. Even as older children, let alone adults, we understand that you must establish the rules ahead of time and stick with them. You don't get to decide after you see the outcome that whatever happened meant that you win.

Adults, however, still do this. Even respected scientists may succumb to the temptation to tweak the rules after the outcomes are known—they just do it in a much more subtle and complex way. They may not even be aware they are doing it.

The rules of science are far more important than any game. They are designed specifically to control for any cheating, whether deliberate or inadvertent. Rigorously adhering to these rules has transformed our understanding of the universe and is largely how we separate what is actually true from wishful thinking. However, much of what passes for scientific research is flawed or preliminary. Some of it is utter crap.

There's so much research out there, with literally millions of new studies being published each year, that you could cite scientific studies to support just about any claim you wish (if you were willing to do a little cherry-picking).

How do we tell the difference between reliable, solid, scientific research and low-quality research? Let's do an SGU "deep dive."

Here's an example of a scientific study gone wrong that we discussed on the show in early 2017: "Field Effects of Consciousness and Reduction in U.S. Urban Murder Rates; Evaluation of a Prospective Quasi-Experiment." This study comes from the Maharishi University of Management.

The idea here is that consciousness is a field and there is a universal field of consciousness of which we are all a part. When individuals engage in Transcendental Meditation (TM) they are not only affecting their own consciousness, they're affecting the entire field.

The point of this and other TM studies is to confirm the belief (they're not *testing* the belief) that if enough people put good vibrations into the universal field of consciousness, society in general will benefit. How many is enough? Well, apparently they have an answer for that. It's the square root of 1 percent of the population (take the population, divide by 100, then take the square root of the result). Why? Because math.

In fact, the number used to be 1 percent of the population. But then (after failing to reach the number needed to affect the world) they "discovered" that the real number was only the square root of 1 percent.

Apparently, there isn't a dose-response effect, there is a threshold effect, and once you get over the magic threshold, the effect kicks in. That threshold has a simple mathematical formula, which makes it sound all sciencey. There's no established theoretical reason for this; it has more in common with a magic ritual and numerology than a scientific process. (The real world is not base ten; that is only a human convention. There's no reason why any natural constant would involve such a round number as 1 percent.)

This is what they claim to have done for the experiment: They tracked the murder rate in the 206 largest US cities where FBI murder statistics were available. They then compared the baseline period from 2002 to 2006 with the intervention period, from 2007 to 2010. The intervention period was when the "Transcendental Meditation and TM-Sidhi group exceeded 1,725 participants beginning in January 2007." They claim that murder rates were increasing before that date, but then when participant #1,725 (because 1,724 was not enough) signed up, the rate started to drop.

What's really funny is the blatant frequentist (basing conclusions solely on statistics) nonsense they spout in the press release. They claim to have "calculated that the probability that the reduced trend in murder rates could simply be due to chance was 1 in 10 million million."

So they think the probability that the murder rate declined from 2007 to 2010 due to chance was one in a trillion.

This is an excellent example of ways to bias the outcome precisely because it tries really hard to have all the form of a serious scientific study, despite the fact that they are essentially studying magic.

The reality is that crime and homicide rates were dropping for the previous forty years. The decrease shown in the study was just a continuation of a long-term trend. In addition, they chose an arbitrary study period (why 2002–2006?) and an arbitrary follow-up period. In fact, during the follow-up period the number of meditators dropped below the magic number at times. There are so many arbitrary choices in this research that you can make the numbers say whatever you want. (You see, this time the robot who loses his head wins.) You can then scrub the resulting write-up to make it seem all perfectly legitimate. This is what we call torturing the data until they confess.

The Problem with *p*-Values

There are many things that can go wrong with a scientific study, but perhaps chief among them is "*p*-hacking." The term derives from the statistical calculation known as the *p*-value. The *p*-value is just one way to look at scientific data. It first assumes a specific null hypothesis (such as: There is no correlation between these two variables) and then asks, What is the probability that the data would be at least as extreme as they are if the null hypothesis were true? A *p*-value of 0.05 (a typical threshold for being considered "significant") indicates a 5 percent probability that the data are due to chance rather than a real effect.

Except—that's not actually true. That is how most people interpret the *p*-value, but that's not what it actually says. *P*-values don't consider many other important variables, like prior probability, effect size, confidence intervals, and alternative hypotheses. For example, if we ask, "What is the probability of a new set of data replicating the results of a study with a *p*-value of 0.05?" we get a very different answer.

In a *Nature* commentary on this problem, Regina Nuzzo reports:

These are sticky concepts, but some statisticians have tried to provide general rule-of-thumb conversions (see "Probable cause"). According to one widely used calculation, a P value of 0.01 corresponds to a false-alarm probability of at least 11%, depending on the underlying probability that there is a true effect; a P value of 0.05 raises that chance to at least 29%. So Motyl's finding had a greater than one in ten chance of being a false alarm. Likewise, the probability of replicating his original result was not 99%, as most would assume, but something closer to 73%—or only 50%, if he wanted another "very significant" result. In other words, his inability to replicate the result was about as surprising as if he had called heads on a coin toss and it had come up tails.

Let me restate that: A study with a *p*-value of 0.01 may have only a 50 percent chance in an exact replication of producing another *p*-value of 0.01 (not the 99 percent chance that most people would assume).

Put another way, people (even experienced scientists) tend to think of the *p*-value as a predictive value, but it isn't. It was never meant to be. It's only a smell test to see if the data are at all interesting or just random noise.

Here is an example from medicine I often use in my lectures. Let's say that 1 percent of forty-year-old women have breast cancer. Let us further say that mammograms are 80 percent sensitive (they will come out positive 80 percent of the time with patients who actually have breast cancer). They are also 90 percent specific (they will be negative 90 percent of the time with patients who do not have breast cancer). These numbers are pretty good for a screening test.

Here is the question: What is the predictive value of a positive mammogram in a forty-year-old woman, or what are the odds that she actually has breast cancer because her mammogram was positive? With a 90 percent specificity you might be tempted to say 90 percent, but that would be wrong. The real answer is 7.5 percent. This is because 99 out of 100 forty-year-old women do not have breast cancer, so with a 10 percent false-positive rate there will be about 10 women out of 100 who test positive but don't have breast cancer. Only about 0.8 of the 1 in 100 women who have breast cancer will test positive.

So, out of 100 women, 9.9 will test positive but be negative, and 0.8 will test positive and be positive. So if you are a forty-year-old woman and test positive, chances are much greater it's a false positive (0.8 true positive / 10.7 total positive x 100 = 7.5%).

This situation is analogous to the p-value. Having a p-value of 0.05 does not mean there is a 95 percent chance the hypothesis is true any more than a positive mammogram in a forty-year-old woman means there is a 90 percent chance she has breast cancer.

Just as with breast cancer, you have to know what the base rate is. We also call this the prior probability. With a scientific hypothesis this often means the scientific plausibility. The lower the plausibility (just like the lower the rate of cancer in the population), the smaller the percentage of positive or statistically significant studies will be true positives.

This means that we can't know the probability that a hypothesis is true just from the p-value of one study. We need to know the plausibility of the hypothesis, and we need to know what all the other relevant studies showed.

We call this a Bayesian approach—you take the new information, you add it to the prior existing information, and you come up with a new probability that the idea is true. While it's possible to debate about how plausible a hypothesis is, one thing is clear: p-values (statistical significance) don't mean nearly as much as most people think they do. Even a significant study doesn't alter the prior probability by that much. It takes a convergence of multiple studies, and multiple independent lines of evidence, before we can be confident that a hypothesis is likely to be true.

If we apply this reasoning to the scientific literature, as statistician and professor of medicine John P. A. Ioannidis has done, we find that most published positive studies are, and should be, wrong. As he demonstrated in his seminal 2005 study, if we assume that 80 percent of new hypotheses in science are wrong (which is a conservative assumption), and we use a p-value of 0.05, then by chance alone 25 percent of positive studies will be false positives. This number increases dramatically as the prior probability goes down.

Prior probability isn't the only factor that worsens the false-positive problem. Eric Loken and Andrew Gelman point out that measurement error also greatly exacerbates the problem. This is why we consider the signal-to-noise ratio in whatever is being measured in a scientific study. In a noisy environment, measurement errors are magnified, and the predictive value of a p-value plummets. "Noisy" data are exactly that—it's like listening to a radio station with so much static (the noise) you can barely make out the announcer's voice (the signal). This could just mean that the natural variability in the data is far greater than the effect you are looking for.

P-Hacking

The problem goes even deeper, and this is where p-hacking comes in. Ioannidis's calculations assume that research protocols are perfectly designed and executed—that everyone is playing by the rules—but we know this isn't the case.

In a 2011 study, Joseph Simmons, Leif Nelson, and Uri Simonsohn revealed the effects of exploiting "researcher degrees of freedom." This means choosing when to stop recording data, what variables to follow, which comparisons to make, and which statistical methods to use—all decisions that researchers have to make about every study. If, however, they monitor the data or the outcomes in any way while making these decisions, they can consciously or unconsciously exploit their "degrees of freedom" to reach the magic p-value of 0.05. In fact, Simmons showed you can reach a p-value of 0.05 60 percent of the time with completely negative data.

Simonsohn points out that p-values in published papers cluster suspiciously around the 0.05 level—implying that researchers were engaging in p-hacking until they reached this minimal publishable threshold.

There is more direct evidence that p-hacking is taking place. A 2009 *PLOS One* review found that 33 percent of surveyed researchers admitted to committing one or more dubious practices. What dubious practices? Those that amount to p-hacking.

It's likely that most p-hacking is innocent, meaning that the researchers don't realize they're essentially cheating. For example, if you survey the data as you collect it, you might decide that once you cross over the $p = 0.05$ threshold you can stop collecting data and publish.

Tracking the data isn't in itself dubious and is often required in medical studies to make sure subjects are not being harmed. But this kind of tracking should be done independently of the primary investigator who is gathering the data to publish. Or, at the very least, the number of subjects needs to be determined ahead of time and not altered by monitoring the data as they come in.

Essentially, anything a researcher changes about the research after looking at data can be p-hacking. That's because it alters the statistics. P-hacking is essentially mining the data or taking multiple throws of the dice and only counting the results you like.

Another way in which much of what is published doesn't reflect reality comes from the problem with replications.

Independent replications really are the ultimate arbiters of what is real in science. Any one study can be a fluke or the product of biased research. Only phenomena that are real, however, should show up in the data no matter who is doing the research.

Exact replications are particularly useful, because by definition they eliminate all the degrees of freedom. All the choices about collecting and analyzing the data have already been made by the study being replicated.

There are many who think there is currently a replication problem in science. In a 2016 study published in *Nature*, 52 percent of scientists surveyed reported this belief based on their own inability to replicate the work of others.

There have been several largely unsuccessful and now famous attempts to replicate seminal studies in psychology and elsewhere. In one 2015 attempt involving a hundred psychological studies, only thirty-nine were deemed to have been successfully replicated.

The problem isn't confined to psychology, but as discussed above, fields like psychology and medicine, which are noisy in terms of measuring outcomes (meaning the data are inherently variable) will tend to have higher false-positive rates.

Remember our friend Daryl Bem? The problem with his psi research is that he was engaging in *p*-hacking. In fact, in a 2017 interview Bem essentially admitted that he engages in research techniques designed to comb through data looking for an effect:

> "I'm all for rigor," he continued, "but I prefer other people do it. I see its importance—it's fun for some people—but I don't have the patience for it." It's been hard for him, he said, to move into a field where the data count for so much. "If you looked at all my past experiments, they were always rhetorical devices. I gathered data to show how my point would be made. I used data as a point of persuasion, and I never really worried about, 'Will this replicate or will this not?'"

Bem could not resist *p*-hacking to rescue his failed psi experiment. He changed the rules specifically to make him the winner—and the scientific community looked at him the same way my daughter looked at her cousin: Nice try, Bem.

The Fix

The problems with both *p*-hacking and overusing the *p*-value are all fixable. One important fix, as discussed above, is valuing exact replications more. Value is easy to determine in science—what gets published, what gets funded, and what gets you academically promoted.

Statistician Andrew Gelman of Columbia University suggests that researchers should do research in several steps. First, collect preliminary data and, if they look promising, design a replication where all the decisions about data collection are predetermined. Then register the study methods before collecting any data. Finally, collect a fresh set of data according to the published methods. At least then we'll have honest *p*-values and eliminate *p*-hacking.

Researchers should not rely only on p-values. They should also report effect sizes and confidence intervals, which are more thorough ways of looking at the data. Tiny effect sizes (a one-week cold was reduced on average by one hour—wow!), no matter how significant, are always dubious because subtle but systematic biases, errors, or unknown factors can influence the results.

Simonsohn advocates researchers disclosing everything they do—all decisions about data collection and analysis. This way at least they cannot hide their p-hacking, and the disclosure will discourage the practice. Nuzzo and others recommend more use of Bayesian analysis, which we discussed earlier—asking, What is the overall probability that this effect is real?

Where Do We Stand?

Looking past all the math, what all of this means for the average science enthusiast or for the science-based practitioner is that you have to look beyond the p-values when evaluating any new scientific study or claim. Also, look for p-hacking when sizing up scientific research—did the researchers have any wiggle room to change the rules as needed to get the result they wanted?

We can still get to a high degree of confidence that a phenomenon is real. This is what it takes for research to be convincing:

1. Rigorous studies that appear to minimize the effect of bias or unrelated variables
2. Results that are not only statistically significant but are significant in effect size as well (reasonable signal-to-noise ratio)
3. A pattern of independent replication consistent with a real phenomenon
4. Evidence proportional to the plausibility of the claim

What we often see from proponents of various pseudosciences or dubious claims is the trumpeting of one or two of these features—but never all four (or even the first three). They showcase the impressive *p*-values but ignore the tiny effect sizes or the lack of replication, for example.

Homeopathy, acupuncture, and ESP research are all plagued by these deficiencies. They haven't produced research results anywhere near the threshold of acceptance. Their studies reek of *p*-hacking, they generally have tiny effect sizes, and there is no consistent pattern of replication, just chasing different quirky results.

Yet there is no clean dichotomy between science and pseudoscience (that pesky demarcation problem). Yes, there are those claims that are far toward the pseudoscience end of the spectrum, but all of these problems plague mainstream science as well.

P-hacking leads to a tremendous waste of scientific resources. It's incredibly inefficient, flooding the literature with spurious results that probably won't be replicated. It's also almost entirely ignored when reporting science to the public. All the public hears is, "Scientists report a significant result," but most of what gets relayed as excited scientific breakthroughs is simply crap cluttering up the scientific journals.

25. Skeptics' Guide Entry: Conspiracy Theories

Section: Science and Pseudoscience
See also: Grand Conspiracies

A conspiracy theory, or more precisely a grand conspiracy, is a belief system that involves at its core the claim that a vastly powerful group is carrying out a deception against the public for their own nefarious ends.

> The government kinda sucks at keeping secrets.　　　　　—Bill Nye

---

Conspiracy thinking is arguably the confluence of many of the logical fallacies and cognitive biases that we have already discussed. In many ways it is the "one ring to rule them all" of faulty thinking.

Conspiracy thinking crops up in just about every topic skeptics deal with. It is an all-purpose get-out-of-jail-free card, played whenever a true believer's back is up against the wall, an attempt to flip the table over when you are losing the game. I have also referred to the conspiracy theory as the last refuge of the intellectual scoundrel.

And yet conspiracy theories remain stubbornly common. Let's take a peek inside to see if we can figure out why.

The Grand Conspiracy

When we talk about conspiracies or conspiracy theories, we're generally referring to what is also called the "grand conspiracy." This term differentiates the big crazy conspiracies from more mundane and plausible conspiracies.

No one doubts the existence of actual conspiracies. Whenever two people get together and agree to work together to perpetrate a crime, that's a conspiracy. There are many conspiracies that exist at the boardroom level—the inner circle of one company or one government agency.

A grand conspiracy is much bigger. This is a conspiracy that by its very nature must span many people and organizations, perhaps even multiple nations and generations.

The grand conspiracy forms a triangle of sorts. First there are the conspirators. This is typically a large, powerful, and shadowy organization with vast resources and control. They need to be powerful in order to fake moon landings, poison the public through jet exhaust, or frame terrorists for 9/11. Then there are the conspiracy theorists, an "Army of Light" that is able to see through the conspiracy (because they are just so clever). Finally, there is everyone else, the dupes or "sheeple" who believe the standard explanation of history and current events.

Conspiracy Thinking

There's a certain flawed logic to conspiracy thinking that makes it both tantalizing and fatally flawed at the same time. The core problem with conspiracy thinking is that it's a closed belief system, specifically designed to insulate itself from external refutation or even the need for internal consistency. Conspiracy theories are often elaborate exercises in special pleading.

Any evidence that can potentially falsify the conspiracy theory is just part of the conspiracy. To the conspiracy theorist, such evidence was obviously fabricated in order to maintain the conspiracy. So, whether it's a scientific study showing that vaccines are safe, high-resolution pictures showing that astronauts were on the moon, or evidence linking terrorists to 9/11—it's all fake.

Any events that seem to contradict a conspiracy narrative were clearly a "false flag" operation: The government engineered the whole thing to throw the sheeple off track. Further, any evidence that we would expect to find to support the conspiracy but is missing is part of a cover-up. The conspirators are really good at covering their tracks. How good? As good as they need to be.

The cover-up is where grand conspiracies tend to get into a death spiral of special pleading. No one has come forward to claim that they were part of the alleged government plan to control the population by releasing "chemtrails" from commercial jets. Why not? Because the government has the power to intimidate or silence any potential whistleblowers.

If George Bush faked the terrorist attacks on 9/11, why didn't Democrats expose the conspiracy when they held the reins of power? They must have been in on it. Why hasn't the media exposed the conspiracy either? They must be in on it too. What about other governments, many of whom are our enemies? Take a wild guess.

Any problems with the conspiracy can be solved by simply expanding it. Before long, you get to the belief that a worldwide shadow government controls everything—the Illuminati, reptilian overlords, the New World Order, and similar grand conspiracies.

Not only does the breadth and scope of the conspiracy have to enlarge, but the power and cunning of the conspirators as well. For example, there's a common belief that "they" have the cure for cancer but are hiding it from the public. Who are "they" exactly? This bit of imprecision hides many assumptions. Big Pharma is a likely answer, but that doesn't cut it.

Individual pharmaceutical companies compete against each other, and new start-ups form all the time. What's stopping a new company from marketing an alleged cure? What about other countries?

Who actually controls research? Most basic cancer research is government funded and carried out by researchers at university labs. How would Big Pharma control this research? Papers are published in peer-reviewed journals, and researchers exchange knowledge routinely. Anything approaching a cure would be the result of dozens of labs, if not more, working over a period of years.

Pharmaceutical companies get involved only in the last step, taking a possible new drug target and developing a workable pharmaceutical.

Perhaps cancer doctors are the ones behind it all? The argument often goes that they want to keep patients coming. Well, the hospitals and clinicians don't control the research either. Since researchers' goals are to be famous and get grants, they don't care if clinicians make money. In any case, curing patients tends to be highly profitable. Allowing your customers to die isn't a good long-term strategy.

Once you begin to see the entire system, involving many people and institutions with different motives, the notion that there is a "they" that can control the whole thing for their own nefarious purposes becomes obviously absurd. No one has the power to hide something as big as a cure for cancer.

While the conspirators are necessarily preternaturally large and powerful to pull off such elaborate conspiracies, they are simultaneously incredibly stupid in the ways they allegedly "expose" themselves to the conspiracy theorists. The conspirators are brilliant when the believers need them to be and careless when they need them to be. So they will stage sending astronauts to the moon and then allegedly allow a fan or open door to blow the flag to give away that it's not in a vacuum. (D'Oh!)

Once the conspiracy narrative is adopted, it becomes a lens through which reality is viewed. Pattern recognition and hyperactive agency detection combine to form a tendency to see disparate events as connected, with an unseen agent behind them. Confirmation bias then kicks in. Every event, no matter how random or innocent, can become evidence for the conspiracy. Anomaly hunting feeds into this as well. Everything even slightly unusual or unfamiliar becomes an anomaly that proves the conspiracy. Every coincidence is part of the pattern.

Conspiracy theorists also commit the fundamental attribution error, ascribing deliberate actions to others and ignoring the quirky external details of everyday life. For example, on November 24, 1963, Jack Ruby shot and killed Lee Harvey Oswald on live television and in the presence of police officers. To conspiracy theorists this was an obvious silencing. Why else would Ruby have done this?

In *Case Closed*, however, Gerald Posner makes a convincing case that Ruby was a loser and a poser (a mobster wannabe) who was deeply disturbed by the killing of President Kennedy and wanted to be the man who killed the man who killed JFK. Yet conspiracy theorists easily impose their own narrative on his motivations.

The ad hominem attack is also a common fallacy employed by the conspiracy theorists. If you question their elaborate conspiracy theory, then you're gullible and lack the vision to see events for what they are. If you point out the factual and logical problems with their case, then you're clearly part of the conspiracy. You are a shill, or part of an "astroturf" campaign, or even perhaps one of the Illuminati.

Conspiracy theories are also often arguments from ignorance. Theorists point to apparent anomalies, coincidences, or things that don't make sense to their limited understanding, and then "just ask questions." If you can't explain everything down to an arbitrary level of detail, there must be a conspiracy. They don't have to provide evidence for the conspiracy, they can just poke holes in the standard version of events.

A 2012 study by Michael Wood, Karen Douglas, and Robbie Sutton showed that believers can accept mutually exclusive conspiracy theories at the same time. Belief that Princess Diana was murdered correlated with belief that she faked her death and was still alive. Belief that Osama bin Laden was already dead when the Navy SEALs got to his compound correlated with belief that he is still alive. Any conspiracy is acceptable, even those that contradict each other.

Finally, the conspiracy theorists make tautological arguments, usually in the form of cui bono—"who benefits?" This is a reasonable question to ask when forming a hypothesis about an actual crime. Conspiracy theorists, however, will use it as evidence for a particular conspiracy.

Every big event will have winners and losers. Those who come to power or may have gained in some way as a result of an historical event didn't necessarily cause it. "Don't be naive," the conspiracy theorist will say. "They gained, so of course they caused it to happen."

Too Big Not to Fail

A 2016 paper in *PLOS One* explores the mathematical probability of a grand conspiracy being revealed from within. The paper, of course, doesn't disprove any particular conspiracy theory, but it does make a compelling argument by putting into rigorous form a frequent case against grand conspiracies, that they're too big not to fail.

The author of this paper, David Robert Grimes, writes about conspiracies as follows:

> Conspiratorial ideation is the tendency of individuals to believe that events and power relations are secretly manipulated by certain clandestine groups and organisations. Many of these ostensibly explanatory conjectures are non-falsifiable, lacking in evidence or demonstrably false, yet public acceptance remains high. Efforts to convince the general public of the validity of medical and scientific findings can be hampered by such narratives, which can create the impression of doubt or disagreement in areas where the science is well established.

When conspiracies get very large, involving large numbers of people across multiple institutions and many years or even decades, this raises questions about the plausibility and logistics of orchestrating and maintaining such a conspiracy. Grand conspiracy theories tend to grow larger and more complex until they collapse under their own weight.

In the *PLOS One* paper, Grimes set out to do a probability failure analysis of grand conspiracies. What's the probability that they will fail from within, meaning that someone who is in on the conspiracy either deliberately or accidentally exposes the conspiracy sufficiently that it fails? He didn't consider extrinsic failure—being exposed by outside investigation.

Grimes used real historical conspiracies as his guide, namely the National Security Agency spying scandal (exposed by Edward Snowden), the Tuskegee syphilis experiments, and the FBI forensics scandal.

In order to make a mathematical model of grand conspiracy failure, he had to plug in several factors. These include: How many people would need to be involved in the conspiracy; how does that number change over time; and how reliable is each individual on average? From the historical examples, he came up with a range of reliability and then used the most reliable end of the spectrum for his calculations to come up with the most conservative estimate of failure probability.

How the number of conspirators changed over time is an interesting variable. For some conspiracy theories, like the alien recovered from a crashed saucer at Roswell, the people involved at the time would die off over the years, reducing the number of people who can expose the conspiracy and decreasing the probability that this would happen.

However, other conspiracies, such as covering up the alleged "real" risks of vaccines, would have a steady roster of people involved. This is because the conspiracy isn't just covering up a single historical event, but ongoing scientific research and the analysis of data.

For his analysis he used four grand conspiracies: the moon landing hoax, that climate change is fake, vaccines, and the hidden cancer cure. For each he had to come up with some way of estimating the number of people necessarily involved in the conspiracy. I think his numbers are conservative, but others might argue that a small number of people could control the process. I disagree, especially whenever scientific data are involved. Any scientists with adequate training can look at the data and do their own analysis.

For the moon landing he used the peak total employment of NASA in 1964, which was 411,000 people. For the climate change hoax he used the membership of scientific organizations that have backed the notion of AGW, resulting in 405,000 people. For vaccines he used the total of those employed by the CDC and the World Health Organization: 22,000. (He could've easily added the members of every pediatrics professional organization.) For the hidden cancer cure he used the employees of the major pharmaceutical companies, who numbered 714,000.

You can, of course, quibble about these numbers. I think they're probably reasonable to an order of magnitude, which is all that matters for his analysis.

Grimes's model predicts these grand conspiracies would intrinsically fail within about four years. Remember, he is using the most conservative estimates for individual reliability and dedication to the conspiracy. For average values, the conspiracies would fail much more quickly. Even if you think he is off by an order of magnitude involving tens of thousands of people, results still showed a high probability of failure within a few years.

You can adjust the variables (reliability, number of people involved, and the change in that number over time) to generate failure curves with his model. If you have several thousand highly reliable conspirators you still get a fairly high probability of failure within decades, with more rapid failure as the conspiracy grows or reliability decreases.

In short, grand conspiracies can only exist in a fantasy world. Author Dean Koontz summarized it this way:

> The sane understand that human beings are incapable of sustaining conspiracies on a grand scale, because some of our most defining qualities as a species are inattention to detail, a tendency to panic, and an inability to keep our mouths shut.

Who Believes in Conspiracies?

While it seems that we each have a little conspiracy theorist inside us, there is definitely a spectrum when it comes to the tendency to engage in conspiracy thinking. There's a conspiracy bias, if you will, and many of us are at least a little prone to it.

Most people are opportunistic conspiracy theorists. We tend to accept conspiracies ad hoc when it suits our beliefs. This is most obvious when it comes to political ideology. For example, a 2016 survey found that 17 percent of Clinton voters believed Clinton's e-mails made reference to a ring inside Washington providing child prostitutes for politicians (the so-called "Pizzagate" conspiracy). Meanwhile, 46 percent of Trump voters believed this. On the other hand, a 2013 survey by Public Policy Polling found that 29 percent of liberals believed or were uncertain that the Bush administration knowingly allowed 9/11 to happen, compared to only 15 percent of conservatives.

Across the board, liberals and conservatives were more likely to accept a conspiracy if it was in line with their ideology. Meanwhile, for conspiracies without clear political implications, such as whether or not the government is covering up a crashed saucer at Roswell (21 percent overall believe this) or whether Paul McCartney was killed in 1966 (5 percent overall), belief is roughly the same across the political spectrum. There's also no political difference when it comes to basic conspiracy ideology, such as the tendency to think there are powerful secret forces at work in the world.

However, the same surveys also find that a subset of the population appears to be general conspiracy theorists—they believe in most conspiracies regardless of the political implications. They believe in conspiracy theories because of a general tendency toward conspiracy thinking.

Here are some common conspiracy theories with the percentage of believers, according to Public Policy Polling:

- 20 percent of voters believe there is a link between childhood vaccines and autism, 46 percent do not.
- 7 percent of voters think the moon landing was faked.
- 13 percent of voters think Barack Obama is the Antichrist, including 20 percent of Republicans.

- Voters are split 44 percent to 45 percent on whether President Bush intentionally misled us about weapons of mass destruction in Iraq. 72 percent of Democrats think Bush lied about WMDs, Independents agree 48–45 percent, but just 13 percent of Republicans think so.
- 29 percent of voters believe aliens exist and are visiting the Earth.
- 14 percent of voters say the CIA was instrumental in creating the crack cocaine epidemic in America's inner cities in the 1980s.
- 9 percent of voters think the government adds fluoride to our water supply for sinister reasons (not just dental health).
- 4 percent of voters say they believe "lizard people" control our societies by gaining political power.
- 51 percent of voters say a conspiracy was at work in the JFK assassination, whereas just 25 percent say Oswald acted alone.

There seems to be a 4–5 percent floor for any conspiracy, no matter how bizarre. Those are the hard-core conspiracy theorists.

There also appears to be some basic psychology at work. We tend to assume that big events must have big causes. It just doesn't sit right with us to think that a major world event, with significant historical implications, was pulled off by some lone nutjob. That's partly why a majority of Americans still believe in a JFK conspiracy, even after fifty years with no real evidence coming to light to support it. There were similar conspiracy theories for decades following the Lincoln assassination. There remain conspiracy theories about the assassination attempt on Ronald Reagan.

The Psychology of Conspiracy

Psychologists Viren Swami and Rebecca Coles reviewed conspiracy research in their 2010 article "The Truth Is Out There." They discuss how early papers on conspiracy theories focused on characterizing the theories themselves rather than the people who hold them. They reference Richard Hofstadter's "seminal" 1966 paper on conspiracy theories, in which he defined them as a belief in a "vast, insidious, pre-ternaturally effective international conspiratorial network designed to perpetrate acts of the most fiendish character."

That sums it up nicely. But the more interesting work came later, when researchers began to explore the psychology of the people who hold conspiracy theories. In this area, ideas followed a typical historical pattern. At first, conspiracy thinking was seen as a form of psychopathology involving paranoid delusional ideation. More recently, conspiracy thinking is seen as fulfilling certain universal psychological needs perhaps triggered by situational factors.

In my view both approaches are correct—there appears to be a spectrum of inherent predisposition to conspiracy thinking. There is at the same time a universal appeal to conspiracy theories and a type of situation in which they are more likely to occur, even among the more rational. For example, Swami and Coles write:

> To the extent that conspiracy theories fill a need for certainty, it is thought they may gain more widespread acceptance in instances when establishment or mainstream explanations contain errone-ous information, discrepancies, or ambiguities (Miller, 2002). A conspiracy theory, in this sense, helps explain those ambiguities and "provides a convenient alternative to living with uncertainty" (Zarefsky, 1984, p.72). Or as Young and colleagues (1990, p.104) have put it, "[T]he human desire for explanations of all natural phenomena—a drive that spurs inquiry on many levels—aids the conspiracist in the quest for public acceptance."

Conspiracy thinking is rooted in a desire for control and under-standing, triggered by a lack of said control or ambiguous and unsat-isfying information. The authors emphasize that the public often lacks access to adequate information to explain historical events (a situational factor). This can be coupled with what has been called a "crippled epistemology"—a tendency to utilize circular reasoning, confirmation bias, and poor logic coupled with this deficiency of information. The result is a popular conspiracy theory that makes sense (even if a perverse sense) of events.

A 2016 study by Damaris Graeupner and Alin Coman found that conspiracy thinking correlates with feelings of isolation and helplessness. This fits nicely with the notion that conspiracy theories are a pathological attempt to feel in control.

The Sandy Hook Conspiracy

A recent grand conspiracy theory that displays many of the features I discuss above is the Sandy Hook conspiracy. In December 2012 a disturbed young man decided it would be a good idea to go into an elementary school and slaughter young children. He killed his mother, twenty children, six teachers, and then himself. It was a horrific event for the families, for the town, and for our country. It was every parent's worst nightmare.

The tragedy took place in Sandy Hook Elementary School, in Newtown, Connecticut. As it happens, my parents and brother (Bob) live in Newtown not far from Sandy Hook. I also have friends who had children in the school (none hurt, thankfully), and one of those friends was at the school when the shooting took place. Her husband (also a friend) was one of the first responders and saw the bodies.

The couple are close friends of my fellow Rogue, Evan, who has known them since childhood. I was introduced to them more recently. I point all this out for a couple of probably obvious reasons. The first is that I have only one degree of separation from direct eyewitnesses. The second is to indicate that this is a small Connecticut town. There are hundreds of families, at least, with similar connections to the shooting.

James Tracy, a professor of communications at Florida Atlantic University until he was dismissed in 2016, runs a conspiracy-mongering blog in which he calls into question the official story of what happened at Sandy Hook. In 2013, soon after the shooting, he was already raising conspiracy questions about its veracity. At the time, he was "just asking questions." He has since progressed to harassing the families of victims.

Tracy believes that the Sandy Hook massacre didn't occur the way the media has reported, and perhaps it didn't occur at all. He makes all the logical errors that conspiracy theorists make.

One of his primary errors is to point to discrepancies in the way the media reported the details. As anyone who has paid attention to the media while a story is breaking has experienced, during a dramatic event the media struggles to get out any tidbit of information as quickly as possible, and some of this preliminary reporting will later prove incorrect. This is evidence of the media fighting for ratings during an unfolding event, not evidence that the media is involved in a conspiracy to fabricate a story.

Tracy's errors are an example of the anomaly hunting I discussed previously—in this case the anomalies are discrepancies in media reporting, plus other extraneous details reported by the media before it was known what they meant. Any event such as this is bound to have many quirky details, and it will be difficult to explain every detail to an acceptable degree without specific knowledge. Strange coincidences will occur—there are just so many details in everyday life, and when a dramatic event suddenly draws attention to all those tiny details, we see all the apparent anomalies.

For example, right after the shooting, when the police were securing the area, several individuals were encountered in the woods around the school. They all turned out to be either reporters hoping to get some photographs or locals just interested in the commotion. But to conspiracy theorists, they are evidence of a broader plot. Who were these people? Why exactly were they there? Why isn't there more detail in the police reports?

In one on-camera interview, Robbie Parker (father of slain six-year-old Emilie Parker) seems to be smiling right before he addresses the camera, when he suddenly looks grief-stricken. To conspiracy theorists this is evidence that he is a "crisis actor" (a term they made up). Those grieving a recent loss, however, can still exhibit a range of emotions, and no one knows what Parker was responding to before he stepped in front of the camera.

The dramatic tragedy of these events in the context of our current media age causes most of the grieving families to protect their privacy. They don't want to be the target of media attention while they're trying to process a horrific loss. Tracy and other conspiracy theorists take this ordinary desire for privacy and interpret it as a cover-up. Ironically, conspiracy theorists can bring extremely negative attention to the families, increasing their desire for privacy, and then they use the privacy they helped to motivate as evidence for a conspiracy.

In addition to all the logical errors, theorists like Tracy who interpret everyday events as if they were evidence of a deep dark conspiracy miss the big picture in such a dramatic way that it's difficult not to conclude that they're divorced from reality. Think for just a minute what it would take to pull off such a conspiracy. The perpetrators would have to fake a mass shooting at an elementary school, pretending to kill twenty children and seven adults who never existed. You would think someone in the town would notice.

The entire community would have to be involved in the conspiracy—in fact, a significantly extended community would have to be involved. Evan and I would both have to be involved, and we live several towns over. The web of direct connections to the families of those killed is massive.

I can't help but point out the flagrant journalistic failure of people like Tracy. He's defending himself as doing the type of investigation that any good journalist should. He is wrong for several reasons. First, as I just pointed out, his theory is absurd, and good journalists don't have to investigate every absurd theory that someone cooks up. Plausibility counts. Further, he's going about it in a clumsy and unethical way. He is harassing grieving families without any probable cause and without a lick of hard evidence to back him up: He has no evidence for a conspiracy (just misinterpreted anomaly hunting).

A competent journalist, if they thought they had a story that needed investigating, would discreetly investigate public records. If they uncovered solid evidence that points to an actual conspiracy, then they might be justified in confronting those involved. Tracy has nothing.

The Sandy Hook conspiracy goes beyond this one former professor (he was fired for this behavior in 2016). Radio host Alex Jones, who has made a name for himself as a conspiracy theorist, hosted a roundtable of Sandy Hook conspiracy theorists on his program. In clarifying his own position, he said:

> I've always said that I'm not sure about what really happened, that there's a lot of anomalies and there has been a cover-up of whatever did happen there... All I know is the official story of Sandy Hook has more holes in it than Swiss cheese.

Alleged anomalies equal some cover-up, but he does not have to commit himself to anything specific. He then appeals to the ad hominem fallacy—we know the government lies, so they are probably lying about this.

Others argue that since the government (under Obama) was using the incident to fight for gun control, they must have staged this incident to that end.

Jones himself, while fighting a custody battle in 2017, used the defense that he is just a "performance artist." He doesn't necessarily believe all the crazy conspiracies that he touts. Of course, that claim can just be another false flag.

Once you're through the looking glass of conspiracy theories, the normal rules of evidence and logic don't apply. All you know is that nothing is what it seems. The only real protection is to understand conspiracy thinking as a phenomenon, to recognize its elements and control against them. This probably has to happen preventively, however. Once you really accept a conspiracy narrative, it's too late.

26. Skeptics' Guide Entry: Witch Hunts

Section: Science and Pseudoscience
See also: Conspiracy Theories

A witch hunt is a dedicated and unjust investigation or prosecution of a person or group in which the extreme and threatening nature of the alleged crimes is used to justify suspending or ignoring the usual rules of evidence.

WOMAN: I'm not a witch! I'm not a witch!

VLADIMIR: Ehh...but you are dressed as one.

WOMAN: They dressed me up like this!

ALL: Naah, no we didn't...no.

WOMAN: And this isn't my nose, it's a false one.
   (*Vladimir lifts up carrot.*)

VLADIMIR: Well?

P1: Well, we did do the nose.

VLADIMIR: The nose?

P1: And the hat. But she is a witch!

(**ALL:** *Yeah, burn her, burn her!*)

VLADIMIR: Did you dress her up like this?

P1: No! (no no...no) Yes. (yes yeah) A bit (a bit bit a bit). But she has got a wart!

Scene from *Monty Python and the Holy Grail*

In the spring of 1692 a number of young girls in Salem Village in the Massachusetts Bay colony claimed that they were possessed by the devil and accused several women in the village of witchcraft. What should have been dismissed as a childhood prank was taken seriously by the authorities.

Eventually over two hundred people were accused of witchcraft and nineteen were executed. But by the following year popular opinion was turning against the trials. The remaining accused were set free and the courts dissolved. By 1711 the judges who presided over the courts apologized publicly, and restitution was paid to the families of those who died.

While this remains a painful episode in American history, it was a pinprick compared to the witch hysteria that gripped Europe for over four centuries. Highly conservative estimates of how many people were killed as witches between 1300 and 1700 in Europe range from fifty to sixty thousand. This is almost certainly an underestimate, however, with some estimates reaching the hundreds of thousands. The time frame is also a bit conservative, with witch hunting beginning before 1300 and ending after 1700.

In 1487 the book that would become the guide to hunting witches was published in Germany, the *Malleus Maleficarum* (Latin for "The Hammer of Witches," *Der Hexenhammer* in German). The book was written by Heinrich Kramer and Jacob Sprenger, members of the Dominican order and inquisitors for the Catholic Church, and was intended to codify knowledge of witches and how to identify them.

The book deals with such critical questions as, "Whether Witches may work some Prestidigatory Illusion so that the Male Organ appears to be entirely removed and separate from the Body." From our modern perspective, this is an almost comical example of the misogyny inherent in the witch hunts.

It's easy to obtain a free electronic version of the *Malleus*, and I suggest that you do, and read at least some samplings of the various chapters. It is stunning. Here's one sample:

And this is when the accused is not convicted of heresy by her own confession or by the evidence of the facts or by the legitimate productions of witnesses, but there are indications, not only light or even strong, but very strong and grave, which render her gravely suspected of the said heresy, and by reason of which she must be judged as one gravely suspected of the said heresy.

In other words, even when there is no evidence that would ordinarily be accepted in court, such as confession, eyewitnesses, or hard evidence, the witch may still be guilty because of other more subtle "indications." This could literally be anything. The book essentially outlines a method for using confirmation bias to prove suspicions.

A "witch hunt" as typified by the *Malleus* and the literal European witch hunts has six parts:

Accusation equals guilt. The very fact of being accused is enough to make someone suspect, and being suspect is enough to justify an investigation. Since the rules of evidence are so fluid and adaptable, the bar so low, any investigation will likely yield a positive result. Effectively, once accused you are guilty.

Suspension of the normal rules of evidence. Even in medieval times people had some sense of what constituted actual compelling evidence. Courts sometimes applied reasonable standards. When the stakes were high, however, such as when an agent of the devil was suspected to be at work, then the normal rules of evidence were jettisoned. Losing one's temper, or staring for a bit too long at someone, would become sufficient evidence to be convicted of witchcraft. The coincidental death of a cow could be enough to justify burning the strange woman who lived at the edge of town.

Allowing spectral evidence. The term "spectral evidence" refers to dreams and visions. There is, of course, nothing tangible to investigate, nor even actual events or other witnesses to corroborate. A vision exists entirely in the mind of a witness, without the possibility of external validation. Spectral evidence was specifically allowed in the Salem trials but was later banned by colony law.

Methods of investigation as bad as the punishment for conviction. The *Malleus* directly recommends torture as an effective method of obtaining a confession. Essentially, once accused, your choice was to confess and be executed or not confess and be tortured, often to death.

Encouraging accusations. Those accused of being a witch would likely know other witches. Therefore you could torture them for names or promise a lighter sentence for turning in others. This led to a chain reaction of accusations.

Accusations used as a weapon. Once the rules of evidence are relaxed to such an extent, it becomes possible, even trivial, to use the Inquisition as a weapon. It can be used against political enemies, against business rivals, and against other "outside" groups. It can become a tool of persecuting marginalized members of society.

This seemingly absurd logic has been translated into modern discourse. In 1996 philosopher Douglas Walton wrote a paper in which he described ten features typical of a witch-hunt-style argument. He wrote:

> In this paper, ten conditions are formulated as a cluster of properties characterizing the witch hunt as a framework in which arguments are used: (1) pressure of social forces, (2) stigmatization, (3) climate of fear, (4) resemblance to a fair trial, (5) use of simulated evidence, (6) simulated expert testimony, (7) nonfalsifiability characteristic of evidence, (8) reversal of polarity, (9) non-openness, and (10) use of the loaded question technique. The witch hunt, as characterized by these criteria, is shown to function as a negative normative structure for evaluating argumentation used in particular cases.

Most of those characteristics are self-explanatory or consistent with the *Malleus Maleficarum*. "Reversal of polarity" refers to the reversal of the burden of proof. In witch trials the accused has the burden of proving themselves innocent, rather than the accuser or prosecutor proving guilt. "Non-openness" refers to being open to the truth of the case. The judge in particular should enter a trial with an open mind, willing to be convinced by the evidence. In a witch trial, they may enter convinced of the guilt of the accused.

Modern Witch Hunts

The witch hunt, and burning witches at the stake, is now a symbol of backward medieval thinking. Of course, it would be a mistake to think that such things don't happen today. We're perhaps just a bit more sophisticated at hiding them.

Most people know about the communist "witch hunts" of Wisconsin senator Joseph McCarthy in the 1950s. Such methods are now referred to as "McCarthyism." McCarthy used fear of communist infiltration into the US government and high society to organize congressional hearings that were little more than modern-day witch hunts.

Accusations of being a communist, or even just a communist sympathizer, were used to defame and destroy many. They were then promised that the devastation to their lives would be minimized if they would just turn in their friends and colleagues. In this climate, accusation was as good as guilt.

The communist hysteria of the 1950s now seems distant. Again, we might be tempted to think that witch hunts are a thing of the past. They aren't. In the 1980s came the satanic panic. This started as an accusation of ritual abuse in the McMartin preschool in Southern California. Despite the long investigation and trial that followed, no convictions were made.

Dan and Fran Keller were not so lucky. They were convicted in 1991 of sexual assault based on the testimony of children who attended their daycare center. The accusation started when a three-year-old claimed that Dan Keller "pooped and peed on my head" and suggested sexual assault with a pen. Of course, all suspicions of child abuse need to be investigated, but just because the idea of such crimes is horrific doesn't mean we can suspend the rules of evidence.

In this and other cases, children were interviewed using dubious leading techniques. When their testimony started to get more and more bizarre—including animal sacrifice, murder, and satanic rituals—that should have called the testimony and the methods into question. Instead the prosecutors assumed that the daycare was the center of ongoing satanic rituals.

The utter lack of physical evidence (it's hard to sacrifice animals and murder people without leaving anything behind) should have also given them pause. Instead the Kellers were convicted and sentenced to forty-eight years in prison. They were released in 2013 after twenty-one years when a judge ruled that their conviction resulted from faulty testimony by an expert witness.

It gets worse. Many people have been sent to prison based on testimony given through facilitated communication (FC). This is the equivalent of spectral evidence, as FC is not a legitimate method, but it still has its true believers who think they can communicate with severely mentally disabled individuals by holding their hand and moving it over a letter board or equivalent. Studies have shown that the facilitator and not the client is doing the communicating. And yet FC testimony has been allowed in court.

There are also many entertaining examples of witch hunts in fiction. It is a powerful plot device, because the audience may know that the accusations are false and see how injustice results from angry crowds and the suspension of the rules of evidence, or the audience may not know, in which case they experience the same uncertainty as the characters they are following.

These modern examples are at the extreme end of the spectrum, but it's important to understand the structure of a witch hunt because witch hunts can manifest in subtle ways. It takes vigilance to consistently apply fair rules of evidence and logic to any accusation, no matter how upsetting.

Perhaps my favorite example is from the video game *Fallout 4*. The synths are artificial humans who are replacing real people with perfect doppelgängers. This means anyone can be a synth, and perhaps all you have to go on is a vague feeling. Better kill them just to be sure.

27. Skeptics' Guide Entry: Placebo Effects

Section: Science and Pseudoscience
See also: Nocebo Effects

Placebo effects refer to an apparent response to a treatment or intervention that is due to something other than a biological response to an active treatment.

> Placebo heals nothing. There are no potentially powerful inner pathways
> by which placebo heals. At best, when lied to, you will feel better.
>
> —Mark Crislip

Radioactive tonics were popular in the early twentieth century, reaching their peak in the 1920s. Radioactivity had recently been discovered and still seemed new and sexy (this was before the atom bomb made radioactivity *very* scary). People extolled the many health benefits of radium or radioactive water, while it was slowly poisoning them.

This isn't an isolated historical incident. Also in the early twentieth century, Albert Abrams, MD, began selling a black box called a Dynamizer that, he claimed, could diagnose any disease by using radio waves. Like radiation, radio was seen as new and high-tech. Abrams followed up his fake diagnostic device with the Oscilloclast, which used those same radio waves to treat illness.

Thousands of practitioners leased and used these machines, and millions of patients swore by their effects. Yet in 1924, *Scientific American* and the American Medical Association published the results of their investigations, concluding: "Analyzed in the cold light of scientific knowledge, the entire Abrams matter is the height of absurdity. The so-called Electronic Reactions of Abrams do not exist…at least objectively. They are merely products of the Abrams' practitioner's mind. At best it is all an illusion. At worst, it is a colossal fraud."

When the FDA finally cracked open some of Abrams's black boxes (soon after his death), they found nothing but simple off-the-shelf electronics, no more complex than a doorbell.

Don't think that such episodes are limited to the quaint past. The recent popularity of bracelets made of rubber and plastic that claim to improve athletic performance is arguably more absurd than these historical examples.

How, then, do we explain how so many people could swear by the health benefits of demonstrably worthless or even harmful treatments? The answer is the placebo effect (or, more precisely, placebo effects).

What Are Placebo Effects?

The operational definition of a placebo effect is any positive health effect measured after an intervention that is something other than a physiological response to a biologically active treatment. (Negative effects from an inert treatment are called "nocebo" effects.) In clinical trials the placebo effect is any measured response in the group of study subjects that received an inert treatment, such as a sugar pill. However, "the placebo effect" is a misnomer and contributes to confusion, because it's not a single effect but the net result of many possible factors.

The various influences that contribute to a measured or perceived placebo effect vary depending upon the situation—what symptoms or outcomes are being observed. Subjective outcomes like pain, fatigue, and an overall sense of well-being are subject to a host of psychological factors. For example, subjects in clinical studies want to get better;

they want to believe they're on the active experimental treatment and that it works; they want to feel that the time and effort they have invested is worthwhile; and they want to make the researchers happy. In turn, the researchers want their treatment to work and want to see their patients get better. So there is often a large reporting bias. In other words, subjects are likely to convince themselves they feel better, and to report that they feel better, even if they don't. Also, those conducting a trial will tend to make biased observations in favor of a positive effect.

It has been clearly demonstrated that subjects who are being studied in a clinical trial objectively do get better. This is precisely because they are in a clinical trial—they're paying closer attention to their overall health, likely taking better care of themselves due to the constant reminders concerning their health and habits provided by the study visits and the attention they are getting, being examined on a regular basis by a physician, and their overall compliance with treatment is higher than usual. Basically, subjects in a trial take better care of themselves and get more medical attention than people not in trials.

For those not in a clinical trial, once they decide to do something about their health by starting a new treatment, they're likely to engage in more healthful behavior in other ways.

There is also a statistical phenomenon called regression to the mean. In any varying system, which could be athletic performance or the waxing and waning symptoms of a chronic illness, any extreme variance is statistically likely to be followed by a more typical variance. This means that when you have bad symptoms, it's likely that your symptoms will eventually become milder, or regress to the mean. This also means that any treatment you take when your symptoms are severe is likely to be followed by a lessening of those symptoms, creating the illusion that the treatment worked.

A similar phenomenon occurs when a person has longstanding symptoms and tries different treatments sequentially. They keep trying treatments until the symptoms improve or resolve, then they credit the most recent treatment for the improvement.

A common belief is that the placebo effect is largely a mind-over-matter effect, but this is a misconception. There's no compelling evidence that the mind can create healing simply through will or belief. However, mood and belief can have a significant effect on the subjective perception of pain. There is no method to directly measure pain as a phenomenon, and studies of pain are dependent upon the subjective reports of those being treated. There is therefore a large potential for perception and reporting bias in pain trials. But there are biological mechanisms by which mental processes can affect pain. For example, increased physical activity can release endorphins that naturally inhibit pain. The perception of pain can also be decreased by simple distraction. Even cursing reduces reported pain intensity. For these reasons the placebo effect for pain is typically high, around 30 percent.

The more concrete and physiological the outcome, the smaller the placebo effect. Survival of serious forms of cancer, for example, has no demonstrable placebo effect. There is a "clinical trial effect," as described above—being a subject in a trial tends to improve care and compliance—but no placebo effect beyond that. There is no compelling evidence that mood or thought alone can help fight off cancer or any similar disease.

Because there's a common and misleading assumption that placebo effects require the belief that a treatment will work (they don't), people further assume there is no placebo effect for animals or babies. However, the illusory effects of regression to the mean and the non-specific effects of getting attention are also in play. Further, someone has to decide if the horse or baby is better, and their assessment can be biased and therefore contribute to a measured placebo effect.

There has been increasing research on placebo effects in recent years. Researchers want to know, for example, how much of what we see as a placebo effect is a biological improvement and how much is just subjective reporting and illusion.

Perhaps the most direct demonstration of this difference was published in the *New England Journal of Medicine* in 2011 by Michael Wechsler et al. They compared albuterol (a drug for asthma), a placebo, sham acupuncture (placing the needles in the "wrong" locations), and no treatment in the management of asthma attacks and measured both subjective and objective response. These two graphs tell the story:

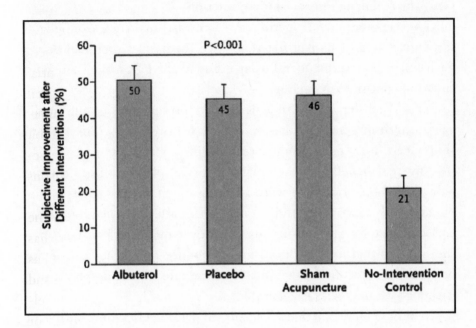

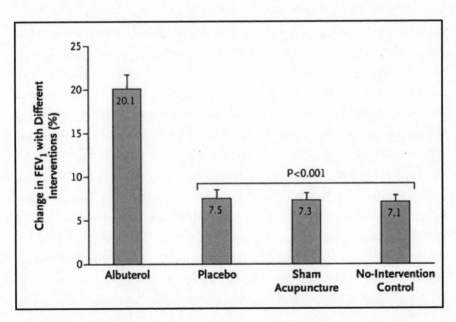

As you can see, subjects reported that they felt better with the placebo treatments (a fake drug or fake acupuncture), but when their breathing function was measured there was no actual improvement. Only the real drug improved lung function.

This is potentially dangerous. An acute asthma attack can be fatal if it's untreated. A patient may convince themselves they feel better with a placebo treatment and delay going to the ER, while their lung function continues to decline.

The Wechsler et al. study is fairly representative of the medical research in general—placebo effects are subjective, illusory, and short-lived. They don't produce real healing.

Any real benefits that contribute to the placebo effect can be gained by more straightforward methods—like healthy habits, compliance with treatment, and good health care. The placebo effect isn't evidence for any mysterious mind-over-matter effect, but since the mind is matter (the brain) and is connected to the rest of the body, there are some known physiological effects that do play a role (although often greatly exaggerated).

In light of all this, it doesn't seem as if placebo effects are sufficient to justify magical, disproven, or highly implausible treatments. Keep in mind, you can get the same placebo effects from science-based treatments. You don't need to believe in magic.

That is exactly what the proponents of all kinds of medical pseudoscience claim, however. They are happy to assert, without evidence, that their treatments have real effects, whether it is homeopathy (which is just water), energy medicine, or coffee enemas. Once scientific studies show their treatments are no better than placebo (meaning they don't work), then the proponents claim that this is okay, because placebos work too.

In a sense, much of so-called "alternative medicine" practice is placebo medicine—using elaborate rituals with fanciful explanations to produce nothing but imaginary placebo effects. This strategy depends on the misperception of placebo effects as being real mind-over-matter effects, when in fact they are mostly illusion and deception.

As cliché as this is—the placebo emperor has no clothes.

THIS SECTION NEEDS WORK -

28. Skeptics' Guide Entry: Anecdote

Section: Science and Pseudoscience
See also: Anecdotal Evidence

An anecdote is a story or experience, often offered as evidence, that was not controlled and is therefore subject to a host of biases and confounding factors. It is considered the weakest form of evidence, useful only for generating, but not testing, hypotheses.

> Anecdotal evidence leads us to conclusions that we wish to be true, not
> conclusions that actually are true.
> —Barry Beyerstein

While attending a lecture by a naturopath at my institution, I had the opportunity to ask the following question: Given the extreme scientific implausibility of homeopathy and the overall negative clinical evidence, why do you continue to prescribe homeopathic remedies? The answer, as much as my question, exposed a core difference between scientific and sectarian health care providers. She said, "Because I have seen it work in my practice."

There it is. She and many other practitioners of dubious modalities are compelled by anecdotal experience, while I am not.

Scientists and skeptics have come to use the word "anecdotal" as a derogatory dismissal of certain kinds of weak evidence—and with good reason. Believers in the paranormal and unscientific healing modalities chafe at this and have rushed to the defense of anecdotal evidence, as it is often the only substance out of which they construct their fantasies and attempt to pass them off as science. Dr. David Katz, who hosted the naturopath I quoted above, thinks that we need to have a "more fluid concept of evidence" in medicine. By "fluid," you can read "weaker anecdotal forms of evidence."

This difference in attitude toward anecdotal evidence is a persistent and critical difference between skeptics and believers, so it's important to understand why skeptics (by which I mean most scientists) don't trust anecdotal evidence. One of the hard-won lessons of the process of scientific discovery is that anecdotal evidence is very unreliable. Anecdotes can be worse than worthless because they can be misleading. This conclusion was arrived at after centuries of being misled by anecdotes. Treatments seem to work and then later are shown to be useless or even harmful. Common experience leads us to believe in many things that more carefully collected evidence later disproves. After being hit on the head so many times by anecdotal evidence throughout history, modern scientists are now appropriately wary.

What we mean, exactly, by anecdotal evidence is the report of an experience by one or more persons that is not objectively documented, or an experience or outcome that occurred outside of a controlled environment. Such evidence is unreliable because it depends upon the accurate perception of the witness(es), often in a situation where the event was unexpected or unusual; it is dependent upon subjective memory, which has overwhelmingly proven to be extremely error prone and subject to a host of flaws; and it cannot account for the random vagaries of a chaotic world.

As we've already learned, our perceptions and memories are constructed narratives loosely based on reality. Memories then change over time, with the details shifting to fit the narrative. There is also confirmation bias, a tendency to selectively remember information that supports the narrative. Casual observations also rarely contain objective measurements or quantification.

Finally, even the most careful observation can't be generalized, because it represents only one or a few cases. The results may be quirky, even a fluke, and not statistically representative. Data are only reliable when they systematically count, measure, and record all outcomes. That is scientific evidence.

At this point in the history of science, with all that's gone before, it is hopelessly naive to put any faith in anecdotal evidence. It's utter folly to use such weak and flawed evidence as a basis for concluding that a new and unusual phenomenon is real and that our science textbooks need to be rewritten. At best, anecdotes can be used as an indication of a possible (not even probable) phenomenon that is deserving of further research or exploration. It can be used to generate a hypothesis, but it shouldn't be used as a basis to confirm one.

This opinion is generally accepted within the scientific community, yet in everyday life we tend to rely heavily on anecdotes, especially our own. If I take a treatment and then feel better, that seems like powerful evidence that the treatment worked. You cannot draw this conclusion, however. Perhaps I would have felt better anyway. Maybe I even would have felt better faster without the treatment. The only way to know is to look at many people taking the treatment versus a fake treatment (placebo) under controlled and blinded conditions and then compare all the outcomes.

We learn to trust our anecdotal experience for a few reasons. For everyday simple things, it's a simple and efficient approach. We don't have the time or resources to conduct a scientific experiment on every little aspect of our lives. If the milk goes sour when I leave it out of the fridge, I conclude that the milk needs to be refrigerated. I wouldn't test this systematically unless I had some reason to doubt this conclusion. The conclusion may even be wrong, but it's not worth the effort to test it.

We are also predisposed to believe stories, especially when they have an emotional theme. Perhaps this is a result of our evolution. Let's say your friend tells you to avoid the valley over the ridge because a large lion is hunting there. We probably descended from those who chose to believe their friend and not those who decided to go to the valley and check it out for themselves. In other words, selective pressures may favor accepting stories as probably true as a default.

Defenders of anecdotal evidence will often claim that since it's reliable for everyday mundane experiences, we can therefore rely upon anecdotes more generally. I think this is both a false premise and a false analogy. Everyday experience can be misleading. There are numerous psychological experiments, mostly designed to look at memory, that have documented the fact that in everyday encounters our anecdotal recollections of events are extremely flawed. And as we've been discussing, our interpretation of events is even more flawed, subject to a host of psychological factors.

Further, we now live in a complex technological civilization. We're asking very sophisticated questions, and anecdotal evidence just isn't up to the task of answering them. The price of accepting what is not true (such as a worthless medical treatment working) can be very high.

Much like our ancient ancestors, we might be tempted to accept an anecdotal report from an alleged "credible witness." But the credible witness is now just as much a myth as Bigfoot. We are all humans. We all have flawed and biased brains. Being an airline pilot, for example, doesn't change the way your brain processes visual information. Some witnesses are certainly better than others, but any witness can be fooled. Even when multiple witnesses insist they saw the same thing, this could be accounted for by memory contamination. It's common for eyewitnesses, especially if they were together at the time, to talk during and after the experience about what they were seeing. The perceptions of one person can thereby easily contaminate the memory of the other witnesses.

Being in an uncompromised "state of mind" is a necessary but insufficient criterion for reliability: Even someone who is wide awake and uncompromised can make profound errors in perception and interpretation. Also, people often underestimate the effects on their mental state of factors such as moderate sleep deprivation. Gullible ghost hunters routinely find that if they stay up all night in an allegedly haunted house, eventually weird things start happening.

The bottom line is this—we know for certain from countless

historical examples that even people who have all the traits of a reliable witness can be profoundly mistaken.

Why Would They Lie?

I'm often asked if I think a purveyor of obvious nonsense is a con artist or a true believer. Of course, I often cannot know for sure (unless they're caught red-handed in fraud), and I suspect that many are a combination of the two to some degree. But it is often argued by believers that if you can't either prove that someone is lying or clearly identify an ulterior motive, they must be telling the truth, or at least are believable.

Clearly, whether or not someone believes what they are reporting says nothing about the accuracy of the information—just that they're not willfully lying. But believers miss some important possible motivations for prevarication. First, people sometimes stretch the truth simply because it makes life more interesting. It breaks the dull routine of our mundane lives for others to believe that we saw a ghost or were abducted by aliens. And that's motivation enough.

Others may commit "pious fraud." They are true believers, but they may embellish their experiences in order to convince those nasty skeptics. This is like a cop planting evidence on a suspect they "know" to be guilty, to make sure justice is done. And finally, people just do weird things for weird reasons. The inability to figure out why someone might lie is no reassurance that they are in fact telling the truth (an argument from ignorance, right?).

The simple fact is that selected memories of uncontrolled quirky observations (anecdotes) are extremely unreliable. They may be suitable for simple aspects of our everyday lives, but not for anything of true importance. They certainly don't suffice as scientific evidence or to establish entirely new phenomena.

ICONIC CAUTIONARY TALES FROM HISTORY

One of the more memorable lectures I heard in medical school involved a story from when the professor was an intern in Israel. He was treating a seriously ill patient who had all manner of laboratory abnormalities, including the fact that most of his electrolytes (ionized substances in the blood, like sodium, chloride, and potassium) were severely out of the normal range. He spent the night carefully adjusting the intravenous fluid and giving electrolytes until the numbers were all in the normal range.

By the morning, despite the professor's best efforts, the patient died. His supervising doctor chastised him, stating, "You corrected the electrolytes but killed the Israelite."

In addition to the pithy rebuke, the story contained an important cautionary lesson: Don't get so focused on fixing the numbers that you miss the big picture. Treat the patient, not just the laboratory findings.

There is a meta-lesson here too—learn from your mistakes, or better yet, learn from other people's mistakes. The professor wanted us to learn from his mistake and not have to lose our own patients to absorb that lesson.

This lesson reflects a more general observation credited to George Santayana: "Those who do not learn history are doomed to repeat it." Part of the duty of activist skeptics is to remember the history of science, pseudoscience, deception, and fraud. The same mistakes, errors in thinking, biases, and cons keep getting recycled. The skeptical lessons of the past are all still relevant today, and studying their history is an easy way to learn from the mistakes of others.

The public may be impressed by claims made for magnetic bracelets, but they are no different than the magnetic snake-oil devices that were popular two hundred years ago. All the pseudosciences we confront today have their roots in the past. The work of scientists and skeptics in previous decades can act like a vaccine, allowing us to mount a more rapid and robust response to fraud and quackery today.

In this final section of core skeptical concepts, we will explore some common deceptions, iconic blunders of the past, and examples of classic pseudoscience. These examples will bring together many of the ideas we have already discussed and allow us to flex our critical thinking muscles. We will see how things can go horribly wrong when people are...less than skeptical. And I promise that no Israelites will be killed in the process.

29. Skeptics' Guide Entry: The "Clever Hans" Effect

Section: Cautionary Tales
See also: Pseudoscience

The Clever Hans effect refers to an unconscious nonverbal communication of information from a researcher or agent to an animal or other subject. This unconscious communication creates the illusion of a cognitive or supernatural ability.

> Fables should be taught as fables, myths as myths, and miracles as poetic fantasies. To teach superstitions as truths is a most terrible thing. The child mind accepts and believes them, and only through great pain and perhaps tragedy can he be in later years relieved of them. —Hypatia

---

It can be fun to think that there's a lot more going on inside the minds of animals than there probably is. We have a tendency to project and anthropomorphize, talking to our pets as if they understand what we are saying (to a limited extent, they do, but not enough to justify our one-sided dialogues).

This tendency might also be a partial explanation for why books like the Doctor Dolittle series (definitely a childhood favorite of mine) are so popular. In the world of Doctor Dolittle, animals have human-level intelligence, we simply don't understand their language.

The first book in the series was published in 1920, at a time when fascination with the concept that animals might have vast unrecognized cognition was near peak level. This in turn is thought to be the result of the rising acceptance of Darwin's theory of evolution. If humans are intimately connected to the rest of the animal kingdom, perhaps animals are more human that we previously thought.

It was against this historical background, in 1904 to be precise, that Wilhelm von Osten introduced his horse Clever Hans to the world. Wilhelm von Osten was a mathematician, but perhaps more importantly, he is described as an amateur horse trainer, a phrenologist (discerning personality from the bumps on the skull), and a mystic. Von Osten spent two years training Hans to read—you read that correctly:...read.

Von Osten created a board with each letter of the alphabet associated with a number. He would then say a letter and hold Hans's hoof as he tapped out the corresponding number of times. Over time he gradually withdrew his hand and let Hans stomp his hoof by himself. In this way Hans apparently learned how to spell out words and eventually entire sentences. Of course, once Hans could communicate in words, then you could ask him questions. Through this technique, von Osten taught Hans to tell time, interpret a calendar, understand music theory, and do arithmetic (take that, Mister Ed).

Hans became an instant sensation in Germany, and then worldwide. He impressed not only the crowds but also the scientific community. Hans was observed by a psychiatrist, Gustav Wolff, who declared himself convinced "that an animal can think in a human way and can express human ideas in human language." Other scholars and scientists were similarly astounded.

While the idea of a horse with apparent human-level intelligence seems implausible, how could they deny the evidence of their own eyes? Hans could tap out the answers to all sorts of questions. The scientists even did a reasonable control: They tested Hans while von Osten was not present, and he was still able to perform. In their minds this ruled out any possibility of fraud.

Of course, the story of Clever Hans does not end there. Oskar Pfungst from the University of Berlin's Psychological Institute did a more detailed investigation of his alleged abilities. Pfungst added a further control that no one else had thought to do: He made sure that no one who knew the answer to the question Hans was asked was in the room with him.

In this situation Hans would slowly tap, looking at the people present, but would just continue to tap without end. It appeared as if Hans was waiting for a signal that never arrived. His amazing abilities vanished entirely when this control was put into place.

Pfungst concluded that during the extensive training of Hans, the horse learned to pick up on subtle cues from his trainer when he hit the target number of stomps on the ground. So he would just keep stomping until he got the signal. There is no indication that von Osten was aware he was giving such cues, and in fact Hans was able to pick up on cues from people other than his trainer.

A Cautionary Tale

The story of Clever Hans contains several lessons for the skeptic. The most narrow lesson is to be especially careful when research- ing animal communication. It's incredibly easy to overinterpret what animals are doing and to impose human intellect on our nonhuman friends. The Clever Hans effect has in fact plagued animal communi- cation research over the last century. It seems every generation has to relearn the lessons of Clever Hans.

More broadly, this episode is evidence for the so-called "observer- expectancy" effect. It's a special case of this more general effect, which simply refers to the influence a researcher or observer can have on what they are observing. Researchers can unwittingly communi- cate their biases or expectations to the subjects of their study, distort- ing the outcome. This is the primary reason for the need for properly blinded experiments. In the case of Clever Hans, observers falsely assumed that only the trainer could be the source of nonverbal cues allowing Hans to simulate having human intelligence. The effect was only detected when all possible sources were eliminated.

Finally, this episode is yet more evidence of why it's critical to consider plausibility when interpreting evidence. Von Osten, who likely did not give deep consideration to scientific plausibility, given that he was a phrenologist and a mystic, obviously began training Hans with the idea that he could learn how to read. This quickly evolved into a theory that Hans could also perform math, grasp complex concepts, and even understand abstract music theory. When a finding becomes more and more and more amazing, it's possible that you are chasing an "artifact" (an illusion caused by bad research methods, rather than a real phenomenon). The apparent abilities of Hans should have made von Osten more skeptical and brought much more caution to the scientists who endorsed the horse's ability.

Similarly, the inability to identify such an artifact doesn't mean that the alleged phenomenon is real (see the argument from ignorance in chapter 10). In cases of extreme scientific implausibility, it's perhaps best to withhold judgment until truly expert investigation can be done and results can be independently replicated. Such a position isn't closed-minded, it is simply realistic. If something seems too fantastical to be true, history tells us it is extremely likely that it's not true, that it's just an illusion, an error, or an artifact of flawed methodology. At the very least, all such possibilities should be ruled out to a degree that is in proportion with the implausibility of the claim.

It is clear that von Osten, Gustav Wolff, and the other scientists who declared Clever Hans legitimate did so prematurely, and now their legacy is being on the wrong side of a cautionary tale (or "tail" if you prefer).

The lesson of Clever Hans is critical to keep in mind the next time a gushing reporter or enthusiastic researcher promotes the apparently impossible. Such stories often come with the challenge to "Explain that, skeptic!" Give it time.

30. Skeptics' Guide Entry: The Hawthorne Effect

Section: Cautionary Tales
See also: The Observer Effect

The Hawthorne or observer effect refers to the fact that simply observing something may alter its behavior, thereby creating an artifact that leads to an incorrect conclusion.

> The universal propensity of humans to learn is a constant threat in almost every experiment on people.　　　　　　　—Stephen Draper

---

The story of the Hawthorne effect—the notion that observers affect the outcome of what they are observing—is a complicated one. It is an excellent demonstration of the fundamental rule that it's always more complicated than you think. If there is one overall lesson I have learned after twenty years as a science communicator, that's it. There is a general bias toward oversimplification, which to some extent is an adaptive behavior. The universe is massive and complicated, and it would be a fool's errand to try to understand every aspect of it down to the tiniest detail. We tend to understand the world through distilled narratives, simple stories that approximate reality (whether we know it or not or whether we intend to or not).

Those distilled narratives can be very useful as long as you understand they are simplified approximations and don't confuse them with a full and complete description of reality. Different levels of expertise can partly be defined as the degree of complexity of the models of reality that you use. It's also interesting to think about the optimal level of complexity for your own purposes. I try to take a deliberate, practical approach here—to what degree of complexity do I need to know?

The Hawthorne Effect

The distilled narrative of the Hawthorne effect is this: The act of observing people's behavior changes that behavior. This is a more general phenomenon, of which the Clever Hans effect is a special case, in which the observer unwittingly communicates information to the observed. The name "Hawthorne effect," coined in 1953 by psychologist J. R. P. French, derives from experiments conducted between 1924 and 1933 in Western Electric's factory at Hawthorne, a suburb of Chicago. The experimenters made various changes to the working environment, like adjusting light levels, and noticed that regardless of the change, performance improved. If they increased light levels, performance improved. If they decreased light levels, performance improved. They eventually concluded that observing the workers was leading to the enhanced performance, and the actual change in working conditions was irrelevant. This is now referred to as an observer effect.

For most people, that one paragraph is a nice bit of knowledge and is probably all they're going to remember long-term. But, of course, it's more complicated than that. Science communicators often try to take that one-paragraph basic knowledge and raise it to a more thorough and nuanced "one-essay" knowledge, the level of a science enthusiast or, in this case, scientific skeptic. But there are deeper levels still, representing various degrees of expertise. The trick is to be essentially correct at each level of approximation.

The problem is that most one-paragraph summaries in the public consciousness are not essentially correct. Sometimes more detail and nuance are needed to get to an acceptable level of correctness. Knowing where that level resides is part of the skill of science communication.

And here we return to Hawthorne. I have heard conflicting statements about the Hawthorne effect and tried to understand the science deeply enough to resolve those apparent conflicts and arrive at an essentially correct and sufficiently nuanced summary of the research. Here is what I found, starting with a recent systematic review of the published literature (always a good place to begin) and the results and conclusion of a 2014 review:

Results

Nineteen purposively designed studies were included, providing quantitative data on the size of the effect in eight randomized controlled trials, five quasiexperimental studies, and six observational evaluations of reporting on one's behavior by answering questions or being directly observed and being aware of being studied. Although all but one study was undertaken within health sciences, study methods, contexts, and findings were highly heterogeneous. Most studies reported some evidence of an effect, although significant biases are judged likely because of the complexity of the evaluation object.

Conclusion

Consequences of research participation for behaviors being investigated do exist, although little can be securely known about the conditions under which they operate, their mechanisms of effects, or their magnitudes. New concepts are needed to guide empirical studies.

From reading this and other reviews, it seems that the evidence over the last century clearly shows that there is an experimenter effect when observing human behavior in social settings, like the workplace. This experimenter effect, however, isn't entirely an observer effect—a response to knowing that one is being observed. The term "Hawthorne effect," therefore, has come to refer to a more general experimenter artifact.

Some more detail on the Hawthorne experiments is interesting and provides further background. They spent five years adjusting light levels for thousands of employees, with control groups in which they didn't adjust light levels. In general, every time they made a change there was an increase in productivity, regardless of the actual change. Both the control and active groups improved without statistical difference. (This was true as long as the actual light levels didn't produce a material problem—when, for example, they dropped the light levels so low it became difficult to see, the workers complained.) They also varied the duration of breaks and the intervals between them, with the same results. They gave shorter workdays, which increased productivity per hour, although if they decreased them too much, daily productivity decreased.

There have been other experiments that help put these results into further context. In a series of teacher expectation experiments, for instance, teachers were told that one set of students performed well on an aptitude test and were expected to perform well in the class. Over the next two years, those students performed better than the other students, even though the groups were actually equal at the onset. Further, the teachers sometimes expressed annoyance at students in the low-expectation group who did very well, exceeding expectations.

There were also the 1900 punch-card experiments of psychologist Joseph Jastrow. One group of workers in the US Census Bureau were told that their quota on the tabulating machines was 550 punch cards per day, which they were able to produce. When asked to increase their quota, they indicated that this would be too difficult. A second group were told nothing, and they produced 2,100 punch cards per day.

How to Interpret All This?

Clearly, there are some interesting psychological effects going on here. To say this is all an observer effect would be a gross oversimplification, to the point of being wrong by missing important other effects. One such possibility is expectation, which is evident in the teacher and punch-card experiments. Teachers have a great influence on the performance of their students, although it's still not clear exactly what the mechanisms are. That influence is biased by expectations, which is exactly why so much attention is paid to gender and race biases in teachers. It's also why we have to be very careful with labels and with using possibly biased standardized tests to establish expectations.

In the punch-card experiments there seems to be a self-expectation bias. People's perception of what is reasonable and even possible was easily established by creating a set point. I think this would now be described as part of the anchoring heuristic—expectations can be anchored to a specific value, and later judgments then use that anchor as a reference point. Setting a quota is a strong manifestation of anchoring expectations.

More recently researchers have been thinking about the social and community aspects of these effects. Essentially, there is a work culture that influences the behavior of those in the group. That culture is altered by many variables, including the anchoring of expectations. Being observed, and feeling as if management is taking an interest in your work conditions or that you have some say in those conditions, will also have an effect. The very notion that someone upstairs cares about details like the light level in your working space may have a profound effect on the work culture. Workers also influence each other. They establish norms of behavior and expectations for standards and productivity. In fact, the expectations of fellow workers may have more of an impact on behavior than management.

Modern corporate culture makes extensive use of consultants. A consulting firm will come in, observe the work culture of a company, then make recommendations based upon some theory or their alleged expertise. Productivity almost always improves after such interventions. It's possible that this corporate self-help industry is essentially selling the Hawthorne effect, even if their specific recommendations are speculative, nonsensical, or counterproductive.

This is all very similar to the placebo effect. Both placebo effects and Hawthorne effects are multiple effects that can be difficult to tease apart. They can also both be used to create the illusion of a specific response to a specific intervention when in fact they are just nonspecific responses to the act of having any intervention and assessing the outcome.

The self-help industry is largely selling placebo/Hawthorne effects as well. If people go on a diet, for example, the details of their diet don't seem to matter (again, like the light levels, as long as they're not so extreme as to create a functional problem). The fact that they are paying attention to what they eat and trying to be more active has an effect.

The bottom line of all this is that any intervention in almost any context will subjectively seem to work. There are a host of observational and psychological factors that will result in a real change in behavior but also a biased assessment of the results. This often leads to the false conclusion that the specific intervention (a treatment, a diet, a self-help strategy, or whatever) has specific efficacy and therefore the underlying philosophy must be valid.

The Hawthorne/placebo effect, interpreted broadly, is extremely far-reaching. The question it almost invariably raises is: Can this type of effect be used in an ethical and positive way? I think the short answer is: It depends. It's more complicated than you probably think. There are many pitfalls and downstream effects, and many things to consider.

My quick summary would be: In the context of medicine, I would not justify any intervention simply because it provokes placebo effects alone. In other contexts, such as making changes in the workplace, I wouldn't do anything that goes against common sense, that seems extreme, or that violates normal ethical conduct just because it seems to create some improved outcome. However, benign and reasonable interventions, even if they have no specific effect, may be a sensible way to make people feel like you are paying attention, that you care, that their voice is being heard, and may therefore improve the culture of the workplace.

Just don't buy into some wacky underlying theory just because the intervention seems to help.

For researchers, the challenge is to separate real specific effects from nonspecific artifacts of doing research or making an intervention. At this point reality is more complex than even the experts have a handle on.

31. Skeptics' Guide Entry: Cold Reading

Section: Cautionary Tales
See also: Mentalism

Cold reading is a collection of mentalism techniques used to create the illusion of having gained specific knowledge about a target through supernatural means. The techniques involve use of vague statements, high-probability guesswork, and feeding back knowledge gained from the subject themselves, while the target makes connections to their own personal experience. Cold reading is used in stage magic for entertainment, but also by a wide variety of less honest practitioners to feign psychic or arcane abilities.

> It is probably a tribute to the creativity of the human mind that a client can, under the right circumstances, make sense out of almost any reading and manage to fit it to his or her own unique situation. All that is necessary is that the reader make out a plausible case for why the reading ought to fit. The client will do the rest.
>
> —Ray Hyman

I am getting a vision...of you. That's right, of you, dear reader. As I write these words I am making a connection through time and space, through the pages of this book, through your fingers and into the very essence of your existence.

I see a person. Someone close to you, an older male figure, and the letter *J* or perhaps *S*. It's hard to see, as I am connecting to more than one reader at a time. I see a uniform of some kind, and there is a strong distinctive odor about him. And why am I seeing a red door? Does this have any meaning to you?

The number 3 is significant. It could be a day, or month, or part of a year. This person wants to connect to you, to let you know they love you and that they are all right.

This is, of course, nothing but a cold reading, a clever stage illusion. Cold reading is at the heart of a long list of practices that pretend to use either supernatural or arcane means to divine special knowledge about a person.

Cold reading is about more than just fake psychics (but I repeat myself). At a medical conference, for example, a pharmaceutical company offered free handwriting analysis—for "entertainment," of course. Always game, I agreed to have the depths of my personality laid bare, betrayed by the sweep of my *s* and the boldness of my *t*. To my skeptical eye, the results were laughably mundane and predictable. The reader knew, of course, that I was a physician, so it was no surprise when he "read" in my handwriting that I like science and have a desire to care for people. Wow!

But others were impressed by the apparent accuracy of their readings. The results aren't dissimilar for many friends and acquaintances who have visited a local psychic, tarot card reader, or astrologer, and the many more who have seen TV psychics like John Edward and Sylvia Browne. "How do you explain this?" people ask, very impressed, convinced that something paranormal must be going on.

Like any magic trick, a good performance can seem amazing, even inexplicable. Once you know how the trick is done, however, it becomes embarrassingly mundane.

So let's have a peek.

The basics of cold reading involve starting with general statements that are likely to be true about anyone: "I see you have financial concerns." "You feel as if no one truly understands you." "Your family has been on your mind a lot recently." Such vague assertions can seem quite specific when someone is applying them to you. As you nod and express amazement, the reader makes other comments, following up on only those statements that garner a good reaction. The psychic says, "Maybe your sister...or an aunt...definitely a woman close to you..."—and he or she watches for your reactions to gauge when the trail is getting warm. Working this way, in a few moments the cold reader is telling you your most intimate secrets.

A cold reader may also start with a general statement and then, once the subject answers positively or negatively, follow up with a more specific statement, pretending that's what they had in mind all the time. For example, they may claim to "see" the letter *J*. A willing participant might then dutifully fill in the detail of "John." The reader will pick up on this and say, "Yes, John. I see a male figure named John who is important in your life."

"John is my father," the subject may volunteer.

"Yes," continues the reader, "because I see that he is an older man, and he was with you as a child." But after the reading, the subject will likely tell others that the psychic knew his father's name was John.

Framing the statements as questions is also very effective. "Do you live on a hill?" If the target or sitter answers yes, then the reader will follow up with, "I was afraid of that, because I see danger involving an accident on that hill." If the sitter answers no, the reader may say, "I didn't think so, because I see a home in a valley or in a flat place." Whatever the target answers, the reader can pretend that's what they saw all along.

Another strategy is to make high-probability guesses. For example, cold readers will commonly see the letters *J* and *M*, or actually guess the names John or Mary, because they are so common. Watch TV psychics carefully—they never see the letter *Q*. They may tell an elderly, affluent New Englander that they see palm trees in the near future. So-called "psychic" detectives will often see water, or a red door, items that seem specific but are common enough that they're likely to be found somewhere near the eventual location of the victim. John Edward is fond of guessing "I see the number 3—it can be a month... or a day...or part of a year." He will keep going until he gets a hit.

These types of guesses are generally unspecific, but they give the illusion of being highly specific once the target of the reading makes the connection to something in their lives. Another common guess is to see a person in uniform. That may sound like a direct specific hit if your father (or whatever person you are trying to contact) was in the army. But a uniform can also refer to a letter carrier, a lab technician, a police officer, or a firefighter. And again, once the target makes the connection, the cold reader will pretend that's what they saw all along.

Going from vague statements to specific examples when connecting a guess to yourself is a known psychological phenomenon. It was first described by the psychologist Bertram R. Forer in 1949 and is now known as the Forer effect, although it's also often called the Barnum effect after the famous carnival showman.

Forer gave his students what he told them was a personality profile specifically created for each individual based on their test results. The students rated the accuracy of the profiles on average as 4.26 out of 5. They were only later told that every student received identical personality profiles.

Follow-up research by other psychologists confirms this effect but also demonstrates that people will rank vague statements as being more accurate than more specific statements. Subjects will even rank vague generic statements more highly than specific statements generated from validated and thorough questionnaires. Other factors that make a fake personality profile seem real are if the subject believes the profile applies to them, they believe in the authority of the source, and the statements are mostly positive.

All of these factors apply to cold readings. Readers pretend the information is specific to the target. They carefully craft their persona and the atmosphere of the reading to create the proper mystique of authority. And they mainly tell people what they want to hear. Strangely, you never hear psychics give clients readings such as "You are never going to find true love or happiness. In your job you will only manage to struggle your way up to middle management, where you will labor away until you retire or die. Your ordinary life will be punctuated by occasional tragedy, but nothing truly exciting will ever occur."

Experienced cold readers are very knowledgeable about statistics—the most common names, jobs, or eye color. That kind of statistical information is useful for other mentalist tricks too. Mentalists may know very specific things. For example, knowing that most airlines have red in their logo allows them to predict, about an airline crash, "I see red on the tail."

Skilled cold readers may also remember information blurted out by their target early in the reading, which they will ignore and then feed back to them later as if they divined the information themselves.

Cold readers are opportunistic and observant. A coworker, skeptical about cold reading as an explanation for alleged psychic performances, once asked me to demonstrate. I told her that she had been very concerned about her love life recently and felt unfulfilled, even desperate.

She was shocked and asked how I knew. It was simple: She was a fortysomething woman without a wedding band. And most people are concerned about their love life. I made a high-probability guess and made it sound specific.

Cold readers may also use information obtained through other means. When this is done it's technically a "hot reading." The most infamous example of this is the faith healer Peter Popoff. He would have his targets fill out "prayer cards" with personal information, including what health issue they wanted healed. He was caught by James Randi (a famous magician and skeptic who uses his knowledge of magic to investigate fraud) having his wife feed him this information through a radio earpiece while pretending to be inspired by angels.

The art of cold reading also involves turning misses into hits. If a statement about the subject of a reading turns out not to be true, well then maybe it's true of the subject's friend who is with them. TV "psychics" who work a crowd use this technique to great advantage, increasing the probability of a random hit by opening up their guess to a room of twenty or more people. A miss can also be turned into a prediction. You haven't found a lost animal recently? Well, keep that one in mind—you will in the near future.

But the most important element of a successful cold reading isn't the psychic but the subject. Most people who visit a psychic, reader, or healer want to believe; they want a successful reading, even if they fancy themselves initially skeptical. Subjects will typically remember all the lucky hits and forget the misses, even egregiously blatant misses, so that the reader's performance will be all the more impressive in the retelling.

A cold reading can take many forms—a straight psychic reading, tarot card reading, palm reading, reading the tea leaves, or even reading one's buttocks. (Apparently, Sylvester Stallone's mother is an "ass reader.") The trappings really don't matter.

You can cold read medical diagnoses. Iridology, for example, is the practice of diagnosing illness by reading the flecks of color in the iris. The techniques are the same, such as giving vague diagnoses like "I see you have an issue with your liver." This isn't a specific diagnosis, just an organ. If the person has no known liver problems, then the iridologist simply says that the person has a "susceptibility" to problems with the liver.

Pulse diagnosis is common in Eastern cultures, feeling the pulse for a long time to discern what is wrong with the body. One not uncommon pulse diagnosis is that "the water is not flowing." This isn't a real diagnosis, and it's intentionally vague. When you think about it, this could relate to many real health issues, from bladder problems to heart blockage.

Even though gaining information from the target is very helpful, especially for the observant reader, cold readings don't have to be done live. The reason for this is simple—it's the target of the cold reading that is doing all the work. They are making the connections and sifting through all the possible correlations in their life. The confirmation bias is all theirs.

Think of the classic *Twilight Zone* episode with a young William

Shatner, where he superstitiously believes that the little devil's bobble-head that's spitting out pieces of paper for a penny has real magical power. He is the one fitting the vague statements to specific events. This process is also referred to as subjective validation, which means people can apply very loose criteria ad hoc to validate a belief.

For a time, phrenology machines were popular. They would measure the bumps on your head with an elaborate helmet, then give a readout diagnosing your ailments. This was nothing more than a prefabricated medical cold reading.

Of course, the best TV portrayal of cold reading was from a *South Park* episode. Stan is trying to convince Kyle that psychics are frauds. Here is his demonstration:

> STAN: All right, look. I'll show you. I just need a volunteer. How about you?
>
> *WOMAN 4: Oh-ho. Me?*
>
> STAN: Okay, I'm gonna pretend that a dead person is talking to me about you, okay?
>
> *WOMAN 4: Okay.*
>
> STAN: Okay, watch, Kyle. Uh, it's an older man, someone very close to you.
>
> *WOMAN 4: My father?*
>
> STAN: Does this month, November, hold a special significance?
>
> *WOMAN 4: My birthday's in November!*
>
> STAN: Right, because he's saying, "Tell her 'Happy Birthday.'"
>
> *WOMAN 4: Oh my God.*
>
> STAN: See, Kyle? I just started with something really vague. I chose "older man" because I'm betting that, based on this woman's age, her father is most likely dead. But if her father wasn't dead, I could still say it was some other older man.
>
> *MAN 2: Well how'd you know her birthday was in November?*
>
> STAN: I didn't. I just asked her if November meant anything. Her father could have died in November, or Thanksgiving could have been really special for them. But I go with the birthday

and validate it now, as if I knew, by saying "He wishes you a
Happy Birthday."

WOMAN 4: *What else does he say?*

STAN: Okay, I'll just use an old standard. He's saying "The
money. Stop worrying about the money."

WOMAN 4: *Oh my God! My sister and I have been fighting over his
inheritance.*

WOMAN 3: *That's amazing.*

STAN: No it isn't! When a father dies, inheritance is usually an
issue, and money is something everyone worries about.

MAN 3: *That sounds a little too coincidental.*

MAN 4: *Yes. There's only one explanation. This kid can communicate
with the dead!*

You should no more believe that someone is psychic because of a
cold reading performance than you should believe that a magician can
really saw a woman in half or catch a bullet in their mouth. These are
all illusions.

Cold reading can be a powerful illusion, however. The core of the
illusion is the fact that most people underestimate the number of possible hits in their life. Our lives are filled with millions of details. Our
brains are also very good at finding connections (remember data mining, pattern recognition, and coincidences).

In this way cold reading is related to the coincidence illusion, the
sense that two events that happen to align could not have done so by
chance. This fails to consider all the possible alignments that didn't
happen.

We likewise underestimate the probability of a guess by a psychic
aligning with something in our lives. The probability, actually, is very
high for any particular guess. So ultimately the illusion is partly based
on our own inability to grapple with the probability of very large
numbers.

Think back to the opening of this chapter—did my cold reading
trigger anything for you? Did your brain find any connections? What

if I told you that your name is Mark and you work in the technology sector? How do you explain that?

(Readers who just happen to be named Mark and work as programmers are really freaking out right now.)

Now that you know how cold readings are done, however, you are less likely to be impressed by the random hits of clever guessing. Even you, Mark.

32. Skeptics' Guide Entry: Free Energy

Section: Cautionary Tales
See also: Perpetual Motion, Overunity

Free energy is the claim that the law of conservation of energy can be broken, that we can create some process that creates more energy than it uses. Such a process would create endless energy seemingly from nothing. However, a free-energy device cannot exist, as it would violate the well-established law of conservation of energy, and no one has been able to successfully demonstrate such a device.

> I fully agree with you about the significance and educational value of methodology as well as history and philosophy of science. So many people today—and even professional scientists—seem to me like somebody who has seen thousands of trees but has never seen a forest.
>
> —Albert Einstein

I have to say I love watching alleged free-energy devices in action. They are often beautiful works of art in their steam-punk complexity, with numerous interacting moving parts. They are also little puzzles to be solved—where in the workings of this device is the deception? Where did they go wrong? Some devices are black boxes. They spit out energy and we're not allowed to peek inside and see how they work. The contraptions are perfect monuments to the pseudoscience that created them.

In December 2006 a Dublin company called Steorn promised to demonstrate its Orbo device, which it claimed was capable of perpetual motion, of generating free energy. In November 2016, ten years and $20 million later, Steorn announced that it was closing up shop and liquidating its assets. The story of this company is a microcosm of the history of free-energy devices and perpetual motion machines.

The poster child for iron-clad laws of science is the conservation of matter and energy—you don't get something from nothing. This means that anyone who claims to have produced a perpetual motion machine, a machine that can generate free energy or that produces more energy than it consumes, is simply wrong. They could be misguided, scientifically illiterate, a hopeless crank, have a tenuous relationship with reality, or simply be a con artist knowingly lying. But chances are they are not, as they would have us believe, a lone brilliant scientist who has succeeded where all others have failed in achieving the seemingly impossible.

There is one characteristic that I have found to be almost ubiquitous among free-energy claimants—an utter lack of humility. This is a general feature of pseudoscience but seems to be epitomized by the type of cranks who think they have surpassed the fundamental laws of physics.

A free-energy machine, also called an overunity machine, is one that produces more energy than it consumes, or has greater than 100 percent efficiency. (Unity means 100 percent efficiency, so overunity means greater than 100 percent.) Unfortunately, the laws of thermodynamics make such claims impossible. The first law of thermodynamics states that you can never get more energy out of a process than you put into it, and the second law states that you cannot even get the same amount of energy out—there will always be some loss in the amount of energy capable of doing work (also referred to as an increase in entropy).

You can think of this in terms of the conservation of mass/energy—energy cannot simply come from nowhere. There has to be a source of the energy.

These laws are so well established that... well, they are laws. Unlike mere theories or guesses, they're as well established as anything in science. Sure, scientific knowledge is always finite. But to claim that the laws of thermodynamics have been broken would be most extraordinary, requiring rock-solid evidence in order to be taken seriously.

That's not what we typically get, however. As with Steorn, we're usually given promises and dubious demonstrations. Steorn gave one failed public demonstration of its Orbo device in 2007. They claimed equipment failure, that they would just turn a few screws and be back in no time. They also did a public demonstration in 2009 of an apparent (black box type) perpetual motion machine, but it provided no actual evidence of perpetual motion and was rightly dismissed as a publicity stunt. THEY GOT PAID FOR 10 YRS

Steorn also challenged the scientific community to directly investigate their device. They did, and in 2009 the expert panel unanimously concluded that they hadn't produced perpetual motion. More excuses, more delays, more promises, and then they folded.

There are various common ways that people convince themselves they have tapped into a source of free energy when they clearly haven't.

Long-Lasting Motion Is Not Free Energy

One way is to create some kind of contraption that will keep itself moving almost indefinitely, such as a rolling ball or spinning flywheel. Such devices try to trick gravity, or they contain magnets that are supposed to give an extra kick and keep things moving around.

These devices, as mesmerizing as they often are, can't possibly work, however. You cannot use some force like gravity or magnetism to create perpetual motion. You have to put in more work to move up the gravity well or away from the magnet than you get back from moving down toward the force. At best, engineers make toys that have minimal friction and can move for a long time, but you cannot get energy out of such systems.

Missing All the Energy Inputs

There are demonstrations of small motors running off an external power supply but with claims that, when the inputs and outputs are all added up, the energy output is greater than the energy input. Sometimes the claimants just get the math wrong. But most often they fail to properly consider or measure all the energies involved. They always seem to take their inventions to some lab or university and claim that their machine works. They typically find earnest but naive "experts" who may have physics or engineering expertise but clearly aren't up to the task of investigating overunity claims (just like the experts who prematurely validated Clever Hans). To date, despite countless claims, no one's free-energy machine has stood up to proper investigation.

Converting Energy from One Form to Another

What continues to amaze me are those who think they've hit upon the secret of free limitless energy just by rigging up some combination of batteries and motors, for example using one battery to run a motor that is then used to charge another battery. This works—but of course at the end of this process you have less energy in the second battery than you had in the first. The motor makes noise and sparks. The light and sound it gives off is therefore carrying energy away from the system. I'm sure it's also heating up a little bit, and waste heat is another source of lost energy. That's why there will inevitably be less energy in the second battery.

It seems that the inventors are simply surprised that such a motor can run for so long on just a battery, especially when you recapture some of the energy by, for instance, charging up another battery. They conclude that they're onto something special, and further conclude that they could scale up their motor and add an arbitrary load and it would still work. When this doesn't work, they assume the failure is due to some technical engineering problem rather than the fundamental laws of physics. So they endlessly tinker with their rig, claiming that it will work as soon as they fix the bugs. But it never does—and the only thing truly perpetual is the endless cycle of over-unity claims.

All We Need to Do Is Scale Up

Another fallacy is to extrapolate from tiny amounts of energy and assume that the whole system can be scaled up. The problem here is that with small amounts of energy subtle errors are easy to miss. When you scale up, those errors become huge and obvious, which is why the devices never work at large scale. That's why I say: Run a house on your free-energy principle and nothing else, and then you'll have my attention.

Fraud

There are, of course, some scams thrown into the mix as well—not just the honest but self-deluded. The promise of free energy seems to be a good way to lure in those with more cash than scientific knowledge. It is a very good rule of thumb to be highly skeptical of any free-energy claims. The person seeking to sell you a device, or often just the chance to invest in the development of a free-energy device, may be sincere or they may be just trying to con you.

Lack of Humility

There's always a certain amount of arrogance with such claims. An ounce of humility would lead one to assume that they were making a mistake rather than the invention that will transform our civilization and has eluded the scientific community for so long.

I find arrogance to be common in pseudoscience in general. If you think you're the one genius who has broken the laws of physics, who will rewrite the textbooks, who has succeeded where countless others have failed before, then you should be really certain.

Instead we get ridiculously grandiose claims based upon intuition and overconfidence.

Perpetual motion machines are the ultimate crank magnets. The promise of free energy is part of the appeal—imagine if it works! Free-energy devices themselves are a tangible metaphor for the crank mind. They're elaborate, often fascinating and even beautiful, and include a great deal of technical detail. But they're missing something essential for real science and progress. They are cut off from reality, removed from the scientific community. They are monuments to arrogance. And most importantly—they don't work.

33. Skeptics' Guide Entry: Quantum Woo

Section: Cautionary Tales
See also: Umm...nothing comes to mind

Quantum pseudoscience or woo or mysticism is the hostile takeover of the success and weirdness of quantum theory to support pseudoscientific beliefs and junk science.

> For all its beauty, honesty, and effectiveness at improving the human condition, science demands a terrible price—that we accept what experiments tell us about the universe, whether we like it or not. It's about consensus and teamwork and respectful critical argument, working with, and through, natural law. It requires that we utter, frequently, those hateful words—"I might be wrong."
>
> —David Brin

Purveyors of pathological science—whether it's for acupuncture, homeopathy, dualism, or whatever—always desire a stamp of approval from..."SCIENCE" (*insert echo-chamber effect*). After all, science undeniably delivers the goods, and having its imprimatur would be a tremendous bonus. Since they have no real scientific support, however, they're content to make their nonsense at least *seem* scientific. After all, even a thin patina of fake science on top of snake oil is often enough to persuade many people.

One effective strategy is to claim that your pseudoscience works because of—whatever the latest scientific discovery is that most people don't understand but sounds really cutting-edge and sexy. A couple hundred years ago the newfangled scientific concept was electromagnetism. This gave birth to countless bogus magnetic devices and healing scams.

The most famous of these was "animal magnetism," promoted by Franz Anton Mesmer. He was ultimately debunked by none other than Benjamin Franklin, but his legacy continues to this day in our language. Animal magnetism, to Mesmer, was an alleged magnetic force created by living things that he could manipulate to cure rich, anxious clients of whatever ailed them. Today the meaning has morphed into "sexual attraction." We also talk of people being "mesmerized" when something draws their rapt attention.

Fast-forward a century, and the new scientific marvel was radiation. This gave birth to all sorts of radioactive tonics and speculation about the magical powers of X-rays and other invisible particles that rip through your body. Believers extolled the virtues of regular radium tonics, and all the while they were slowly killing themselves with radiation poisoning.

The Bailey Radium Laboratories of East Orange, New Jersey, sold a radioactive tonic they called Radithor, which they guaranteed contained "at least 1 microcurie each of Ra-226 and Ra-228." They also guaranteed their product was entirely safe, and "certified radioactive." The market for such products didn't go away until they were specifically banned by the FDA in the 1950s.

But fear not, there are always new scientific ideas to fill the void left by the older ones, those that have become too familiar to be ideal for shining up pseudoscience (although magnetism remains popular for this). Today one of the most popular legitimate scientific ideas used to justify nonsense is quantum mechanics.

Quantum theory is uniquely perched on the top of the heap of all sciences that could reasonably be expected to support wild paranormal theories persuasively. One reason is its unmitigated success. It has been so successful and so impactful that quantum theory and relativity are considered the premier scientific theories of the entire twentieth century. This theory isn't just some obscure observations or laboratory curiosity. A significant portion of the world's gross domestic product is completely based on quantum mechanics. Transistors, computers, lasers, even the World Wide Web would not exist without quantum mechanics. You are bathed in its influence from the time you get up to the time you go to sleep.

More importantly, in its study of the forces and particles and behaviors of the atomic world, quantum theory has revealed a fundamental reality that is as wildly counterintuitive and bizarre as it is beneficial and innovative.

My three favorite manifestations of this wackiness are superposition, entanglement, and tunneling.

Superposition is the fundamental tenet of quantum theory that turns the most heads. Imagine all the distinct attributes that particles can have, such as spin, mass, and velocity. Some of these attributes can be polar opposites, like "spin up" and "spin down," but particles actually exist at times in a freakish state called a superposition, in which they're both spin up and spin down at the same time. This state is fragile, however; it persists only until the particle interacts with the environment and settles into one specific spin orientation, a process called decoherence or wave-function collapse.

The concept of entanglement takes superposition and puts it on steroids. Imagine a quantum system with no inherent spin. If two particles emerge from that system, they will have opposite spins (such as an up spin and a down spin). Since the original system that spawned these particles had no inherent spin, they can't both be up or both down (due to conservation laws). They can both be in a superposition of up and down, and they can also potentially travel millions of light years through the relative vacuum of space in this state. If one of these particles is hit by another atom or is measured by an alien scientist, it will decohere into either spin up or spin down. When that happens, however, the other particle, quintillions of miles away, will necessarily have to be the opposite spin. How could it know instantly, on the other side of the visible universe perhaps, which spin its partner "chose"? This is unknown. Yet the fact remains that there appears to be some type of superluminal connection between such particles.

If that's not enough head-spinning atomic behavior, let's look at quantum tunneling. This refers to a phenomenon in which a particle moves through a barrier that classic physics tells us it shouldn't be able to. One way to describe how this happens is that it is due to quantum theory's wave-particle duality. This means that what we think of as point-like particles can also be described as extended wavelike entities. Particles are therefore often in, you guessed it, a superposition of both wave and particle at the same time until they interact with the environment. An electron wave therefore can have a tiny part of its wavelike leading edge extend just past a barrier. Therefore, a small percentage of the particles will decohere on the other side of the barrier.

If scientists and lay people can accept such a fundamentally weird reality, is it really too much to ask that this wild theory could also support mind reading, the fundamental interconnectedness of all living things, or any of the host of paranormal ideas attributed to quantum mechanics?

In a word, Yes.

The overall allure of quantum theory for the mystic is that it is so counterintuitive. It's only counterintuitive to us, however, because we evolved in a macroscopic world where quantum and relativistic effects are not apparent.

Quantum mechanics is so counterintuitive, in fact, that even Albert Einstein had trouble accepting it. He made his career, among other things, from having the vision to realize that relativity theory was not just a mathematical trick—the world actually worked that way. When you get closer to the speed of light, time and distance actually change. Space and time are the variables, while relative velocity is the constant. Very weird to someone evolved in an environment where nothing moves that fast, but Einstein realized it was a fundamental aspect of how nature works.

Then, toward the end of his career, while other physicists were essentially saying the same thing about quantum theory, that things like superposition and entanglement were how the world actually works at that scale (in this case very small rather than very fast), Einstein balked. He figured we just didn't understand what was really going on. *outside the author's area of expertise*

So we can be forgiving when non-geniuses today have a hard time wrapping their brains around relativity and quantum theory. But this leads to the potential for pseudoscientists to say, "Hey, I know my claims are weird and fantastical, but quantum theory is weird and fantastical, so therefore I may be onto something." That's when common misconceptions about quantum theory can cross the line into quantum woo.

Author and alternative medicine guru Deepak Chopra is perhaps the most well-known proponent of quantum woo. His books and frequent lectures often invoke the word "quantum" to explain a host of New Age beliefs.

The role of consciousness in the universe seems to be especially interesting to Chopra. He has said the following:

> It was only with the advent of quantum physics that scientists began to consider again the old question of the possibility of comprehending the world as a form of mind...Indeed, the quantum theory implies that consciousness must exist, and that the content of the mind is the ultimate reality. If we do not look at it, the moon is gone. In this world, only an act of observation can confer shape and form to reality.

This very common misconception stems from the idea that it takes a conscious measurement or an intelligent observation to collapse a superposition (wave-function) and force reality to choose its...reality. This is wrong in multiple fundamental ways. As I described earlier, it's not specifically the act of observation that removes a superposition but interaction with the environment—mindless decoherence. One atom randomly bouncing off another will therefore do just as much as the most conscious mind experimenting in the atomic realm.

Chopra's moon comment exemplifies a common theme in quantum woo, that the bizarre phenomena of the quantum realm bubble up easily into our macroscopic realm of everyday life. This is simply untrue. Decoherence is so universal that once you start looking at anything much bigger than an atom, the weirdness starts disappearing. That's why quantum computers and similar experiments are so hard to pull off. The high level of isolation or extreme cold required to demonstrate quantum behavior is exceedingly difficult to maintain. By the time you start dealing with macroscale objects with millions to quintillions of atoms, all interacting in some way, any quantum behavior is long gone, replaced by more classical Newtonian mechanics. The objects of everyday life are essentially decoherence machines.

This is what modern quantum mechanics is telling us. For Chopra to think that the moon is in some superposition that requires the regard of a conscious mind to be fully actualized shows that he has left modern quantum theory and is well into Quantum Woo Land. (Actually, I think he is the mayor of Quantum Woo Land. Someone give him a sash.)

In short, quantum theory doesn't prove that consciousness is fundamental to the universe, that the universe itself is conscious, or that we control reality with our minds. And entanglement does not mean that all things are interconnected in such a way that we can use it to explain telepathy or the distant effects of planets and stars required for astrology. Entanglement is indeed weird, and we don't fully understand it, but it's a fragile state that only occurs in special circumstances. Our brains, macroscopic objects that they are, aren't entangled with each other.

By now, you may have recognized that quantum woo is often used as a giant argument from ignorance. It's often invoked as something we don't fully understand and therefore my magic is real. I say, if you don't understand quantum mechanisms, then just don't invoke it as an explanation. You might embarrass yourself to the dozen or so people in the world who actually do understand it.

Extra Geeky Aside: Theoretically, even large objects haven't lost

100 percent of any quantum weirdness. The wave/particle duality applies to objects like the moon or even people. Objects have a quantum wavelength, called the de Broglie wavelength, which can be thought of as a measure of their quantum effects. Your corresponding wavelength, however, would be orders of magnitude smaller than an atom, meaning that your wavelength is for all intents and purposes zero and can be safely and completely ignored. In other words—no quantum weirdness at the macroscopic level.

POOR REASONING IN A BOOK ON SKEPTICISM

34. Skeptics' Guide Entry: Homunculus Theory

Section: Cautionary Tales
See also: Pseudoscience

Homunculus theory is a class of medical philosophy that assumes one part of the body contains a functional map of the entire organism.

> The claim of alternative practitioners to not treat disease labels but the whole patient...allows alternative practitioners to live in a fool's paradise of quackery where they believe themselves to be protected from any challenges and demands for evidence. —Edzard Ernst

———

Before the advent of science, medical systems were philosophy-based. This means that they were built on an idea or philosophy that wasn't established by science. That idea or philosophy was an organizing principle used to explain and make sense of health and illness. Such notions, however, were isolated ideas, not based on evidence or woven into a deeper understanding of how the universe works.

For example, most cultures believed in a type of life spirit or force that animates living things and makes them different from nonliving things. Health and illness were explained using the idea of a life force, with illness resulting from the force being weak, blocked, or out of balance. In the West, the concept of four bodily humors was popular for almost two thousand years without being based on anything real. Sympathetic magic was also common—the notion that substances contain an essence reflecting what they look like. Powdered rhino horn, therefore, should make an effective aphrodisiac.

One philosophy that seems particularly quaint to modern ears is the notion that one part of the body "maps" out to the rest of the body. For example, the foot contains one of these so-called "maps." The attempt to diagnose health by reading the map of the foot is known as reflexology. Iridology measures well-being through studying the iris, and the touching of the features and forms on the palm's surface to determine salubriousness is known as palmistry. These are just three examples of a homunculus theory of medicine.

The word "homunculus" originated in the seventeenth century, when the practice of alchemy was still in its heyday. Alchemy is the attempt to transform matter—turning lead into gold, for example—by using "magical" concoctions, incantations, and ceremonies. Contained within the belief system of alchemy is the idea that a single sperm contains a miniature person—fully developed, merely scaled down to sperm size. The mini-me inside a sperm is capable of "birth" without the requirement of fertilizing an egg.

How? According to one of the more interesting ideas of the famous philosopher and alchemist Paracelsus (1493–1541), the process is as follows:

> Let the semen of a man putrefy by itself in a sealed cucurbit with the highest putrefaction of the venter equinus [horse manure] for forty days, or until it begins at last to live, move, and be agitated, which can easily be seen...If now, after this, it be everyday nourished and fed cautiously and prudently with [an] arcanum of human blood...it becomes, thenceforth, a true and living infant, having all the members of a child that is born from a woman, but much smaller.

Despite the grotesque description, the word "homunculus" is in this context synonymous with "little man," and it is from this the term is borrowed to describe the various practices of mapping an entire body through one of the body's features. For reflexologists, the foot is the "little man" of the body. For iridologists, the eye is the little man. For palmists, the little man is the palm.

Homunculus philosophies of health and medicine lack both evidence and plausibility to be understood as anything other than a belief system. There is no underlying anatomy or physiology that can explain how the iris, the foot, or the palm could reflect the state of function of any other part of the body. Despite this, these ideas have endured the test of time (hundreds of years in some cases) and are still in practice today, with eager practitioners and willing customers.

Homunculus theory is mainly a desire for simplicity, to subsume all the complexities of biology in one nice neat little system. It would be wonderful if we had a map of the whole body, somewhere we could just read in order to make detailed diagnoses, or massage or poke with needles to treat illness. This theory also reflects prescientific and predigital thinking. Before advanced technology, people understood the notion of templates. Templates were used to make clothes, and molds were used to make precisely identical items. It's an easy idea to understand.

Therefore, when trying to imagine how a baby is made, it makes sense to reach for the familiar, to think that there must be a little template of a person in the gametes.

Furthermore, there are real homunculi in your brain. For example, there is a motor homunculus, a motor strip in your cortex that actually does map to your body. These are real connections. Those brain cells give off axons that actually go to all parts of the body through the spinal cord and nerves. Developmentally, this is called somatotopic mapping. When the brain is forming, it maps itself to neurons connecting to other parts of the body, along both the motor and sensory strip. This also works for the retina, which maps to the visual cortex. We can see the brain's homunculi when we map the brain's function, and we can follow their connections to the body.

There is no such evidence, however, for the alleged homunculi in the foot, iris, or palm. There's no connection between one particular spot on the sole of your foot and your liver.

There are other sorts of pseudomapping systems, such as acupuncture. Acupuncture points allegedly map to different parts of your body, including the organs. And even chiropractic—straight chiropractic, specifically—postulates that the spinal cord and the vertebrae map to different organs in your body, even when those nerves don't actually go to or provide any kind of nerve supply to those organs.

All these different kinds of "mapping" systems—and, therefore, homunculus-based theories of health—are pure pseudoscience.

That, however, didn't stop the 2010 Jerusalem International Conference on Integrative Medicine from accepting an abstract based on a newly invented homunculus system. The abstract was submitted by Dr. John McLachlan, who later published the results of his prank in the *BMJ*. He presented case reports and testimonials, which, he claimed, established the usefulness of "butt reflexology." That's right, he just superimposed the real homunculi from the brain onto a schematic of human buttocks and invented a new "alternative" medical system. His abstract was gleefully accepted for presentation at the conference.

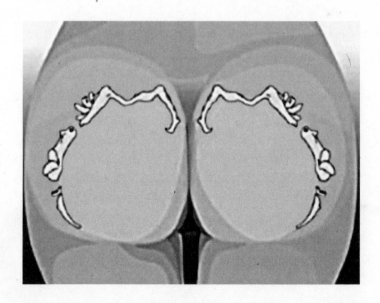

There are some general lessons we can derive from the history of the homunculus. First, beware of simplistic systems that purport to explain complex phenomena. Usually the world turns out to be more complex than we think, not simpler. Also, if you can't think of a possible mechanism by which something can work, then be skeptical. This may just be a deficit of knowledge on your part, but an "expert" should be able to give you a reasonable explanation that you can understand.

And be suspicious of ancient prescientific ideas. They really had no clue, so unless their ideas have been reasonably demonstrated and explained by modern science, they probably are as nonsensical as they sound. Think of Paracelsus and his putrefaction of the sperm. He was what passed for a scientific genius back in his day.

35. Skeptics' Guide Entry: Intelligent Design

Section: Cautionary Tales
See also: Creationism, Fundamentalism

Intelligent design (ID) is the notion that the complexity of the universe can only be explained as the result of an intelligent force, a designer. Critics maintain the idea isn't a proper scientific theory because it is fundamentally unfalsifiable.

> Design in nature is but a concatenation of accidents, culled by natural selection until the result is so beautiful or effective as to seem a miracle of purpose.
>
> —Michael Pollan

In 1996 Michael Behe published *Darwin's Black Box*, a book he hoped would take the scientific community by storm, posing a serious challenge to the theory of evolution and solidly establishing intelligent design as a legitimate scientific alternative. He failed miserably on both counts (as have all his ID colleagues), because he failed to realize the core problems with ID—not the least of which is that it simply is not science.

The core belief of ID, according to the Discovery Institute (the main organization promoting this idea), is as follows:

> Intelligent design holds that certain features of the universe and of living things are best explained by an intelligent cause, not an undirected process such as natural selection.

The primary scientific criticism of ID is that it isn't a legitimate scientific theory but rather a transparent attempt to recast religious faith (creationism) in scientific-sounding jargon. ID lacks the minimal criterion to be considered science: falsifiability. ID proponents, of course, reject this argument, because the entire purpose of ID is to disguise creationism as a scientific theory.

Much of the discussion on this topic focuses on the specific point of whether or not ID can be falsified: Can it theoretically be proven false by scientific evidence? If it can, then it is at least potentially scientific. If it cannot, then it is forever outside the realm of science.

This creates a dilemma for ID proponents. They desperately want ID to be considered legitimate science—that is in fact the entire reason for its existence. But they cannot allow for it to be actually falsified by scientific data, for that again would defeat the purpose.

As an example of this strategy, let's take a look at an essay from Jonathan Wells of the Discovery Institute. He is responding to an essay by Francis Collins, a geneticist who accepts the science of evolution and is highly critical of ID but is also a devout Christian. Wells wrote:

> More surprising is the fact that Collins is here citing experimental evidence against a theory he maintains is unscientific because it is not open to experimental testing. In claiming that evidence from gene duplication disproves ID, Collins is demonstrating that ID can be tested with scientific evidence. Either ID is unscientific, in which case evidence is irrelevant; or evidence can be cited against it, in which case ID is scientific. Collins can't have it both ways.

It's actually Wells who is trying to have it both ways, citing evidence against ID to demonstrate that it is falsifiable without acknowledging that it has been falsified (nice trick). This is the game: pretending ID can be falsified, but then always keeping it just out of reach of scientific evidence so that in practice it can never be falsified.

There are actually several problems with ID that render it unscientific.

Asking the Wrong Question

ID is based upon the false dichotomy that design in nature is necessarily intelligent and evolution is an "undirected process" incapable of producing design. ID proponents have carefully crafted their premise: design = intelligence, evolution = random. Then all they have to do is show the appearance of design in nature and claim that ID is verified.

The fatal flaw in this strategy is the failure to recognize that evolution is capable of producing design, because it's not a random process. Natural selection is the nonrandom survival of organisms based upon their inherited traits. Mutation and variation are random, and the long-term path of evolutionary change is best described as chaos, but natural selection allows for the nonrandom accumulation of favorable changes.

So the ID proponents are asking the wrong question—always a fatal problem in science. The question is not whether or not there is design in nature, but what the nature of that design is. Evolution is a bottom-up process whereby design and complexity emerge out of blind but nonrandom processes. "Intelligent" design, by contrast, is a top-down process where the final result is known ahead of time by the designer and is achieved with purpose.

There are many analogies we can draw to illuminate this difference. For example, a city that grew over decades without any central planning, based rather upon the decisions and actions of individuals acting in their own interest, is like an evolved city. An ID city, however, is one planned and mapped out ahead of time by a committee, corporation, or some other body. In the evolved city there will still be design—streets and utilities will follow residences and business, for example, and shops will tend to pop up and survive if they meet a demand. But it will be messy, with lots of redundancy, and with abandoned buildings where neighborhoods collapsed or businesses failed. Streets would likely not be optimally arranged. A planned city, however, would look vastly different—more clean, purposeful, and direct. The streets would be laid out in a deliberate way, one that probably would not have emerged spontaneously with use.

The difference between evolved and top-down designed life would be even more stark. Buildings and entire city sections can be torn down and rebuilt, so some top-down design can always be imposed later. But biological systems are far more constrained. Bottom-up evolutionary systems can only work with the raw material at hand. They can't start from scratch, develop new organs or limbs wholesale from nothing, or completely eliminate unneeded bits of anatomy.

If we ask the correct question—does life display bottom-up or top-down design?—the answer is obvious to anyone with sufficient knowledge of biology and an unbiased mind. Life is overwhelmed by the features of bottom-up design, from the vestigial eyes of cave salamanders to the bits of viral DNA junking up our genome. But none of this can falsify ID, because they're still asking the wrong question.

Common Descent

It's important to recognize in any such discussion that evolutionary theory actually has multiple components: the fact that life's diversity arose through evolutionary branching descent over time, the specific mechanism(s) of evolution, and the particular pathway of evolution taken by specific branches of the tree of life. One important line of evidence for the first claim of evolution—that life arose through branching descent—are the many biological similarities among various species that demonstrate not only common descent but also a branching pattern of descent predicted by evolutionary theory.

For example, there is evidence from gene duplication that over time one gene can become duplicated as a mistake of replication. Descendants will therefore have two copies of a gene instead of one. The duplicate copy is then free to change over time through mutations, essentially free to experiment with variations of function, because the original gene copy is able to carry out the original function of the gene. The duplicate gene may therefore hit upon a new function that helps the original gene carry out its purpose, or perhaps it may become co-opted for an entirely different purpose.

Biologists can examine the sequence of base pairs in genes to map out their relationships with other genes, and in that way they can build a detailed map of which genes evolved from which other genes. What we find when we compare such maps among species is that they fit into a nice pattern of branching common descent. There are multiple other independent lines of evidence that also demonstrate not only branching descent but a reasonable overlap—the different lines of evidence generally agree about which species evolved into what, when.

These lines of evidence could have completely falsified evolution. It's possible, for example, that we could have found patterns of gene variation that were incompatible with the theory of common descent. If we bring this back to the question of whether or not ID is falsifiable, we have to ask: Is there a pattern of gene variation we could have found that could potentially falsify ID? The answer, apparently, is no. At least, I'm not aware of any ID proponent making such a prediction that was open to falsification.

This is where we also get into the "it depends" answer. As I stated above, evolution has various components, only one of which is common descent. Some ID proponents, like Michael Behe, actually accept the fact of common descent. They think that life did change over time through branching descent. They just don't think this process was due to natural selection acting on variation. They think it was guided top-down by an intelligent designer. So evidence of common descent isn't evidence against this form of ID.

What about those who do not accept common descent, those who, like young-earth creationists, think that life was created more complete, that the "designer" didn't spend millions of years making slow changes over time but *poof*ed life into existence pretty much as it is? Evidence for common descent does falsify such claims.

Or does it? In theory, yes it does. And those like Wells are now using this to argue that ID is science because it can be falsified. But in practice, it doesn't, because ID proponents and creationists who reject common descent make the argument that the intelligent designer could have chosen, for their own unfathomable reasons, to make life so that it has the appearance of branching descent. The "God can make life to look like whatever he wants" defense renders the beliefs of anyone who makes it unfalsifiable. So, either way, this line of argument does not make ID falsifiable science.

Irreducible Complexity

This is the primary line of argument by its proponents for the falsifiability of ID, and it's the core concept of Behe's book—that we can look for design in biological nature by looking for structures (bits of anatomy) and biochemical pathways (reactions necessary for life) that are irreducibly complex, meaning that they could not function if they were any simpler.

The biggest problem with this argument is that, once again, it's based upon a false premise and one that was chosen to achieve the desired result. The premise is that if a structure could not function for its current purpose if it were any simpler (if any complexity were removed), then such a structure couldn't have evolved because it couldn't have passed through simpler forms to get to its irreducibly complex state, because evolution requires that in order to be selected, a structure would have to provide an adaptive advantage every step of the way.

The iconic example is the mousetrap. Take away the spring, or the base, or the lever, and it cannot function as a mousetrap; therefore the mousetrap is irreducibly complex.

Superficially this sounds reasonable, but it has been shown to be a false premise (and yet the ID crowd won't abandon it). Specifically, the premise ignores the possibility that an irreducibly complex structure could have evolved from a simpler structure that served a different purpose. A favorite example of ID proponents, first proposed by Behe, is the bacterial flagellum. The bacterial flagellum is fairly complex, and Behe argued it could not function without all this complexity and so could not have evolved. However, the flagellum could have evolved from a simpler tube that couldn't move but was used to inject substances into another cell. This simple syringe could have been an evolutionary stepping-stone to the more complex flagellum—and in fact evidence now supports this hypothesis.

The premise is also false because natural selection isn't the only force operating in evolution. There is also genetic drift, which is random variation not under selective pressures. Remember the gene duplication example above—a redundant copy of a gene is free to drift, producing changes that can at some point stumble upon a crude function. Even the slightest reproductive advantage from the most rudimentary of structures would provide a toehold for selective pressures to take over, fine-tuning the new structure.

In practice the notion of irreducible complexity contains two strategies. The first is to argue that a biological entity could not be simpler even in theory, which we've proven invalid. So ID proponents fall back to their second strategy, arguing that evolutionists haven't fleshed out the actual evolutionary history of an apparently irreducibly complex structure or pathway. But this second strategy is nothing more than an argument from ignorance. It relies upon our current state of knowledge and assumes that "currently unknown" equals "unknowable," and further, that unknowable means impossible—impossible for evolution, therefore we need to invoke an intelligent designer.

An argument from ignorance—basing a conclusion on what is not known—is always a weak argument, because it doesn't require any positive evidence for a theory: It's just knocking down a competing theory. What has happened since Behe made his initial claims for irreducible complexity is that scientific progress has continued and many of the holes in current knowledge that Behe relied upon have been filled in, like the bacterial flagellum example above.

What implications does this have for the question of whether or not ID is falsifiable? Well, it establishes that specific claims used to support ID are falsifiable. We have now falsified the claim that the bacterial flagellum has no simpler evolutionary antecedents. Wells and others use this to say that ID is therefore falsifiable. But once again, in practice it isn't, as is evidenced by the very fact that Wells, Behe, and others have not abandoned ID because the flagellum argument was proven wrong. This is because when one argument falls, they simply migrate to another. They even state that in order to falsify ID, evolutionary scientists would have to flesh out the complete evolutionary history of every biological component. This, of course, is an impossible and absurd standard.

What they're admitting, without meaning to, is that ID is ultimately a "god of the gaps" belief, and the only way to falsify it is to close every single last gap.

The False Dichotomy

The notion of ID falsifiability has a deeper logical problem: ID is defined entirely by what it isn't—namely, evolution. ID is based upon the claim that evolution cannot explain life. This is a false claim, but even if true it wouldn't be evidence for ID. So ID proponents spend their time trying unsuccessfully to poke holes in evolutionary theory and point out gaps, all the time pretending that current gaps are somehow evidence for ID, when they aren't.

If ID proponents had an actual theory, they should be able to say something about the features of intelligently designed life (predict what we should find)—and these features can be looked for to see if their design predictions pan out. So far, they have not been able to do this. Their answer to this is irreducible complexity, but as I explained, this doesn't cut it.

What they won't do is make any statements about the intelligent designer. What marks would such an intelligence leave upon creation? No one appears to know. When pressed, they often play the "it's a mystery" card. How can we possibly fathom the intelligence necessary to design and create life? Evolution, on the other hand, does make many specific predictions about what evolved life should look like—predictions that have been validated. Evolution's successful positive predictions are a problem for ID, but they deal with them by arguing that the designer could have arbitrarily and inexplicably decided to make life look as if it evolved.

To put it another way, if your theory is compatible with every possible observation, then by definition it is not falsifiable, and it is therefore not a scientific theory. Since the intelligent designer (by which, let's face it, they mean God) can, according to ID proponents, create life to look like anything, no possible observation of life can falsify such a designer.

The challenge that remains open for the ID community is to state a specific prediction about what positive evidence should be present if life were top-down intelligently designed. They cannot do this.

Perhaps the Discovery Institute (if they were overcome by an uncharacteristic spasm of intellectual honesty) should change the definition of ID to:

Intelligent design holds that certain features of the universe and of living things were caused by an intelligent agent who, for reasons we do not care to get into, chose to make the world look exactly as if it were the product of random variation and natural selection.

36. Skeptics' Guide Entry: Vitalism and Dualism

Section: Cautionary Tales
See also: Life Force, Essence

Vitalism is the belief that living things are animated by a life force, or "vital" force, that gives them not only the quality of being alive but also their essence. Dualism is the same concept applied to consciousness—that the mind is more than or separate from the physical brain.

> Science does not aim at establishing immutable truths and eternal dogmas; its aim is to approach the truth by successive approximations, without claiming that at any stage final and complete accuracy has been achieved.
>
> —Bertrand Russell

At a recent skeptical conference there was a kiosk in the hotel selling energy bands. These are cheap rubber and plastic bracelets that are claimed to improve your health or athletic performance by balancing or supporting your "energy." We asked the vendors how they worked, and they gave us the usual line about vibrations and life force. (I don't think they really cared; they were just trying to make a living.)

It's amazing how persistent the idea of life energy is, even in our modern world. If you were trying to make sense of the world thousands of years ago, prior to any modern scientific knowledge, it makes sense that you would assume there is some fundamental difference between living and nonliving things. Living things are active, they appear to have agency, they can use energy to do stuff, and they can die. Inanimate objects have none of these properties.

Our brains are in fact organized to divide the world into these two categories—the animate (which it operationally defines as moving in a way that suggests agency) and inanimate. Our brains assign direct emotional significance to animate things but not to inanimate things.

It's understandable, then, that pretty much every human culture independently developed some version of the belief that a special animating force makes living things alive. The Romans called it *spiritus* and the Greeks *pneuma* (both words also meaning "breath"). In China it is called *chi*, which also translates to "blood" because they felt the life force was carried in the blood. In Japan it is *ki*, in India *prana*, in Polynesia *mana*, and in Arabic it is *baraka*.

The notion that our thoughts, memories, feelings, ambitions, and all our mental functions are the result of the activity of cells in our brains is a very modern concept. Prior to modern neuroscience, even our best thinkers assumed that our minds were something more than physical. How could a clump of matter have feelings? It simply doesn't fit with our intuitions.

Interestingly, studies of young children find that their naive reasoning about biology includes the notion of life energy and essence. Further, they think that organs have agency and intentionality and that they transfer energy as part of their function. This was found in both Japanese and American children and is therefore not specific to any one culture. Children use the assumption of vitalism until they gain a more scientific understanding of life.

Up to the nineteenth century, the notion of vital energy or essence was a placeholder for our collective ignorance about biology. At no point did we prove, necessarily, that the vital force doesn't exist; it was simply edged out by scientific explanations. Vitalism became worse than wrong—it was simply unnecessary. It stands as an excellent historical example of how ideas emerge and fade in science, and how important it is to recognize when a scientific notion is supported by evidence and when it is just a guess we use to explain the unknown.

The more we learned about biochemical, physiological, genetic, and neurological functions, the less there was for the vital force to do. Each step of the way, the vital force was retained to explain whatever it was we didn't yet understand, a vitalism of the gaps per se. Eventually there was simply nothing left for the vital force to do, and by the end of the nineteenth century vitalism had simply vanished from scientific thought.

The beginning of the end came perhaps with Friedrich Wöhler. Many vitalists argued that organic compounds couldn't be synthesized from inorganic material, because living matter was fundamentally different from nonliving matter. But in 1828 Wöhler made urea from inorganic material, definitively disproving this notion. This discovery later gave rise to the myth that vitalism ended when Wöhler made urea, but this isn't supported by history. The tendrils of vitalism continued into the twentieth century, but only on the fringes of science. More and more, vitalistic thinking became the arena of pseudoscience and the paranormal.

In our modern world I think it's fair to say vitalism is entirely relegated to these areas. It has morphed into a vague New Age concept, something to do with energy and vibrations. What kind of energy? Don't ask—just energy.

In a 2013 study looking at vitalistic thinking in adults, researcher Stuart Wilson found that belief in a vital force corresponded strongly with beliefs in the paranormal and New Age philosophy.

Many alternative medicine modalities are explained with hand-waving references to "energy" or the life force. Chiropractic (at least the purist form, still practiced by about 30 percent of chiropractors) is based upon so-called "innate intelligence," which is their term for the life force. D. D. Palmer, who made up chiropractic, claimed that innate intelligence would descend from heaven into your brain and then down through the spinal cord to the rest of the body. Subtle subluxations, or misalignments of the spine, cause health problems because they are blocking the flow of innate. I have never heard an explanation for how someone with a spinal cord injury can live at all with such a complete blockage of their innate.

Another life energy–based treatment is therapeutic touch, which consists of waving the hands over a person, fluffing up and unblocking their "human energy field." Reiki is essentially the same thing, just using the Japanese term. Traditionally, acupuncture is explained as placing needles into special points that represent critical junctures in the flow of chi, unblocking and rebalancing this life force.

No one, of course, has been able to demonstrate that anything like life energy exists. There doesn't appear to be any chi, human energy field, or innate. It can't be detected by any instrument. Practitioners cannot demonstrate that they can detect it. There doesn't appear to be anything for the life force to do. And no practice based upon such a notion has been proven to be effective.

Now before you say, "But wait, what about Kirlian photography?" well, that is just one pseudoscience used to prop up another. It's like saying, "I know aliens exist because I can communicate with them telepathically."

Kirlian photography was made popular by parapsychologist Thelma Moss in her book *The Body Electric* (1979). She thought she had discovered a window into the world of "bioenergy" that would bring things like acupuncture into the scientific mainstream. However, her techniques led to nothing useful. The images generated by these photographs are simply a manifestation of barometric pressure, moisture, electrical fields, and other known physical phenomena.

Despite its continued popularity, the vital life force remains a pre-scientific placeholder displaced by our knowledge of biology. It stands, however, as a good reminder that just because a belief feels right, that doesn't make it real. There is clearly a primal appeal to the notion of life energy, but we can't ignore a couple centuries of biological science that tell us it doesn't exist. But I didn't explain all this to the woman at the conference kiosk. She seemed to have her hands full with a hotel full of skeptics.

Historically our path out of belief in a vital force was mirrored by our path beyond belief in a noncorporeal mind inhabiting our physical bodies. Before modern neuroscience, dualists, like vitalists, had a point, and their position does seem to resonate with how we feel. We all want to believe Yoda when he says, "Luminous beings are we, not this crude matter." It sounds like wisdom.

Now we know that the truth is closer to what psychologist Susan Blackmore said:

When I say that consciousness is an illusion I do not mean that consciousness does not exist. I mean that consciousness is not what it appears to be. If it seems to be a continuous stream of rich and detailed experiences, happening one after the other to a conscious person, this is the illusion.

It has now been established beyond any scientific doubt that the mind is what the brain does. The functioning of the brain is consciousness. In a sense, we are our brains.

We know, for example, that if the brain is not functioning, then you are not conscious. If we put the brain to sleep, you go to sleep. In fact, recent studies have shown that you need about 40 percent of your cortex to be healthy and active to maintain wakeful consciousness. If more than that is damaged, you will be in a coma or vegetative state.

If we change the brain, then we also change the mind. Drugs that alter brain function alter consciousness (you may have experienced this yourself).

Different brain states correlate to different states of consciousness. For example, when you're asleep your brain is still functioning, but in a different way than when you are awake. When you dream, your consciousness is altered in a way that correlates nicely with alterations in brain function. As the brain develops, our mental states also develop. Children are different from adults because their brains are different.

We can now use technology to turn different parts of the brain on or off, and the corresponding mental state also turns on or off. We've been able to do this for a long time during brain surgery by stimulating different parts of the brain. Now we can do it less invasively with transcranial magnetic stimulation.

And finally, when your brain dies, you die. There is no evidence of any consciousness without brain function or after death. As an aside, dualists often invoke near-death experiences as evidence for mind separate from body, but near-death experiences do not require such an explanation. They are simply experiences the mind may produce during trauma or when recovering from trauma or severe stress.

If you look at the hypothesis that the brain causes consciousness, and then you make predictions based on that hypothesis, it turns out every one of those predictions is true. To deny this connection is to deny all of modern neuroscience. In that way modern dualists are neuroscience deniers in the same exact way that creationists are evolution deniers. They deny the current model of biological neuroscience in order to manufacture a gap, and then they try to slip their dualism—their "ghost in the machine"—into it.

Mind of the Gaps

To explain where most dualists go wrong in their logic, let me start with a quote from Jerry Fodor, who stated in a book published in 2001 that "So far, what our cognitive science has found out about the mind is mostly that we don't know how it works."

At this point you may be thinking that we still don't fully understand how the brain works and creates the phenomenon of mind, and you would be correct. But that is the misdirection—confusing the question "Does A cause B?" with "How does A cause B?" We can know with a high degree of certainty (based on the lines of evidence I outlined above) that the brain causes the mind, without fully understanding exactly how it does so.

Let me illustrate this denial tactic by referring again to creationism (or evolution denial).

Evolution deniers will often take quotes from biologists that refer to either the mechanisms of evolution or the details of the historical pattern of evolution and present them as if they are questioning the fact of evolution. For example, one of the classic icons of evolution is the American Museum of Natural History display of the evolution of horses—from eohippus (now tragically renamed as *Hyracotherium*) to the modern horse. As our understanding of the pattern of evolution developed, it was realized that such linear progressive representations are fictions. Evolution with speciation creates a branching bush of descent, without any linear trend toward an apparent goal. Cherry-picking horse relatives from this bush and placing them in a line toward one branch—modern horses—completely misrepresents the true pattern of speciation over time in the horse lineage. Creationists still delight in taking quotes to that effect completely out of context, as if they call into question the fact of evolution itself.

Debates over the details of evolution don't decrease our confidence that evolution happened, just as research into the details of genetic complexity doesn't call into question the basic facts of genetics or the fact that DNA is the molecule of inheritance. Science progresses largely by creating a more and more detailed picture of nature, digging deeper and deeper into how the world works. Generally, lower-level or more detailed questions don't affect higher-level ones. Even if they reveal an anomaly that cannot be explained by the overall paradigm, this usually results in a modification of the paradigm, not a dismissal of it.

Returning to consciousness and the brain, all the evidence we have suggests that the mind is a product of the brain. The evidence that the brain creates mind is, in my opinion, overwhelming, while the evidence for a mind separate from the body is dubious and questionable—certainly insufficient to disprove the brain-mind hypothesis.

Dualists have therefore adopted the strategy of creationists by requiring that neuroscientists explain, in detail, exactly how the brain creates the subjective experience of mind. There are preliminary answers to this question. The mind is an emergent property of the brain and cannot be reduced to any single component of brain function. This is, admittedly, just a partial answer, merely describing the type of phenomenon we are dealing with and not really explaining it.

We have also identified many of the components of consciousness, localized them to specific brain regions. For example, we've pinpointed brain regions that create the sense we are inside our bodies, that we exist as an entity separate from the rest of the universe, and that allow us to direct our attention and form memories. We long ago determined those parts of the brain that see, feel, plan and execute movement, and generate emotional reactions. With functional MRI studies neuroscience has accelerated, and we are quickly reverse engineering the brain piece by piece.

This brings up another aspect of the dualist strategy: focusing on a snapshot of a science rather than its progress over time. As mentioned previously, there will always be gaps in our scientific knowledge, and their presence doesn't really tell us anything about the power and success of a scientific explanatory paradigm. A better method of assessment is to look at how it has progressed. Pseudosciences like homeopathy and ESP have not progressed—they're chasing their tails, going around in circles. Evolution, by contrast, has been remarkably successful as a means of scientific explanation.

Likewise, the materialist paradigm of mind and consciousness—the notion that the brain is the cause of mind—has been and continues to be a very successful model. One manifestation of this is that neuroscience, as a discipline, has grown and advanced. As new tools come online, our ability to explore the brain and to explain the phenomenon of mind has increased. The dualist paradigm, by contrast, has not produced anything tangible or reliable.

The handwriting is on the wall for the dualists, just as it was for the creationists. Scientists follow what works. Evolution works as an explanation for the complexity of life on earth and has vanquished the need to invoke magical creation. Biology works as an explanation for why living things are different from nonliving things, and we no longer have to invoke a life force. Neuroscience works as an explanation of mind and consciousness, and we no longer have to invoke a spirit to fill in the gaps. We will never have all the answers, never fill in all the gaps, but as long as these paradigms continue to flourish and succeed, scientists and the scientific community will follow them. And the deniers will be further and further marginalized to griping on the sideline, peevishly pointing at the shrinking gaps and desperately trying to prop up false anomalies.

Cartesian Dualists

There are different flavors of dualism. Cartesian dualists (named after René Descartes) believe that consciousness is fundamentally nonphysical, or spirit, in nature and that it communicates with the body through the brain. Humans, therefore, are both physical and spirit.

This relates to the second strategy dualists often use to argue for dualism. The first is the "mind of the gaps" described above. The second (which is really just an argument from ignorance against neuroscience) is the claim that the mind doesn't have the same properties as physical things and therefore cannot be physical. Here's what they say: Physical things have a location. You can say where your house is. They have size, mass, and shape—the properties of material things. Then the dualists ask, "What is the size of an emotion, the shape of a memory, or the mass of your dreams?"

Consciousness, they argue, doesn't have the properties of material things, therefore it is not a material thing. Therefore it is something else. Let's call it spiritual. This may sound superficially interesting, but it is utter nonsense. They are making what we call a category mistake. They are assuming that consciousness is a thing, and they're wrong. The brain is a thing, and it has all the properties of a thing. The mind is not a thing, it is a process. The mind is not the brain, it is what the brain does—it is the brain in action.

Let me give you what may at first seem like a strange analogy—capitalism. Capitalism itself does not have concrete physical properties. There is currency, there are goods, and there are consumers. These are like the brain: They are the physical substrate of an economy. Capitalism, however, is what all of these things do. It only exists when things are happening, when people are spending money on goods or services. We must no more conclude that capitalism is an interaction between the physical and the spiritual than that consciousness is.

Light Fairies

Another way that dualists attempt to rescue their belief from the avalanche of neuroscientific evidence is to appeal to the "correlation is not necessarily causation" fallacy. They say, "Sure, the brain correlates with the mind in all those ways you point out. But that doesn't mean the brain causes the mind."

Perhaps, they argue, the brain is simply a receiver for the mind, which lies outside of the physical. If you change the channel on your TV (the TV being the brain in this analogy), that changes the program, but the TV didn't cause the program.

This analogy fails, however. I might be able to change the channel, but I cannot change what the actors do on a television show by adjusting the TV. I cannot turn a drama into a comedy.

This line of reasoning also violates Occam's razor. I might argue that when you turn on the light switch, a light fairy goes from the switch to the light and activates the light with their own fairy light energy. When you turn off the switch, the fairy returns and the light turns off. The fairy, of course, is invisible, incorporeal, and can travel at the speed of light.

There is no way you can prove that my light fairy doesn't exist. Sure, you can argue that opening and closing a circuit correlates with the light turning off and on, but that correlation doesn't prove causation. When the electricity is off, the fairy goes to sleep. Electricity doesn't cause the light, it is just a medium for the fairy.

The problem with my fairy hypothesis is not that it's wrong, it's that it is not even wrong. It is simply unnecessary. It is an added supernatural step that is entirely superfluous and doesn't actually explain anything. Saying that the mind is caused by spirit is as unnecessary. Occam slices it away.

The Hard Problem

So how do we explain how the brain causes consciousness (or how brain function *is* consciousness)? We don't yet know all the elements of brain function that are essential for consciousness or know how they work together, but we are making steady progress. It does seem, however, that every part of the brain contributes its bit to consciousness. There is no "seat" of consciousness. Neuroscientists thought there might be one central control center, which they called the global workspace, but recent research has shown this is not the case.

Consciousness, therefore, is like a committee. No one person is the committee, every person contributes. You also need a certain quorum (a minimum number of people) before you have a committee, and you need a certain amount of brain function in order to be awake and conscious. The brain's committee seems to be a bit chaotic, with everyone shouting for attention and talking over one another. They may even fight, but in the end a collective decision is made.

In this analogy the thoughts of the individuals are your subconscious mind. "Subconscious" basically refers to brain processes that occur below the level of awareness. You can even make decisions without being aware that you did, and act on them before you're aware the decision was made.

Some modern dualists, like David Chalmers, acknowledge all these advances in neuroscience but refer to them as the easy problems. Chalmers argues that we can figure out how the brain moves your arm or sees something or does mathematical calculations. What we can't figure out, however, is the hard problem: Why do you experience your own existence?

Chalmers isn't a Cartesian dualist, he's a property dualist. He feels that consciousness is a manifestation of the material world, just a phenomenon currently unknown to science. Chalmers talks about what he calls philosophical zombies (or "p-zombies" for short). He argues that it's possible to imagine a being that does everything a human does but is not aware of its own experience. Subjective experiences of things, such as the experience of seeing red, are called qualia. So, Chalmers asks, why are there qualia?

The other side of this argument is perhaps best articulated by Daniel Dennett. He says, essentially, that there is no hard problem. It's easy problems all the way down. Once you have solved all the easy problems, then you've solved the hard problem by extension.

To explain further, he encourages you to simply ask, "And then what happens?" The brain sees an image; that information goes to the association cortex, which gives it meaning as a thing; and then that information goes to the amygdala for the assignment of emotion to the thing you are seeing (if it is living). And then what? Well, then that information goes to some other part of the brain where memories are stored. And then what? Well, that provokes a pattern recognition in your memory of a previous event involving the object. And then what? Then that information goes again to the emotional centers to provoke the emotions you have associated with that memory.

And the chain keeps going. It never stops. All these components of brain function that we've identified just keep going around in circles. The brain talks to itself, it receives new information, it activates circuits and networks, which provokes other networks, on and on until you die. That endless chain of activity (which pauses when you sleep but is actually replaced by other types of brain activity that just don't produce wakefulness) is your stream of consciousness. It doesn't ever have to go anywhere or report to anything. It's simply consciousness.

Therefore there is no hard problem. We understand consciousness to the degree that we have explained and understand the easy problems.

What about Chalmers's p-zombies? That's an interesting question, but I challenge his central premise that qualia are not necessary to do everything we do. While this is a difficult question to answer definitively, here are some possibilities:

We know that even the most basic bacteria have a system of attraction and repulsion. They'll swim up a chemical gradient toward food and away from poisons. This is also the most basic function of nervous systems, to provide reward for adaptive behavior and pain for harmful behavior. Pain and reward are the essence of function. For this system to work, however, pain has to "feel" bad and reward circuitry has to "feel" good on some level. As creatures become more sophisticated, their reward and aversion circuitry also become more sophisticated. This evolves into complex emotions. Think, too, about motivation. Fear is a great motivator to get you to focus and marshal all your energy to run really hard away from that predator.

Others have speculated that, because a memory of an experience is essentially identical in the brain to the experience itself but organisms need to know if they're remembering something or experiencing it in real time, the two things have to feel different in some way.

And finally, there needs to be some mechanism for allocating attention. Our brains cannot pay attention to everything at the same time, nor can they perform every type of processing at the same time. In fact we exert a significant amount of our mental energy just focusing our attention on the important stuff and filtering out the rest. Attention certainly implies a conscious agent.

Even if it's possible to design a system that mimics all these things without actually paying attention or feeling things, that doesn't mean that evolution would not have hit upon qualia as an easy solution. The fact is, we do experience our own existence and it works.

These are all understandably difficult concepts. In a way it's the brain trying to understand itself, the single most complex thing we know of in the universe. For this reason, we need to be especially careful and especially humble. We can't trust simple reasoning or our naive intuitions.

Essentially, I think that dualists are invoking magic to explain why their intuitions can't easily grasp consciousness as a phenomenon. It is quite possible that, because our brains evolved to give us the very useful illusion of a continuous stream of seamless consciousness, we evolved to not easily grasp our own consciousness. That would require piercing the illusion.

Science can accomplish understanding consciousness with decades of careful thought and research. Neuroscience is working out pretty damn well. I see no reason to lose our nerve, to start invoking light fairies and essentially give up trying to scientifically understand one of the greatest puzzles of the universe.

37. Skeptics' Guide Entry: N-Rays

Section: Cautionary Tales
See also: Pseudoscience

The story of N-rays is a classic tale of how bias can affect research, how an appealing fad can take hold, and how a little bit of skepticism can serve as an antidote.

> The scientific method consists of the use of procedures designed to show not that our predictions and hypotheses are right, *but that they might be wrong.* Scientific reasoning is useful to anyone in any job because it makes us face the possibility, even the dire reality, that we were mistaken.
>
> —Carol Tavris

Prosper-René Blondlot (1849–1930) was an accomplished French physicist who received three prizes from the Académie des Sciences for his work in electromagnetism. Despite this, he is best known for his role in one of the classic scientific blunders of modern times.

In 1895 German scientist Wilhelm Röntgen discovered X-rays, electromagnetic radiation with a wavelength ranging from 0.01 to 10 nanometers. This discovery made Röntgen famous and brought significant prestige to the German physics community.

The discovery also created the expectation that other forms of electromagnetic radiation would soon be discovered, and many physicists, including Blondlot, sought to do just that. It was therefore not surprising when, in 1903, Blondlot announced the discovery of another source of radiation that he named N-rays (after his birthplace, Nancy, France, and his university).

Between 1903 and 1906 no less than three hundred papers on N-rays were published in the scientific literature, involving about a hundred scientists and medical researchers. At least forty scientists claimed to see these N-rays.

Blondlot reported that N-rays were produced by many types of matter, both living and inert (everything except green wood and some treated metals), and that their strength increased with the "psychic activity" of the source. His experimental setup included a hot wire inside an iron tube to generate the N-rays, which were then refracted through a 60-degree angle prism of aluminum, and in turn were detected by a calcium sulfide thread that would glow slightly. The subtle glow could only be perceived in the dark by those with sharp vision.

Other researchers discovered much about the properties of N-rays: They could apparently penetrate metal and wood but were blocked by water. Medical scientists discovered they were emitted by muscles and the human brain and hoped they would be as useful in medical diagnosis as X-rays were turning out to be. It was discovered that N-rays could be transmitted over a wire.

The science of N-rays was taking off, but there was one slight problem. N-rays do not exist. All of this research was nothing more than self-deception.

The first clue that there was something wrong with N-rays came from the fact that English and German physicists couldn't replicate the French research. The laws of physics don't follow national borders, and so this failure to replicate generated considerable skepticism (as it should).

Scientific reputations and French pride were on the line. The French physics community would certainly have wanted their own electromagnetic radiation to match the Germans' X-rays, so it was easy for them to tell themselves that the German and English researchers just weren't doing it right: They didn't have the sensitive eye necessary to see the subtle glow of the thread that was the only evidence N-rays existed.

N-rays had other curious properties that served only to increase skepticism (among non-French physicists). Blondlot claimed that N-rays had a long wavelength like infrared light but that they passed through substances that should only admit much shorter wavelengths. N-rays had no direct effect on photographic plates and so could only be detected through the subjective perception of the researchers (although in some experiments the glowing detector was photographed to document its intensity).

Blondlot also found that he continued to detect N-rays even after the source was removed, leading him to conclude that certain substances, such as the quartz lens used in some setups, could store N-rays and later emit them. He even "discovered" that the vitreous humor of the human eye could store and emit N-rays.

The more Blondlot and his colleagues discovered about N-rays, the more their confidence grew, while the skepticism of German and English scientists grew equally large.

To settle this growing dispute, Robert W. Wood of Johns Hopkins University was sent by the journal *Nature* to observe Blondlot's experimental procedure. Wood had been unable to detect N-rays himself and wanted to see if he could resolve the discrepancy. As he later wrote: "I went, I must confess, in a doubting frame of mind, but with the hope that I might be convinced of the reality of the phenomena, the accounts of which have been read with so much scepticism."

Wood was perhaps the perfect person to undertake this investigation. He had a reputation for being a prankster. He was also a science popularizer and took it upon himself to investigate and debunk the pseudoscience of his day.

Blondlot and his assistants set up an experiment in which an aluminum prism was used to show the spectrum of N-rays. Blondlot insisted that he was able to see the faint glow of the detector in a completely dark room, while Wood could see nothing.

Wood spent hours in Blondlot's lab conducting several experiments to see if the N-rays could be reliably demonstrated. In the first setup, Blondlot and his assistants claimed they could see the glowing thread, while Wood could not. This was explained as Wood not having sufficiently sensitive vision.

So Wood did a simple control. The presence of his hand in the pathway of the N-rays should have been enough to block them, so, in the darkened room, he asked his colleagues to view the thread and tell him when his hand was blocking the N-rays. They couldn't reliably do so, stating that the thread was dimming or brightening even when his hand was still.

He next observed their attempts to show differences in how the experimental setup changed the brightness of the detector as documented on film. However, he observed that the cathode used to generate the N-rays naturally fluctuated by about 25 percent—so changes in the light source could explain apparent changes in the brightness of alleged N-rays. There was also the possibility that the precise techniques used by the experimenters—the angle and duration of exposure to the film—could affect the outcome.

Finally, Wood performed the control for which he is most famous. In the third experimental setup, an aluminum prism was used to bend the N-rays and spread them out into a spectrum. As he recounted in his eventual letter to *Nature* detailing his observations:

> I was unable to see any change whatever in the brilliancy of the phosphorescent line as I moved it along, and I subsequently found that the removal of the prism (we were in a dark room) did not seem to interfere in any way with the location of the maxima and minima in the deviated (!) ray bundle.
>
> I then suggested that an attempt be made to determine by means of the phosphorescent screen whether I had placed the prism with its refracting edge to the right or the left, but neither the experimenter nor his assistant determined the position correctly in a single case (three trials were made). This failure was attributed to fatigue.

Wood, with a few simple controlled experiments, single-handedly demolished the entire field of N-rays.

Lessons from N-rays

The story of N-rays is a classic cautionary tale of pathological science, bias, and self-deception among otherwise accomplished scientists. And yet the story is not nearly as well known as it should be.

The French scientists ignored many red flags that should have stoked their own skepticism. In science it pays to be one's own most vehement critic and greatest skeptic. This means trying earnestly to prove your own hypothesis wrong, and tentatively accepting it only to the degree that it survives dedicated attempts at doing so.

Blondlot displayed the opposite of self-skepticism. He too casually proffered ad hoc explanations for the curious aspects of N-rays, such as concluding that only certain people could see them. Even when blinded evaluation failed, the failure was attributed to fatigue rather than the far simpler explanation, that N-rays do not exist.

This is why Occam's razor is so valuable to scientific inquiry. We can always invent clever explanations for why our hypothesis appears to fail. But we must take very seriously the straightforward possibility that our hypothesis is simply wrong.

History Repeats Itself

Lest you think this is a quaint example from the turn of the nineteenth century, not relevant today, the research of Jacques Benveniste will provide a more recent example. In 1988 Benveniste published a paper in *Nature* magazine claiming that water retains the memory of substances that had been diluted in it—even through extreme dilutions.

This research was obviously done in the context of providing a mechanism for homeopathy, which uses fanciful treatments that are diluted often to such an extreme degree that none of the original substance remains (beyond background levels). Benveniste's claim was that even when the original substance is gone, the water retains its chemical signature through its own structure.

Benveniste's lab worked primarily with antibodies that activated basophils, which are immune cells, causing them to release histamine, which can be detected microscopically (Benveniste was an immunologist by training). He claimed that water that once had antibodies in it could still activate basophils after the antibodies were diluted away.

John Maddox, then editor of *Nature*, published the results even though he was skeptical of their validity, because the paper had survived peer review. But, in keeping with the history of his prestigious journal, he went to inspect the experimental procedure for himself. He took along Walter Stewart of the National Institutes of Health, who was an investigator of scientific fraud, and noted skeptic and magician James Randi. Like Wood before them, they published their findings in a follow-up article in *Nature*, detailing a number of failings of Benveniste's lab.

During the investigation, the team did what Wood did: They blinded the evaluation. When properly blinded, the replicated experiments were negative. The only debate among the investigators was whether the original results were due to bias or fraud. It was noted that one co-researcher critical to carrying out the experiments, Elisabeth Davenas, was the only one who could consistently produce positive results. Randi, in his talks on the experience, recounted observing her read a plate, check to see whether or not the sample was a test or a control, and then recount the plate to obtain a number more to her liking.

Randi also recalled that she was alone among her team in not being nervous when the results were about to be revealed. He interpreted this to mean that she knew the results, properly blinded, would turn out negative. However, a team of observers representing the journal *Nature* concluded that unwitting bias—not fraud—on the part of Davenas was a major explanation for the previously reported results. Like Blondlot, Benveniste ignored multiple anomalies in the data that should have prompted skepticism. He further refused to adhere to the blinded results, instead dismissing the investigation as a "witch hunt."

These historical cases, while perhaps among the most dramatic, aren't unique. More subtle manifestations of self-deception are common, even among mainstream respected researchers.

I routinely encounter rationalizations for why therapeutic trials turn out negative. Perhaps, it is suggested, the dose wasn't high enough or the treatment duration long enough. Perhaps the treatment would work on a subset of patients if they could be identified. Perhaps a better outcome measure should be used.

It's certainly possible that one or more of these explanations could be true. It is also possible, and a lot more likely, that the treatment simply doesn't work. What makes these explanations unconvincing is their ad hoc nature.

Humans in general are great at coming up with reasons to maintain their desired beliefs in the face of contradictory data. More intelligent and educated people aren't necessarily better at critical thinking, but they are likely to be more clever and creative in coming up with such excuses—and scientists are no exception.

38. Skeptics' Guide Entry: Positive Thinking

Section: Cautionary Tales
See also: Self-Help Pseudoscience

The power-of-positive-thinking movement is the cornerstone upon which self-help empires are built. From Chopra to Oprah, self-proclaimed gurus of the movement have made large sums of money via traditional consumer-based strategies, like television programs, books, seminars, and retreats. The underlying rhetoric is simple: "If you think positively, positive results will come."

> The correspondence between reality and my beliefs comes from reality controlling my beliefs, not the other way around.
>
> —Eliezer S. Yudkowsky

Wouldn't it be nice if just imagining something—thinking about it really, really hard—made it come to pass? This is a type of magical thinking that has broad appeal: It gives us the illusion of power and control. But it's just that—magical thinking.

It is also what many self-help authors and gurus would have you believe. It falls under many names—mind over matter, the law of attraction, manifestation, magnetism—but all its iterations have one thing in common: Absolutely no scientific evidence backs up such ideas.

From Rhonda Byrne's *The Secret* to Wallace Wattles's *The Science of Getting Rich*, self-help authors—whether con artists or sincere believers—are cashing in on the power-of-positive-thinking movement. Search any bookseller's website for "law of attraction," and you'll get thousands of hits for books touting a foolproof scheme to make money, find love, improve your memory, and even cure cancer. And if that worries you, it should.

Not only can we not change the universe just by wishing, doing so seems to be counterproductive. Psychologist (and friend of the SGU) Richard Wiseman points out that the research clearly shows no benefit from positive thinking; in fact there is a negative effect.

Wiseman notes that in a 1999 study, Lien Pham and Shelley Taylor compared "effects of process- versus outcome-based mental simulations on performance." In other words, they had students think about getting a good grade on an upcoming exam (outcome-based), compared them to students who thought about the process of studying and taking the exam (process-based), and compared both to a control. The students who imagined the process did better than the control, but the students who imagined getting good grades actually did worse.

The lesson is clear: Positive thinking by itself is not only worthless, it's a waste of time and effort that detracts from more practical activity. However, imagining the process you need to go through in order to achieve a goal is useful, because it leads to actual action. It's the action (in this case studying) that has the positive outcome.

In fact, research shows there are many benefits of a negative or pessimistic outlook. Pessimism correlates with higher earnings, fewer marital problems, more effective communication, greater generosity, and less disappointment. It is apparently helpful to worry, at least to some extent. Excessive optimism can make us careless and set us up for failure.

An excessively positive outlook can also complicate dying. Psychologist James Coyne has focused his career on end-of-life attitudes in patients with terminal cancer. He points out that dying in a culture obsessed with positive thinking can have devastating psychological consequences for the person facing death. Dying is difficult. Everyone copes and grieves in different ways. But one thing is for certain: If you think you can will your way out of a terminal illness, you will be faced with profound disappointment.

Individuals swept up in the positive-thinking movement may delay meaningful, evidence-based treatment (or neglect it altogether), instead clinging to so-called "manifestation" practices in the hope of curing disease. Unfortunately, this approach will most often lead to tragedy. In perhaps one of the largest investigations on the topic to date, Dr. Coyne found that there is simply no relationship between emotional well-being and mortality in the terminally ill (see James Coyne, Howard Tennen, and Adelita Ranchor, 2010). Not only will positive thinking do nothing to delay the inevitable; it may make what little time is left more difficult.

People die in different ways, and quality of life can be heavily affected by external societal pressures. If an individual feels angry or sad but continues to bear the burden of friends', loved ones', and even medical professionals' expectations to "keep a brave face" or "stay positive," such tension can significantly diminish quality of life in one's final days.

And it's not just the sick and dying who are negatively impacted by positive-thinking pseudoscience. By its very design, it preys on the weak, the poor, the needy, the down-and-out. Preaching a gospel of abundance through mental power sets society as a whole up for failure. Instead of doing the required work or taking stock of the harsh realities we often face, individuals find themselves hoping, wishing, and praying for that love, money, or fame that will likely never come. This in turn has the potential to set off a feedback loop of despair and failure.

Don't get me wrong. I'm not saying that being a (reasonably) positive person is a bad thing. Nor am I advocating that one actively squelch a sunny disposition. But don't expect to think good thoughts and be showered with some sort of cosmic blessing. Unfortunately, the universe doesn't work that way.

39. Skeptics' Guide Entry: Pyramid Scheme

Section: Cautionary Tales
See also: Multilevel Marketing

A pyramid scheme is a type of business that lures in new recruits by promising them an income from those they recruit themselves. Typically, people who engage in these businesses need to spend money to buy products or other materials related to the business. The training they receive is geared toward finding new recruits. If you imagine a pyramid, new recruits must find another bigger layer of recruits below them, and that new layer will go on to find yet another, even larger layer. Money is paid up the pyramid to those above.

> I think scientists have a valid point when they bemoan the fact that it's socially acceptable in our culture to be utterly ignorant of math, whereas it is a shameful thing to be illiterate.
>
> —Jennifer Ouellette

A friend of yours calls you up, excited to talk to you, and requests that you come to a meeting to learn about a life-changing opportunity. They explain that they're running their own business now and that you can too. You are going to make a great deal of money, but the first thing you have to do is go to a meeting. You are apprehensive, but your friend won't take no for an answer.

At the meeting there are a few other people you don't know. The friend who invited you is happy to see you and reassures you that this is going to be great. You're introduced to Tony. Tony will be running the meeting and soon will explain everything. Before long, Tony addresses everyone in the meeting:

> Welcome, everyone. I'm glad to see new faces tonight. Take a look around you at everyone in the room, because all of you are going to have your lives changed together tonight. Not long ago a friend of mine came to me and introduced me to Vita-Root multivitamin and meal replacement. I thought, "What the hell is this?" but I sat through the meeting. By the end of that meeting, I was sold. I realized that this was the opportunity I've been waiting for. My goal is to show you how smart and easy this business is. I want to show you that all you have to do is work a few hours a week and you will start making money. You see this check? It's my latest check from Vita-Root. It's for $5,270. This was what I made last month by running meetings just like this one. I'm also my own boss and I live a zero-stress life now.

Interesting, huh? Admit it, a part of you wants to know more. That's because basic psychology is at play here. Tony knows that everyone wants to feel like they're a part of something, and everyone wants to have control over their lives, make money, and be happy. All of these things were promised in the first minute of the meeting. Tony's goal is to get you to join the business, which means in one way or another you have to pay. Persuading people to buy the company's vitamins to sell, pay a membership fee, and find new recruits is how Tony and the friend who recruited you make money. On top of that, the person who recruited Tony makes money too, and the person above them. Anyone you recruit—including their recruits—is called your downline. You make money from the multiple levels below you, which is why businesses like this are called multilevel marketing.

Some multilevel marketing companies (MLMs) that have their agents sell products to the public are legitimate, well-established businesses. A multilevel marketing business is considered a pyramid scheme if the money a recruit makes comes mostly from the people in their downline (the people they recruit). In a very simple example, you ask four of your friends to give you $50 each. You make a profit of $200. Now each of your four friends has to go and find four more people to give them $50, at which point they will make a net profit of $150. This pattern keeps going, and as it goes on, the number of people involved increases exponentially. Hence the name "pyramid scheme," because the base of the pyramid is wider than the top. Pyramid schemes can be much more complicated than this, but that's the essence.

The FTC (Federal Trade Commission) warns the public to look out for three basic signs that you're dealing with a pyramid scheme: Most of your income comes from your recruits, not from direct sales; you're required to buy and maintain product, whether or not you sell it; you may be required to buy other things (like seminars) or pay other fees.

Often the products being sold are themselves dubious. They may be services that seem too good to be true (can you really cut your electricity bill in half?), questionable medical products, or everyday products with extraordinary claims. You can ask yourself, "If these products are so good, why aren't they sold in stores?"

Often the person trying to recruit you will insist that the company is "not a pyramid scheme" and that it is "totally legal." It's a safe bet that any time someone tells you their company isn't a pyramid scheme, it is.

Studies involving hundreds of MLMs have shown that 99.6 percent of people involved actually lose money. Only the people at the very top of the pyramid make any money; everyone else in the "downline" serves only as the ultimate source of their profit.

MLMs may also be dubious from a marketing perspective. Let's take our supplement example again. The company itself may not make direct claims to consumers that violate regulations by, for example, declaring that the supplement cures a disease. You won't find such an assertion in any official company information.

However, in person at MLM recruiting meetings, they do make

or imply such claims. The marketers themselves are customers who are purchasing the supplement for their own use. They then try to sell it with their own anecdotes and claims. They're solely responsible for these claims, and the company has a comfortable buffer against the resulting fraud.

The next obvious question is: Are MLMs legal? Well, yes and no. Pyramid schemes where no product is involved are illegal in most countries. The gray zone is when an actual product is being sold through an MLM model. In some cases the product is just a shield for what is ultimately a pyramid scheme.

In many cases where people do actually sell the product, that is sufficient to render the company technically legal. But again, even if legal, most people will still lose money because they're forced to buy and maintain product, they may be forced to spend money in other ways, and there isn't enough product being sold to keep the pyramid going.

Even with a legitimate MLM and product, the ultimate enemy is math. The structure of an MLM is inherently self-defeating. If, for example, the company starts with just six people, and each level has to recruit six people of their own, in just twelve levels you have over two billion people.

Even without complete recruitment, you quickly saturate a population. Further, every person you recruit is now a competitor. They may share your social network, and you'll both be recruiting from the same pool.

Any business model that depends on an ever-expanding sales force is simply unsustainable. Your best bet is probably just to stay far and away from any MLM. The chances are overwhelming that you'll end up at the bottom of that pyramid.

SECTION 2

Adventures in Skepticism

Now that we've filled your bags with critical thinking gear, let's put those tools to work.

Often on the show we do what we like to call a "deep dive" on a topic. There is something interesting in the news, or we get a question, and in order to discuss the topic meaningfully, we need to do a fairly thorough examination of all the relevant claims and the evidence. It's easier than ever to do this because of the internet.

In this case the promise of the internet and the web were largely fulfilled—we now have a readily accessible and searchable reservoir of collected human knowledge. In April of 2017, the estimated size of the World Wide Web was 47.5 billion web pages. Of course, much of this is misinformation, cat videos, pornography, and clickbait ads, but there's some useful knowledge tucked in there as well.

The really valuable skills today are in knowing how to find information and (more important) how to critically assess what you find. Here is a guide to navigating the jungle of information out there in the digital wilds:

What is the source of the information? There is no single unimpeachable source of information (that's right, not even the SGU). Some sources are better than others, however. Academic sites have been found, when checked for quality, to be the most reliable. Other institutions may also be reliable. If accuracy is important to their reputation, there is likely a team of people working on content, some sort of editorial process or peer review, and they don't have an extreme agenda.

Group blogs by professionals are also usually good sources of information. They're more credible than individual blogs or web pages, which are highly variable depending on the individual. Commercial sites are generally less trustworthy. The worst sites, however, are extreme ideological sites that exist to promote a specific narrative. Sometimes the goal is to promote a narrative and sell you stuff, as on the infamous medical conspiracy site Natural News.

But, as I said, no one site is perfect. So **you need to look at a variety of sources**. Further, it's important to find out where that source got its information from—what the primary source is. It's common on the web for sites to reproduce or simply link to other sources of information (that's what the web means). You have to **track information back to the original source**.

Sometimes that original source isn't reliable, even though it is being repeated by mainstream outlets. It's also possible that the original source, which may be a published scientific study or primary journalism, has been misrepresented by secondary sources. At other times one primary source gets repeated so many times that you may think it is many sources, but all roads lead back to the same primary.

It's extremely important that you specifically **look for a variety of sources and opinions**. Before I accept any claim, I always want to know who disagrees with that claim and what the reasons are for their rejection of it. If possible, I then want to find out how the first side answers their critics, and how the critics answer these responses. In essence **it's important to follow a discussion through to the end**. Sometimes you don't know who has the better position until you've heard all the point-counterpoints and find out who has the last word.

You also need to **independently examine the arguments of each side to see who has the better arguments**. Which side has better sources of evidence? Which side uses more valid logic? If one side has to consistently resort to logical fallacies and distortions, and never really answers their critics, they probably have a weak position.

It's possible that neither side is completely right or completely wrong. That doesn't mean there is always a balance. Sometimes one side is completely wrong (like anti-vaxxers). Sometimes the answer is pretty much in between two extremes, and sometimes we just don't have the answer and you have to reserve judgment. It's perfectly okay to say that you don't have an opinion about a topic if you don't feel you've done adequate research.

It is also extremely useful to **check yourself against other people you respect and who have the appropriate expertise.** I always check my understanding of a topic against the experts when I can. If my take is different from theirs, I need to find out why. Usually it means I'm missing something.

You need to make a concerted effort not to cherry-pick—**don't just accept the first answer, or the answer you want.** Don't seek out an expert who agrees with you and then conclude that you must be correct. Seek out experts who disagree with you. Those are the people you need to listen to.

The process is never over. All conclusions are tentative, and you should not stake them out forever or make them part of your identity. **If you learn new information, happily incorporate it into your assessment.** Take pride in the ability to change your mind. This doesn't mean you never have strong opinions, just that they should be only as strong as the evidence and logic support, and open to revision.

Here are some personal adventures we have had in using skeptical scientific thinking to learn about, and sometimes debunk, topics over the years. Grab your towel, put on your slippers, and join us as we venture forth into the wilderness. It is often frustrating, sometimes scary, but always rewarding.

40. Motivated Reasoning About Genetically Modified Organisms

Steve's Adventure

There doesn't seem to be any other way of creating the next green revolution without GMOs.
—E. O. Wilson

When we started our skeptical careers in the 1990s, GMOs (genetically modified organisms) weren't on our radar at all. Crops and other useful organisms that result from modern technology directly altering their genetic makeup were just being introduced, and they had not yet entered the public consciousness. They were on the radar of the organic food lobby and organizations like Greenpeace, however, who started a campaign to turn the public against this technology.

By the time I knew that anti-GMO was a thing, the campaign was in full swing, GMO opponents had a long list of scary claims about GMOs, and public opinion was already solidly against this food technology. Even many of my fellow skeptics felt that Monsanto was an evil corporation and that at least the regulation of GMOs needed an overhaul.

I knew I had to roll up my sleeves and delve into the science and the many claims being made against genetically modifying our food. I also knew I had to do it as objectively as possible, to find out where the truth actually lay. I had no stake in the outcome—the only thing at stake was my reputation as a skeptic and science communicator. I had to get it right.

So I followed the process we have been laying out—what are all sides saying, what are the objective facts, who has the better arguments, and where does it all shake out? Getting started was challenging because most of what you will find on the topic is already biased by an agenda, either pro or con. I had to dig down to original scientific studies as much as possible and see what objective experts say about the evidence.

What I found is that the anti-GMO side is built entirely on sand. The more you dig, the more their claims utterly collapse. In the end I had to conclude that the anti-GMO position was nothing but an elaborate and well-funded propaganda campaign. I try really hard to be fair to all sides, especially sides with which I disagree. But even giving anti-GMO arguments the most charitable interpretation I possibly can, they just don't hold up to critical analysis.

I also checked my own analysis against the experts. I found that every major scientific organization in the world that has reviewed the scientific literature on GMOs has come to the same conclusion—GMOs are safe and don't represent any unique risk to health or the environment.

Despite this solid scientific consensus, the topic of GMOs has the greatest disconnect between public opinion and scientific opinion. In a 2015 Pew survey, 37 percent of US adults said they thought it was safe to eat GMOs. Meanwhile, 88 percent of members of the American Association for the Advancement of Science (AAAS) thought that GMOs are safe—that's a 51 percent gap, the largest for all the issues they surveyed.

Why such a big difference? This is probably the result of a deliberate propaganda campaign over the last two decades by the organic food lobby and groups like Greenpeace. It's an excellent example of how effective a good narrative can be. In this story there are good guys and villains, and there is a moral to the story that seems to empower those who accept it. Unfortunately, the narrative isn't based in reality.

I think it's also true that the skeptical community was slow to recognize the situation. The anti-GMO side had about a fifteen-year head start, where their narrative was spread unopposed. We had a lot of catching up to do.

The story of GMOs is also a lesson in how to recognize a position that is ideological rather than science-based. Ideological positions tend to be rigid. They settle on their conclusion and then search for justification, and they will change their justification as needed but never question their conclusion (which is actually their starting point).

As you'll see, those who are against GMOs are against GMOs all the time. They will cite one reason, and if you ever definitively knock it down, they migrate over to another.

Journalist Mark Lynas went on a similar journey to mine, but he started out as anti-GMO until he took a close look at the actual evidence. He wrote in the *New York Times* in 2015:

> I, too, was once in that activist camp. A lifelong environmentalist, I opposed genetically modified foods in the past. Fifteen years ago, I even participated in vandalizing field trials in Britain. Then I changed my mind.
>
> After writing two books on the science of climate change, I decided I could no longer continue taking a pro-science position on global warming and an anti-science position on GMOs.
>
> There is an equivalent level of scientific consensus on both issues, I realized, that climate change is real and genetically modified foods are safe. I could not defend the expert consensus on one issue while opposing it on the other.

Another journalist, William Saletan, also looked deeply into the GMO issue and came to a similar conclusion:

> I've spent much of the past year digging into the evidence. Here's what I've learned. First, it's true that the issue is complicated. But the deeper you dig, the more fraud you find in the case against

GMOs. It's full of errors, fallacies, misconceptions, misrepresentations, and lies.

So, let's take a look at what GMOs are and then examine the arguments for and against them.

A Brief History of GMOs

GMO advocates are quick to point out that pretty much all the food consumed by humans has already been extensively modified by human activity. Corn, for example, was cultivated from teosinte, which looks nothing like modern corn. In fact it took some deep detective work to figure out that they're essentially the same species.

Cultivation is mostly about artificial selection—saving the best plants from one year's crop to provide the seeds for the following year. Repeat that a few thousand times and you have the development of agriculture and all the food you recognize today.

Cultivation can also involve cross-pollination, creating a hybrid species in an attempt to get the best traits from closely related species. Using a combination of cross-pollination and artificial selection, breeders have created countless varieties of common plants. The black or purple tomato, for example, of which there are now about fifty varieties, is high in flavonoids, which give them their color. Orange carrots were developed by a fortuitous mutation resulting in high levels of beta-carotene. This turned carrots into an important staple crop as a source of vitamin A.

Essentially, almost none of the foods you eat look much like they did in their evolved state prior to human tinkering. Raspberries are one of the rare exceptions.

Of course, cultivation isn't the same as direct genetic alteration, and no one is saying it is. The point is that genetic alteration itself is apparently not a problem, and we would all starve without food that resulted from massive genetic change. It seems to be *how* we make the genetic changes that makes some people queasy. But let's keep going.

Eventually, breeders who were too impatient to wait for a fortuitous mutation to occur developed what is called mutation breeding—exposing plants to radiation or chemicals that increase the mutation rate. Between 1930 and 2007, 2,540 mutagenic plant varietals have been released. Greenpeace, for some reason, has not protested chemically mutated food.

There's also now a technique known as forced hybridization, where plants that would not crossbreed in the wild are forced to mix their genes. Making a hybrid of any kind between two species or cultivars results in changes to hundreds of genes in unpredictable ways.

"Genetic modification" refers to several techniques for changing organisms to suit our wants and needs. The technology involves various methods for inserting one or more specific genes directly into a target organism, or selectively altering or silencing a gene that is already there. There are two basic types of gene insertions: transgenic and cisgenic. Cisgenic insertion involves inserting genes from closely related species, ones that could potentially crossbreed with the target species. Transgenic insertion involves genes from distant species—even from different kingdoms of life, such as putting a gene from a bacterium into a plant.

There are several types of GM plant traits currently approved for use: herbicide tolerance, insecticide production, altered fatty acid composition (for canola oil), non-browning (for apples), fungus resistance (for chestnut trees), and virus resistance. Many other potential applications are in various stages of development.

While GMO critics often refer to transgenic plants as the source of their anxiety, opposition to "GMOs" lumps them in with cisgenic plants and those that don't even insert new genes, just silence or alter existing ones.

Opposition to transgenic technology often takes the form of "This couldn't happen in nature." This is not valid logic (it is an example of both the appeal-to-nature fallacy and the genetic fallacy), and it's also factually incorrect. There is something called horizontal gene transfer. Genes can move between unrelated organisms. For example, it was discovered in 2014 that cultivated sweet potatoes contain a transgene from a soil bacterium (*Agrobacterium*), a completely natural transgene.

Objection to transgenes seems to be based on the notion that genes from one organism are inherently different from genes from another organism. But this is untrue. A gene doesn't know it's a fish gene or a tomato gene or a person gene. They are just genes. In fact, fish and tomatoes share about 60 percent of their genes. It's true that different kingdoms use different promoters, which are gene regulators, but these are easily swapped out to make them compatible with the target species.

Still, many people feel there is something icky about a "fishmato" simply because a gene from a fish was inserted into a tomato. This wouldn't confer any fish characteristics on the tomato, however, except for the one gene being inserted (in this case a gene for cold tolerance, which never made it to market).

Genetic engineering technology is part of a spectrum of technology used to alter living things so we can eat them. There is no clear or obvious dividing line among the various techniques used, and so the term "GMO" is arbitrary to some degree. But this is part of the anti-

GMO narrative, to create a false dichotomy and declare everything on one side of the divide to be suspect and not safe. You could just as easily, for example, demonize mutation breeding—yet most GMO critics I talk to are completely unaware that such a thing even exists. Why are plants derived from parents that were mutated with radiation considered organic, while a plant with a silenced gene is condemned as "unnatural"?

It's not my position or the scientific position that all GMOs are automatically safe. That would be just as irrational as blanket hostility. Rather, it is simply clear that the label "GMO" is arbitrary and you can't treat all cultivars labeled as GMO as a group. They should be evaluated individually for their safety and environmental effects. Not all plants made from traditional breeding are safe, and not all GMOs are harmful.

Let's take a look at the primary arguments against GMOs and see how they fare.

Health Effects

There are various specific controversies surrounding GMOs. Perhaps the most emotionally laden concerns the potential health effects of GMO food. Critics have coined the term "Frankenfood," which is a politically useful slogan but not very useful as a concept. There is a legitimate concern, however, that introducing new proteins into human food might lead to allergic reactions or unforeseen health consequences.

For this reason, the safety of GMO food has been researched, and the bottom line is that the research shows that existing GMOs are safe for human consumption and as animal feed.

A 2012 statement by the AAAS concluded:

> Contrary to popular misconceptions, GM crops are the most extensively tested crops ever added to our food supply. There are occasional

claims that feeding GM foods to animals causes aberrations ranging from digestive disorders, to sterility, tumors and premature death. Although such claims are often sensationalized and receive a great deal of media attention, none have stood up to rigorous scientific scrutiny. Indeed, a recent review of a dozen well-designed long-term animal feeding studies comparing GM and non-GM potatoes, soy, rice, corn and triticale found that the GM and their non-GM counterparts are nutritionally equivalent.

The National Academy of Sciences agrees:

To date, no adverse health effects attributed to genetic engineering have been documented in the human population.

The World Health Organization also agrees:

GM foods currently traded on the international market have passed risk assessments in several countries and are not likely, nor have been shown, to present risks for human health.

As I noted above, essentially every major scientific or medical organization that has reviewed the evidence has come to the same conclusion: There is nothing inherently unsafe about the processes used to make GMOs, and no GMO on the market is unsafe. In fact they're arguably safer than the products of hybridization or mutation farming. For example, GMOs are specifically tested for proteins that have the potential to cause allergies—such proteins share features in common that can be screened for.

What does the published evidence show? As referred to by the AAAS, reviews of animal feed studies have concluded that GM food is safe:

Results obtained from testing GM food and feed in rodents indicate that large (at least 100-fold) "safety" margins exist between

animal exposure levels without observed adverse effects and estimated human daily intake. Results of feeding studies with feed
derived from GM plants with improved agronomic properties,
carried out in a wide range of livestock species…did not show
any biologically relevant differences in the parameters tested
between control and test animals.

Multiple systematic reviews have come to the same conclusion—
no evidence of any safety risk. Critics claim there hasn't been enough
testing. It is easy, however, to simply ask for more testing and make
that seem as if it's the rational position (moving the goalpost). This
is the same strategy used by anti-vaccinationists—no testing is ever
enough, and the precautionary principle is endlessly invoked.

A 2014 study was particularly impressive. The authors looked at
statistics for negative health outcomes for livestock before and after
GMO feed was introduced. This is a great natural experiment, since
animal feed was changed rapidly and almost completely to GMO.
The authors found the following:

In all industries, there were no obvious perturbations in production parameters over time. The available health parameters,
somatic cell count (an indicator of mastitis and inflammation in
the udder) in the dairy data set (Fig. 1), postmortem condemnation rates in cattle (Fig. 1), and postmortem condemnation rates
and mortality in the poultry industry (Fig. 2), all decreased (i.e.,
improved) over time.

So, multiple health parameters for multiple animals—billions of
animals over about fifteen years—showed no adverse effects from the
rapid introduction of GMO animal feed. If there were any significant
adverse effects from GMO, it seems reasonable that it would show up
clearly in the data.

Critics also have their studies to cherry-pick from, like the infamous
Séralini study of 2012 (which was retracted by the original publisher

after a year of criticism and controversy, although the study was then republished in a more friendly journal). Gilles-Eric Séralini is a French researcher who is decidedly anti-GMO and has published a few fatally flawed studies that were largely trashed by the scientific community.

In general, that's what we have: a few small or flawed studies suggesting GMOs are unsafe against a mountain of studies showing they are safe. When scientists look at all the data, they generally agree that GMOs are safe.

It can also be pointed out that while the safety of GMOs has been rigorously tested, the plants that are produced through hybridization, which can chaotically mix in hundreds of genes, and plants resulting from mutagenic breeding do not require the same safety testing currently required of GMOs.

There is, for example, a GMO black tomato that takes two specific genes from the snapdragon and inserts them into a tomato to produce higher levels of flavonoids. This species must go through regulatory hoops, while the fifty varieties of black tomatoes made by cultivation, with much more unpredictable results, require no testing.

As with most extreme positions, if the evidence doesn't go the anti-GMOers' way, it is common to dismiss it as part of some conspiracy. In this case the casual dismissal of the safety data is justified by saying that the studies are all conducted by the industry. If this were true, that would be a legitimate concern. When a company conducts studies of its own products, those studies tend to be highly favorable for the company. That's pretty clear.

However, there are two things to consider. When the research itself is tightly regulated and there are quality standards, it's much harder to fudge the results. More importantly, scientists, not companies producing GMOs, independently conducted the majority of research showing that GMOs are safe. So the claim is simply factually incorrect.

Safety and health effects are clearly the number one concern for those who are anti-GMO. However, when I point out that the evidence is sufficient to conclude that currently available GMOs are safe, a common response is to simply say, "Well, the real reason I oppose GMOs is not safety, but X." They have a position looking for a justification and are willing to be flexible.

Let's take a look at those other reasons.

Environmental Effects

While the safety of GMOs is rather straightforward scientifically, the net environmental impact of specific GMOs is a horrifically complex topic.

Herbicide-tolerant plants, such as so-called "Roundup Ready" plants, are engineered to be resistant to the herbicide glyphosate. This allows farmers to control weeds by spraying their entire crop, even after the crop has sprouted. The advantages of herbicide-tolerant crops are that they're less labor intensive and save money. They also reduce the use of soil tillage, which is bad for the soil and releases carbon dioxide (CO_2) into the environment.

The disadvantage of herbicide-tolerant crops is that they increase herbicide use, which gets into the environment and encourages the development of resistant weeds. However, glyphosate is much safer than the herbicides it has largely replaced.

So what's the net effect? That all depends on how these crops are used. Relying solely on glyphosate and glyphosate-resistant crops is turning out to be a bad idea, mainly due to the development of resistance among weeds. But as one tool among many, glyphosate resistance can be a net advantage. Farmers are better off using minimal, rather than no, tillage farming and using a variety of herbicides, not just glyphosate.

The real issue (a theme that will keep cropping up—pun intended) is that the bottom line doesn't rest with GMO versus no-GMO, but how glyphosate-resistant GMO and other crops are used as part of overall farming practice.

This is also true of insecticide-producing GMOs, specifically Bt crops. Bt is an insecticide produced by a bacterium. It's a popular insecticide among organic farmers because it is environmentally safe. Bt GMOs have the gene from the bacterium inserted so they make their own Bt.

The advantage of Bt crops is that they're pest resistant and they reduce insecticide spraying. The disadvantage is that overreliance on this one strategy results in pests becoming resistant to Bt. This is worsened by cross-pollination spreading the Bt trait to wild plants. There are also concerns about the effects of Bt on friendly insects (I would point out that Bt is already used as a pesticide).

Again, the bottom line is that the Bt trait can be a useful addition to the farmer's bag of tricks. But farmers should mix Bt crops with non-Bt crops, to reduce the evolution of resistance, and use other insecticides.

There's a practice called integrated pest management (IPM). The goal is to use a variety of methods to control pests while minimizing the development of resistance and environmental harm. Since the same risks exist for all pesticides, not just GMO-compatible ones, GMOs can be a very effective part of IPM, and there is no advantage to avoiding them as a category.

Some raise the environmental concern that GMOs may get out into the wild. While we certainly should make attempts to keep crops in the field and minimize unwanted contamination (and there are techniques for doing this), we don't have to worry about killer GMOs getting loose. Crops are fragile things. We've bred them to be fragile. Plants make natural pesticides to defend themselves. In fact, most of the pesticides you'll consume are made by plants on their own for self-defense. But crops have far-reduced levels of natural pesticides, because they can be bitter and make us sick. That is why we have to defend our crops from pests.

Further, crops are not optimized for survival. They are optimized to use a lot of energy to make edible parts for humans to eat. Anyone who has grown a garden knows this to be true on some level. Crop plants have to be babied, while you can't keep wild weeds from growing. Crops out in the wild would have about as much chance as a dachshund on the Serengeti.

Objections to GMOs are not limited to health and environmental risk, however. It seems that no potential propaganda opportunity has been missed.

Indian Farmer Suicide

The claim, popularized by environmental activist Vandana Shiva, Al Jazeera, and the 2011 movie *Bitter Seeds*, is that 270,000 Indian farmers have committed suicide as a result of expensive seeds and crop failure among GMO cotton in India. The myth seems to have been invented out of whole cloth.

In reality, Indian farmer suicides were on the rise prior to the introduction of GMO cotton in 2002, they stabilized after the introduction, and there is essentially no correlation between planting GMO cotton and risk of suicide. In fact, Indian farmers using GMO cotton are making more profit, and overall cotton production has increased significantly in India.

Suicides correlate with risky business decisions, lack of irrigation, lack of government subsidies and lending support—but not with the use of GMO cotton. This one is quite simply pure BS, but it persists nonetheless.

Terminator Seeds

The claim is that Monsanto developed terminator seeds that will grow for one generation, but the seeds from the resulting crop are sterile. The truth is that Monsanto acquired a company that had a patent on a terminator seed, but they never developed it further or brought it to market. And Monsanto promise they will never market a terminator seed.

This complaint always struck me as odd. Sometimes it's argued that GMOs are dangerous because their traits might get out into the wild. The terminator seed could potentially address this fear, because the seeds are sterile. However, the terminator was presented as a greedy attempt to force farmers to buy seeds every year. As we'll see, this claim falls flat also.

Saving Seeds

Forcing farmers to buy seeds every year is now perhaps the most common complaint against the business practices of Monsanto (the poster child for Big Agro). The claim is that farmers for thousands of years would save seeds from one year to replant the next. This is presented as if it were a natural right. Seed companies, however, through their GMO monopoly, now force farmers to buy new seeds every year.

I'm not going to discuss whether it is better for farmers to save seeds or buy new ones each year. I will simply point out that this issue is not unique to GMOs and, when put into perspective, is really a nonissue.

First, many seeds on the market are hybrids. According to the USDA, 95 percent of US corn was grown from hybrid seeds prior to the introduction of GMO varieties. The situation is similar with wheat, soybeans, grain sorghum, cotton, peanuts, and many other crops. So basically, most crops are hybrids. This is critically important, because you can't replant the seeds from hybrid plants. Because of the nature of genetics, hybrid traits don't breed true, so the mix of dominant and recessive traits in the next generation will be unpredictable.

Therefore, farmers already cannot save their seeds from one year for the next for the vast majority of crops. Yet I never hear anti-GMO activists railing against hybrid crops because they force farmers to buy new seeds every year. Hybrids have been popular since the 1930s and are "natural," so I guess it's okay.

Even without the hybrid issue, many farmers choose to buy seeds each year rather than save their seeds because it can be time-consuming and isn't cost-effective. Some small farms save their seeds and cultivate their own heirloom varieties, but of course they aren't buying GMO varieties from big seed companies anyway.

At least in the US and Europe, the whole issue of saving seeds and GMOs is simply a nonissue. It may be different for some third world farmers, and I've read conflicting information on this question. If this is your concern (and again, this is not about GMOs but about big seed companies in general), then advocate for better regulations for third world farmers, not for a ban on GMOs.

Suing Farmers for Contamination

It just doesn't happen. In fact, organic farmers sued Monsanto in order to protect themselves from the possibility that Monsanto might sue them in the future for contamination, but they couldn't cite a single case in which this had already happened. Ironically, they proved in court that Monsanto never sues for this and their claim was baseless.

The notion that Monsanto will sue for contamination is based on a misrepresentation of a few cases in which farmers tried to nullify Monsanto's patent for a particular GMO (by claiming patent exhaustion, for example). In every case, the farmers deliberately stole seed from Monsanto or tried to violate the patent and their contract. These were not cases of simple accidental contamination, but that's how anti-GMO activists have spun these cases.

Monsanto and Agent Orange

To prove that Monsanto is "evil," some opponents point out that Monsanto produced Agent Orange for the US government in the 1960s and '70s. This is true but—who cares? They, along with many other companies, took a government contract to produce a chemical. This has absolutely nothing to do with GMOs and is a transparent attempt to poison the well.

Patenting Life

Without any evidence of actual risk or harm caused by GMOs, some anti-GMO activists will resort to the subjective argument that life shouldn't be patented. We can debate the ethics of allowing organisms to be patented and the relative benefits and risks of using patents to defend intellectual property. However, this issue isn't directly related to GMOs and does not constitute a valid reason to oppose the technology.

First, not all GMOs are patented. Some are open source, although opposition to GMOs has made it so expensive to bring them to market that it has become more difficult to develop open-source GMOs.

Further, most patented seeds are not GMOs. Hybrid seeds can also be patented. So for decades prior to GMOs, farmers were largely planting patented seeds they purchased every year and could not save seeds for replanting. However, when GMOs were introduced this suddenly became a problem. The merits of allowing companies to patent their seeds is simply a tangential issue to GMO technology. You can oppose one without the other.

Also note that seeds eventually come off patent. Roundup Ready soybeans came off patent in 2015. Farmers can plant generic seeds and save them for replanting if they want (they still can't save hybrid seeds, however, as discussed above). So ironically, shifting from hybrid to GMO crops will lead to more seeds being replantable by farmers, even though the farmers themselves don't seem to really care about that.

GMO Research

This is an issue that actually does deserve attention, but I rarely hear it raised by anti-GMO protestors. It's a good example of letting propaganda overshadow genuine concerns.

The big seed companies control who can do independent research on their seeds, and they have been accused of blocking any unflattering research. In 2009 twenty-six seed researchers wrote an anonymous complaint to the Environmental Protection Agency about such restrictions. The result was a roundtable with the researchers and the big seed companies. The seed companies aired their worries about piracy of their technology, and researchers argued the need to be able to do independent research on safety and environmental effects. They came to a resolution that led to research agreements with many universities, and the situation is now much better.

Future Potential of GMO—Golden Rice and Beyond

Golden rice, for me, is a touchstone issue. I can't see any legitimate reason to oppose the use of golden rice, which is rice that has had genes for beta-carotene inserted. Since rice is a staple food in many parts of the world where blindness and death from vitamin A deficiency in children is a huge problem, the introduction of golden rice seems like a no-brainer.

This technology is ready for field-testing and therefore could be saving lives very soon. I get the sense that GMO critics oppose golden rice because such a home-run success for GMOs would destroy their narrative (that it's all bad all the time). GMWatch, for example, argues that golden rice has not been adequately tested. However, they ignore a 2012 study (and their article has been updated to reflect that recently) showing that beta-carotene from golden rice is converted to vitamin A in children at the same rate as pure beta-carotene from oil. The researchers estimated that one bowl of rice would provide 60 percent of the daily vitamin A requirement. (This study was retracted for what I consider to be trumped-up ethical concerns, even though the science and the findings are valid.)

GMWatch and other critics argue that there are better solutions than this "high tech" fix, for example fortifying food, vitamin supplements, and growing carrots and other vegetables high in vitamin A. In addition to this argument being a Nirvana fallacy and a false choice, such efforts are already underway and, while they have produced good results, are a long way from solving the problem of vitamin A deficiency. Golden rice is being advocated as an additional method alongside (not instead of) these other methods. In fact the issue has caused Patrick Moore, ex-president of Greenpeace Canada, to criticize Greenpeace for their opposition to golden rice.

There are many other potential applications of GM technology. Increasing the nutritional content of food is just one. Another is nitrogen fixation. Some plants use bacteria to fix nitrogen from the atmosphere. Others, including the cereals that account for most human calories, need to get their nitrogen from the soil, which means heavy fertilization. This has a huge environmental impact, a huge cost, and is one of the primary limiting factors in big agriculture. Imagine if corn and wheat could fix their own nitrogen. This capability will probably never be developed by traditional breeding techniques—we need GM technology for this. *POTENTIAL ENVIRONMENTAL DISASTER*

Another very promising goal is enhancing photosynthesis. Some plants have more efficient photosynthetic pathways than others. If we could get the optimal photosynthetic process into our major crops, that could increase yields significantly (estimates are by 20 percent). Other potential applications to increase yields include drought resistance, pathogen resistance, and cold tolerance.

Ultimately, I found that GMOs are neither a panacea nor a menace. Genetic modification is simply a powerful technology, and its impact will depend entirely on how it is used. In fact it's difficult to talk about GMOs as if they are simply one thing. Each individual GMO needs to be assessed on its own risks and merits.

As with many technologies, what matters most is how it is used. Safely feeding the growing population of the world in a sustainable way without having a major negative impact on the environment is a great challenge for our civilization. We shouldn't uncritically accept the hype and spin of companies offering simple answers (that involve buying their product), but nor should we reject an entire technology based upon fear and misinformation.

In the end, I think the conversation can be a healthy one—exploring all the complex issues of the use of GM technology can lead to better practices and solutions. However, we need to reject the false dichotomy of GMO versus non-GMO. We need to reject the fearmongering about Frankenfoods. And we need to clearly identify the actual issues and not confuse them with a simplistic villain.

But my adventure was not over. Once I felt I had sufficiently wrapped my head around the topic, I started discussing it in my talks to skeptical groups. I received much more pushback than on any other issue I have discussed with fellow skeptics. Clearly we were in a deep hole on GMOs.

Even though I had a thorough understanding of the issue, I had to figure out the best way to communicate my information to the public. It was very easy to trigger emotional reactions and to come off as dismissive. Of course, anyone pointing out the factual errors in the anti-GMO position immediately gets accused of being a shill.

In the last few years, however, I do feel this has changed significantly, at least among skeptics and science enthusiasts. Changing general public opinion, however, is going to be a steep hill to climb.

WE'RE "IN THE KNOW", NOT JUST PLAIN OLD PEOPLE

A POOR EXAMPLE, IN A BOOK ON SKEPTICISM AND A FAILURE TO ADHERE TO THE AUTHOR'S OWN METHODOLOGY.

BTW, I'M NOT AGAINST GMOS - I'M PRESENTING A WARNING AGAINST FLAWS IN LOGIC & REASONING - EXACTLY WHAT THIS BOOK IS SUPPISED TO DO.

41. Dennis Lee and Free Energy

Perry's Adventure

> Dennis Lee has broken a lot of laws, but he hasn't broken the laws of
> thermodynamics.
> —Robert L. Park

Anyone who is even vaguely familiar with the laws of thermodynamics and conservation of energy is aware that perpetual motion machines, more generally referred to as "free-energy devices," are impossible. There is no way to construct a machine that creates energy from nothing. Energy always has to come from somewhere, and at least a little bit of energy has to be wasted in any process.

As we discussed in the chapter on free energy, the allure of such devices has ensnared people throughout the centuries. Endless tinkerers have imagined, designed, and built devices that appear to generate a theoretically endless supply of energy. They never work, but that doesn't stop some inventors from clinging to their fantasy. The promise of being the person to end the energy problem and transform our civilization in a single stroke is just too great.

Then there are those who aren't just tinkering in their garage or making YouTube videos—they're soliciting investors with the promise of free energy. Whether they actually believe their claims or not is an interesting question but one that's difficult to prove one way or the other. The most infamous modern-day example of a perpetual-motion-machine seller is Dennis Lee. By the time the New England Skeptical Society (NESS) decided to confront Dennis Lee back in 2001, he had already built up quite an impressive résumé:

- Dennis Lee was arrested eight times between 1974 and 1979 in New Jersey on fraud, forgery, and drug-related charges.
- In 1975 Lee pleaded guilty in Bergen County, New Jersey, to five counts of passing bad checks and taking money under false pretenses. He was given a one-year suspended sentence and three years' probation.
- In 1978 Lee persuaded Pat Robertson to give him $150,000. Two months later, Robertson accused Lee of false advertising, operating a pyramid sales scheme, and unauthorized sales of securities. Robertson never got his money back.
- Lee was arrested in New York in 1982 for again passing bad checks.
- He was accused by the Washington State attorney general in a 1985 civil action of violating the state's Consumer Protection Act. Lee agreed to a stipulated judgment of $31,000 but left the state without paying the fines.
- In 1990, he pleaded guilty to two felony counts of consumer fraud filed by the Ventura County, California, district attorney and six counts of violating the state's Seller Assisted Marketing Plan Act. Lee served two years in a California state prison.
- In 1999 the Washington State Department of Financial Institutions served a cease-and-desist order accusing Dennis Lee of marketing unregistered securities and defrauding potential investors. It labeled Lee's activities "a clear and present danger to the investing public."
- In 1999 Jim Murray, an electrical engineer employed by Dennis Lee, was asked by Lee to rig Lee's system to make it appear 200 percent efficient, instead of the actual 20 percent. Murray refused and resigned.

And that's just through the year 1999. His broken promises and unlawful conduct continued well into the twenty-first century.

I suppose it's conceivable that someone whose career was marked by fraud and deception might actually be the one person to crack the apparently impossible challenge of creating a free-energy device, but I wouldn't bet on it. Many people, however, were ostensibly willing to do just that.

Dennis was on tour in the summer of 2001, and his travels brought him to our home state of Connecticut. Being the peaceful activists that we were, we decided to go to the event and talk directly to the attendees in an effort to persuade a few who might be persuadable. Perry DeAngelis organized our efforts and documented the details of our adventure from that day:

On July 23, 2001, the NESS decided "To Be a Part of History!" At least that's what we and hundreds of others were invited to be by Dennis Lee's ever-traveling huckster show of technological magic. This is the opening line of the advertisement announcing the beginning of Better Home Technologies' (one of Lee's current front companies) new nationwide tour. It goes on to promise kits to convert your present car or furnace to run on water, a heat "paint" that will eliminate the need to shovel snow, and, of course, free electricity for life.

The NESS put together its own flyer to hand out at this very first show of the new tour, which was held at the Waterbury Sheraton Hotel. The show was to begin at seven p.m. and four NESS members arrived at six thirty with hundreds of these flyers in hand, warning attendees not to be fooled by Dennis Lee. The crowd was beginning to build, and the NESS faction placed itself strategically down the hall and around a corner from the registration table for the show. With two members on each side of the hall, we were well placed to put a flyer in the hand of almost every would-be victim of Lee's chicanery. Many of the folks going to Lee's show were elderly, unfortunately—an often-exploited segment of our society. One gentleman sat down on a nearby bench with our flyer, replete with its multiple warnings and a chronology of Lee's criminal past, and then asked us, "So you believe in all this?" We took a moment to correct that confusion.

This process went on happily for about fifteen minutes wherein approximately one hundred NESS flyers were distributed, until a security guard came up to us, asked for a handout, and whisked it away. Moments later another fellow came up holding one of our flyers and began loudly berating us for not having tried Lee's miracle machines and yet decrying them. We tried to explain to him that others had indeed tested Lee's devices that defied the known laws of the universe and their findings were clear and damning. We also pointed out Lee's multiple convictions for fraud, which only led this fellow to accuse the NESS of having a hidden criminal past.

Before we had a chance to explore this inanity, the security guard, having gotten his marching orders from the MOD (manager on duty), reappeared and informed us we would have to leave. The NESS asked if we could continue our endeavors if we purchased a room, but this would only allow us to stay on the premises without distributing the flyers. We said that we did not wish to break the law and began walking out, still handing out flyers to passersby, several of which were angrily yanked right out of the people's hands by the Lee comrade. Once we had been escorted to the parking lot, the guard informed us that we would have to leave the property completely. Again, not wishing to break the law by trespassing, we got into our cars and departed.

We can only hope that the flyers the NESS was able to distribute got the warning out about Lee and kept a few extra wallets safely in pockets.

The tactics used by Dennis Lee and others is to pack as many people into a conference room as possible and then dazzle them rapid-fire with one fantastic device and claim after another. Over the course of several hours, those with more sense slowly drift away, leaving the most susceptible behind. Mental fatigue also begins to take its toll.

After the demonstration, the audience is left with the sense that even if only one of these amazing devices turns out to work, they will become rich by investing in them. And that is what Lee sells—the opportunity to invest in his company. People are not asked to purchase a product, because Lee has no legitimate products to sell. Instead he sells an investment or a dealership.

Lee uses a generous helping of appeal to patriotism and religious faith to sell "opportunity." In a firsthand report, a former owner of one of Lee's dealerships relates how Lee keeps people on the hook, sucking as much money out of them as possible, and then cuts them loose when they are finally onto the scam.

Lee's attempts to milk the public for investments in his useless technologies did not, unfortunately, end with our humble efforts. As is common with such careers, Lee just kept moving on to new scams, occasionally getting slapped on the wrist by the authorities.

His next scam was a device to increase the efficiency of internal combustion engines through a so-called "Hydro-Assist Fuel Cell (HAFC)." This is an example of the common free-energy claim that they can make cars burn gasoline more completely or more efficiently. Of course, car companies have been working to optimize efficiency for decades and probably haven't missed anything simple and dramatically effective.

The FTC named Lee as a defendant in a 2009 investigation of these claims, but unfortunately, he has been able to maintain a career selling pure pseudoscience for decades. He occasionally gets banned from an individual state or has to retract a claim, but he is largely left to continue his scams.

He sells his amazing claims by appealing to those who tend to accept conspiracies. Conspiracies are the best friend of con artists, because they can do a lot of heavy lifting when it comes to explaining away inconvenient facts. Scientists say free-energy devices are impossible because they're part of a conspiracy. Anyone trying to debunk his claims, or warn of his previous fraud convictions, is just trying to suppress his message. They are probably shills for Big Oil. If the government tries to shut him down, well, what more proof do you need?

Lee also weaves in belief in God and American patriotism, creating a trifecta of emotional appeal. Even being directly warned that Lee has a long history of felonies is apparently not enough to dissuade some from thinking a bad-check writer from New Jersey has revolutionized physics.

The existence of Lee reminds us that not everyone out there making paranormal or pseudoscientific claims is a true believer. There are also sharks in the water.

42. Holly-woo

Cara's Adventure

If you are searching for sacred knowledge and not just a palliative for your fears, then you will train yourself to be a good skeptic.

—Ann Druyan

Texans get a bad rap. Sure, some have a tenuous relationship with climate science and evolution. And yeah, the state consistently falls toward the bottom of national educational rankings. But when this Texan found herself living in Los Angeles, she was in for a rude anti-science awakening.

People in the South generally eat terrible food and have the obesity statistics to show it. I'm reminded of this every time I go home to visit. "Dunlap disease" is epidemic. (You've never heard that one? "My gut done lapped over my belt!") And I fall victim too. When I head back to LA after visiting family, my luggage has accumulated an extra five pounds, and I'm carrying the same weight in fried chicken and doughnuts.

But are the meat-and-potatoes practices of my Texas brethren really less healthy than those of image-obsessed Angelenos? (That's what people who live in Los Angeles are really called.) Sure, it's easy to stay coiffed, plucked, and toned in the City of Angels. But beneath the plastic veneer of perfection lies an entrenched world of pseudoscientific woo.

Walking down a street in my neighborhood of Los Feliz, I'm likely to see a juice bar specializing in cleansing tonics, a day spa offering

Reiki and Ayurveda, a boutique with crystals and singing bowls in the window, and a "medical" facility advertising B-vitamin infusions and colon hydrotherapy. This is not an atypical stretch of storefronts. Amid the CrossFit gyms and spray tan salons, the alternative medicine business is booming.

Southern California is home to psychiatry-denying Scientology, vaccine-denying Jenny McCarthy, science-denying Gwyneth Paltrow, and reality-denying Deepak Chopra. Cult leaders and celebrities alike hold a flummoxing amount of power over the residents of the country's second-largest city. Perhaps LA is simply a microcosm of a greater scourge infecting the American people, nay the world. Or maybe there's just something in the air out here. (Smog? Cannabis? Desperation?) But what strikes me as most odd is the age-old quandary of why otherwise brilliant people fall victim to such hucksterism.

As of this writing, I've been living in Los Angeles for nearly a decade. I've met many, many different people—haves and have-nots, young and old, intelligent and, well, less so. In an effort not to destroy previous or current relationships (and to avoid throwing anyone under a massive bus), I've decided to amalgamate my experiences into one composite human, whom I'll call Aubrey. I don't personally know any Aubreys, and the name is "genderful," as will be my stories. Yet from here forward, for ease of reading, Aubrey will be referred to as "she." She is anyone and everyone I've encountered in the woo-tastic land of Hollyweird.

Out here I often observe a phenomenon I like to call "celebrity syndrome," a gradual cordoning off from reality among the Hollywood elite. Aubrey exemplifies it beautifully. Because celebrities are usually hounded in public spaces, they begin to avoid them (much like an agoraphobic gradually reduces their own list of safe spaces in an effort to avoid severe anxiety symptoms). They have assistants who buy their clothing and groceries. Tailors, personal trainers, and beauticians come to them, instead of the other way around. And eventually they develop deep, singular relationships with alternative medicine practitioners, yoga instructors, life coaches, and other boutique wellness "experts." They often forgo legitimate medical advice for the polished, intuitive, captain-of-your-own-ship rhetoric of their gurus.

Aubrey once told me that "mosquitos need a swamp to survive. If we treat our bodies like a swamp, disease will flourish. If we remain clean from the inside out, we won't ever get sick." Let's shelve the fact that even the basic argument is flawed (mosquitos don't need swamps to live; they often require water to breed). More important is the fact that, unfortunately, infection, cancer, and genetic disease simply don't work that way. Sure, your immune system has the power to ward off weak invaders, but with a high enough dose of bacteria, virus, or parasites, you're going to get sick, regardless of your smoothie regimen. And avoiding carcinogens is always a good idea, but for many, cancer (and other heritable diseases) is pretty much written in their DNA.

In her efforts not to become a swamp, Aubrey has become obsessed with wellness. Symptomatic of the stereotypical LA lifestyle, she gloms on to any trendy, Pinterest-inspired diet she can find. Cayenne pepper and lemon juice for days on end? Sure! Gluten-free today, dairy-free tomorrow? Why not? And don't even try to offer her anything with GMOs, animal products, high fructose corn syrup, nightshades, artificial colors, flavors, or preservatives. Basically, if the "Food Babe" Vani Hari has railed against it, Aubrey won't touch it. And although multiple studies show fad dieting (especially of the yo-yo variety) to be potentially harmful, when it comes to Aubrey, there are definitely bigger fish to fry. Take her reliance on vitamin infusions. Although she swears she feels energized and sharper after a B-vitamin drip, studies don't support such claims. In fact they consistently show no better efficacy than placebo. And they aren't without risks—when performed outside of a hospital setting (as most of these so-called "medi-spas" are), there is no guarantee of purity or sterility. And although intravenous vitamin delivery may be recommended for a small percentage of individuals with a medically verifiable deficiency, most physicians agree that oral administration is the proper course.

But Aubrey doesn't stop there. She also swears by her colonics.

Aubrey sincerely believes that a blast of water up the rectum is the only way to remove the "hardened toxins" she's accumulated after years of eating…well, kale and quinoa. She's so dependent on this so-called "therapy" that she even has a home hydrotherapy kit—a kind of DIY water torture. Now, this would all be well and fine if a colonic had the same effects as a bidet. But unfortunately, it can be very dangerous. Let's ignore the fact that Aubrey's holistic "doctor" recommends this bizarre practice for a moment and break down what actually happens when you turn your large intestine into a water slide.

Methods vary, but colon irrigation generally involves inserting a tube into the rectum and flushing gallons of water against gravity using a fancy machine or passive pressure. Enthusiasts tout its ability to clear unnamed toxins from the walls of the intestinal tract, but there's no evidence that such toxins even exist or that hydrotherapy does anything but make you feel full and then poop yourself. And the risks are severe. Individuals with kidney or heart problems could throw off their electrolyte balance, leading to organ failure. Those with previous surgeries, hemorrhoids, diverticulitis, or other GI problems risk perforating the bowel, a serious complication that can lead to death. And, of course, since colonics are rarely performed in legitimate medical settings (especially DIY administration), there's a risk of using improperly sterilized equipment, which may have the opposite effect of its intended benefit: a raging infection. I don't know about you, but I'm not willing to risk dying on a poop machine in an effort to rid my body of vague toxins that my own biology is pretty good at taking care of on its own.

Aubrey's holistic "doctor" also uses a diagnostic tool known as iridology. I hadn't heard of such a thing until I met Aubrey, and for good reason. Iridology was debunked almost immediately after it came to be, in 1893. As discussed in the chapter on pseudoscience and the demarcation problem, this so-called "method" was formulated based on a single observation and has never produced an iota of evidentiary support. It fully mimics phrenology, except instead of bumps on the head, practitioners equate markings in the iris with all manner of medical complaints. It's pure pseudoscience, plain and simple, yet Aubrey swears by it.

Her obsession with fitness borders on mania. She once exerted herself so hard in a CrossFit gym that she developed rhabdomyolysis. Dubbed "rhabdo" for short, this is a life-threatening medical condition characterized by cola-colored urine, compartment syndrome (dangerously high-pressure buildup impeding blood flow), and muscle pain and weakness. Rhabdo occurs when skeletal muscle is damaged so severely, it breaks down inside the body, sending myoglobin and other muscle constituents throughout the bloodstream. Because the body isn't accustomed to such large molecules floating around, they can clog up the kidneys, leading to severe dehydration, kidney failure, paralysis, and even death. When your obsession with health is in and of itself unhealthy, it's time to take a step back and rethink your approach.

Perhaps one of the worst offenders in this category is Gwyneth Paltrow (although there sure are others). Her lifestyle website, Goop, is a regular source for Aubrey's nonsense. She advocates for vaginal steaming and vaginal eggs made of jade (yes, you are to wear these things inside yourself—a practice gynecologists have vehemently condemned). She sells an entire line of unproven supplements and she regularly promotes detox cleanses, colon cleansing, and homeopathy. She erroneously claims (through her "experts") that underwire bras cause breast cancer and that chemical-based sunscreens (isn't everything chemical based?) cause hormone disruptions (they do not). Gwyneth Paltrow's advice is not only misguided, it's outright dangerous, as Timothy Caulfield, author of *Is Gwyneth Paltrow Wrong About Everything?* points out. Yet Aubrey and others can't get enough.

The list of Aubrey's wooey approaches to health and wellness is extensive—chiropractic, acupuncture, chelation therapy, vaccine denialism, etc., etc.—but it all stems from a similar place: My friend Aubrey does not trust the "mainstream medical establishment." Aubrey is smart. Aubrey is educated. Aubrey is well read (although I often question her echo-chamber reading material). Yet Aubrey thinks that the health care system doesn't have our best interests at heart. She doesn't trust allopathic doctors or the Western scientific establishment as a whole, yet she cherry-picks when and how she chooses to apply established, evidence-based reasoning to her life. You'd better believe she took antibiotics when she got a raging urinary tract infection, but the yeast infection that followed was "treated" with a vaginal mask made of yogurt and kefir. I once pleaded with her that if she ever got cancer, she seek chemotherapy or any treatment her oncologist recommends. She said she couldn't promise she would. I was devastated.

Aubrey displays an awful lot of motivated reasoning, and truth be told, it's hard not to get caught up in this kind of thinking in Los Angeles. This sprawling metropolis has a long and rich history of healers, gurus, and even cult leaders duping the populace for financial gain and personal glory. The cognitive biases are so overwhelming in many of the city's residents, they are more likely to listen to Dr. Oz than their own licensed, board-certified physicians. And instead of a $25 copay, they'll shell out hundreds for bogus therapies promising a longer life and more youthful appearance. Maybe it's the same motivation that brings millions of young hopefuls to the city to pursue their dreams that ultimately victimizes them at the hands of duplicitous hucksters. Of course, not everyone in Los Angeles is rich and famous and credulous. We have the largest population of unsheltered homeless people in the country. Our residents are extremely diverse, ethnically, financially, and cognitively. Many Angelenos live a perfectly normal life, but I'm not talking about those people. I'm talking about Hollyweird, land of the privileged, pretty, and credulous.

Narcissism is a real bitch. *so are you, Cara, making up a character in a poorly written script*

43. The Singularity

Bob's Adventure

Clearly, unless thinking beings inevitably wipe themselves out soon after developing technology, extraterrestrial intelligence could often be millions or billions of years in advance of us. We're the galaxy's noodling newbies.
— Seth Shostak

I am an ardent technophile. Technology is the most fascinating by-product of human culture and cognition to me. In a lot of ways I'm also a recovering techno-optimist. The possibilities and allure of future technology can leave me giddy with anticipation. Techno-optimism can also make it easy to expect major advances unrealistically soon or to dismiss the complexity and time frames of interacting variables involved in bringing us these new tools and devices. It's a constant struggle to temper my enthusiasm enough to allow for a less biased, more realistic assessment of when we can expect today's science fiction to become tomorrow's science.

So what are the common missteps that many people take when crystal-balling the rise of revolutionary future science and technology?

People overestimating near-future technology and underestimating the deeper future is a truism that rarely disappoints. Even as few as twenty years ago, I confidently imagined that we'd surely have commuter jetpacks and flying cars by today's date. Yet almost no one going back many decades foresaw the emergence and transformative power of the internet.

Prognosticators are also fond of underestimating the full sweep of changes that can be wrought by new technology and just time. Early twentieth-century depictions of the future show reasonably extrapolated flying cars, but they're being ridden by quaint people wearing spats and holding canes and feathered parasols. Today we see a near-term future of autonomous cars being shared by everyone, since cars are often idle in garages and parking lots. Seems reasonable, but the notion of sharing cars may turn out to be as much of an anachronism as wearing a monocle while flying a jetpack. Perhaps people will love spending time in cars once they're free to do anything from sleeping to socializing to Netflix binging to having sex. They may then want to own their own cars so they can customize them to their unique tastes and hobbies. This could lead to cars ballooning in size beyond all current expectations.

In the 2002 movie *Minority Report*, for example (a compelling vision of the near future), they imagined we would have tiny cell phones. This was a reasonable extrapolation from the trend in the 1990s. They completely missed the smartphone revolution, however, that led to bigger and bigger phones to maximize screen size.

Perhaps the most pernicious pitfall for predictors is called wish-casting. Meteorologists came up with this term after they noticed that weather forecasters predict with an unrealistic frequency that the Fourth of July will be sunny and Christmas Day will be snowy. People have a strong tendency to let conscious or unconscious desires influence their predictions. Anticipating and expecting anthropomorphic robotic maids and butlers like Rosie from *The Jetsons* has a history and an undeniable appeal. The appeal is so great that many people just assume it will happen and then build their prediction around that assumption. That metal man in a tux may never actually clean your house, however. Today we have disk-shaped Roomba doing the vacuuming. In the future, cooking may be done by robotic arms attached to the ceiling, with no legs or torso required. Beds may be self-making (please please please). There are no foregone conclusions when it comes to the future.

Most often, imagining the actual bit of futuristic technology itself is the easy part. The specific implementation of that technology and, more importantly, how people decide to use it are so contingent and unpredictable that they are the downfall of many visions of the future. If you told people in 2000 that we would be texting rather than video chatting, you would have likely been ridiculed.

With those caveats firmly in mind, let's look at one of the more controversial and game-changing predictions about the future, the technological singularity.

The term "singularity" has many different meanings in many fields of science. In mathematics, it's a point that is undefined or behaves erratically. A gravitational singularity is a point of mind-numbing density producing infinite tidal forces. A mechanical singularity is a mechanism that has inherently unpredictable future states. The technological singularity is similar—a disruptive and unpredictable change of massive proportions.

The technological singularity is described with different nuances by different people, but many would agree that it is a hypothetical future point in human history in which an intelligence explosion caused by an artificial superintelligence (ASI) or augmented human minds changes civilization so rapidly that prediction beyond that point becomes inherently infeasible.

One of the first hints of the concept of such a singularity came in 1958 from famous mathematician John von Neumann, as quoted by his friend Stanislaw Ulam:

> One conversation centered on the ever accelerating progress of technology and changes in the mode of human life, which gives the appearance of approaching some essential singularity in the history of the race beyond which human affairs, as we know them, could not continue.

Another conceptual pioneer was I. J. Good. He was a British mathematician and cryptologist who also worked with mythic computer scientist Alan Turing. It is Good who originated the concept of an intelligence explosion with the following quote:

> Let an ultraintelligent machine be defined as a machine that can far surpass all the intellectual activities of any man however clever. Since the design of machines is one of these intellectual activities, an ultraintelligent machine could design even better machines; there would then unquestionably be an "intelligence explosion," and the intelligence of man would be left far behind. Thus the first ultraintelligent machine is the last invention that man need ever make, provided that the machine is docile enough to tell us how to keep it under control.

This outcome stems from an initial critical juncture in this entire scenario: the creation of an artificial general intelligence. It doesn't have to be smarter than a human, but being in the ballpark of humanity is probably best. From there, simply speeding up the computational substrate is all that would be needed to accomplish superhuman feats. Imagine a human-level artificial intelligence (AI) being able to spend ten years on a single task. Now imagine if those ten years of work can be accomplished in one workweek.

What if that task was the improvement of its own mind, to rewrite or reorganize the foundation of its very cognition? Many more options would obviously be available to such a digital mind beyond those available to a PhD-earning human. We've already discussed overclocking the brain, but what about adding the equivalent of millions more digital neurons to the artificial prefrontal cortex? What about doubling the number of connections in the connectome, the map describing the hundred trillion neuronal interconnections that make up the human biological brain? Either of these strategies—or others—could reasonably produce an intelligence smarter than any human could possibly be. The moment we start down that road, the fuse to an intelligence explosion has been lit.

Once this occurs, the process of recursive self-improvement could quickly produce minds twice as smart as the cleverest human, then ten times smarter. As this happens and unanticipated insights are forged, and paradigm shifts start…shifting, the "prediction horizon" of our culture starts shrinking. At this moment in 2018, we can confidently predict advances that are likely or possible in the near future, say in fifteen or twenty years. It's certainly hazy, but I'm pretty certain that many cars on the road could be autonomous in that time. However, we certainly won't be getting into our cars and scooting to the moon in twenty years.

The smarter these ever-improving AIs become, the closer our prediction horizon gets. Within a short period of time, looking beyond one year into the future will be utterly hopeless, then six months, then perhaps…mere hours, and finally…

When we've lost all hope of predicting any of the future, the technological singularity will have arrived.

I always liked Ray Kurzweil's description of the singularity as "technological change so rapid and profound it represents a rupture in the fabric of human history." This of course makes sense, because when a mind hundreds or thousands of times more intelligent than anyone starts running the show, how can mere humans anticipate anything that it would do or could discover? To reassert our former power of prediction, one would have to be as intelligent as the AI.

You may be wondering—then what? Of course, by definition, we can't predict the answer to that, but we can go over all the extreme possibilities that our meager meat minds can muster. Our artificial superintelligence could conceivably care for and respect all life. It could work with humanity to improve our condition in innumerable ways: curing disease, vastly extending life-spans, making our entire solar system habitable, or even uplifting us to superintelligence as well, if we so desire. This is the benevolent artificial superintelligence scenario imagined in Neal Asher's Polity science fiction universe. (Highly recommended, BTW.)

The other end of the spectrum is…not fun. An entity a million times smarter than us could see humanity as V'Ger saw Captain Kirk and his crew in *Star Trek: The Motion Picture*: as a carbon infestation that needs to be ignored if we're lucky or, more likely, wiped out. Listing the other examples of this dystopian ASI scenario in culture would exceed the recommended word count for this book.

So what do you think of this technological singularity now? Is this all nonsense, merely the Rapture of the Nerds, as it's been called? Or is the clamor that's been rising up around this idea becoming increasingly justified? Is even that a false dichotomy?

Some disagree with the very premise that a sentient intelligence can be created artificially, digitally...nonbiologically. I call this squishy bio-chauvinism. There's no reason to think that wetware or electrochemistry are the only mediums that could support any of the innumerable types of intelligences that could potentially exist. Many theorists of mind subscribe to a functionalist view in which the organization and self-interaction of the brain are more important than the material substrate itself. Many different nonbiological substrates could potentially support what we would describe as a mind.

Consider this: What if we had a process to slowly convert each of the neurons in your brain into an electronic digital device that precisely mimics their inputs and outputs. After a period of time, your brain would be completely nonbiological. Assuming the technology functions as described, would there come a time during the transformation when self-awareness started fading away or stopped completely? Would you become a philosophical p-zombie that behaves normally but without experiencing the rich inner life, the qualia each of us enjoys and assumes others enjoy as well? Is there something completely undiscovered or unhinted at in neuroscience that allows only biological neurons and nothing else to produce the epiphenomenon of mind?

Aside from whether it is possible to replicate sentience, there is also the question of whether we should even try. Now we're aware that ASIs could present an existential risk, perhaps we won't let that genie out of the bottle. We've seen in the past decade that narrowly focused, non-self-aware AI can still achieve a great many goals, and we've only scratched the surface of what even this limited AI can do. We don't necessarily need robots and appliances that have fully formed minds. It's not ethical to create minds anywhere near human level or higher to essentially just be our slaves. Any research creating self-aware synthetic minds can be done safely in isolated labs that have no networks, using air-gapped computers to prevent escape. There may be no real need to replicate these types of minds in thousands of robots just waiting to be pissed off and go all Terminator on us.

On the other hand, business models in the last ten years have enthusiastically embraced the fruits and potential advances that AI offers. Indeed, investment in AI worldwide has surged to unprecedented levels, with no sign of abatement. The power of an artificial superintelligence to improve the bottom line beyond all imagination will likely become only more widely appreciated and increasingly insistent as time goes by. Keeping it safe, locked away in a sterile lab, may eventually be seen as hopelessly naive and unrealistic in a world of intense global competition and seemingly intractable problems.

Perhaps building an artificial intelligence is eminently possible but so exquisitely difficult that it's essentially beyond humans. In that catch-22 scenario we'd need a superhuman intelligence just to create an artificial intelligence. That does make sense when you consider that the human mind is the most complex object we know of in the entire universe. A worse realization is the possibility that to build a fake brain, the human brain might need to fully understand itself first. Is that even possible? Wouldn't that break some fundamental law of logic? Fortunately, that's most likely a straw man argument. Coding a brain from scratch would be one thing, but copying an existing brain may well be the most reasonable approach to this task.

Whole brain emulation is the popular concept of scanning a brain with such fidelity and resolution that we could then emulate it on an appropriate computer platform. This could potentially create a human-equivalent digital mind without our having to understand the brain significantly better than we do now. Sure, this would be monumentally complex, and we don't have the scanning technology or the computing hardware to accomplish it. But the question then becomes: Can we eventually create that sort of supporting technology? There is little doubt about it. We're far from the limits of what scanning and computing can do.

For example, computronium is a hypothetical substrate that is calculated as the densest and most efficient computing material allowed by the laws of physics. Physicist Seth Lloyd calculates that such a mythical material could compute on the order of 5×10^{50} (500 quinquadecillion) calculations per second. That's thirty-five orders of magnitude greater than the fastest supercomputer on the planet as of 2017. Granted, the idea of computronium is a little tongue in cheek. In practice it probably could never match 10^{50} calculations per second. Additionally, the implication that one arrangement of matter can be optimized for all types of computing (parallel versus serial computing, for example) is just silly. Yet, even a tiny percentage of that theoretical maximum would be vastly greater than what we have now and more than enough to emulate a human brain.

As you can see, the techno-optimist in me sees a lot of potential in future AI, as long as it's harnessed properly. But I also need to do my skeptical due diligence (which Steve never lets me forget). So where could all this talk of a singularity go wrong?

First, we haven't made a self-aware general AI yet. It's still hypothetical. Fifty years ago, making such an AI was thought to be fifty years away, and yet here we are and experts are still predicting we may be fifty years away. It has turned out to be a lot more difficult than we anticipated, and it may be harder still. We're not close enough that we can say how close we are. Also, as I suggested above, we may never really need a self-aware AI. More focused AIs are doing just fine, and we may simply continue in that manner.

If we're going to talk about intelligence, we also have to mention that this is a far more complex topic than was previously understood. There are many types of intelligence, and it's likely that AI will excel in some respects and lag behind meat brains in others (as is currently the case). Our AIs may all have the equivalent of Asperger's syndrome, and we will not know why or how to change it. They may lack true creativity.

Furthermore, we have no idea if there are physical or practical limits to complex self-aware entities. As we try to make our AIs faster, better, smarter, they may all become psychotic or suffer from some inherent limitation in self-communication. And, of course, we may run into limits of computing power itself.

So we cannot just simplistically extrapolate such a complex technology into the future and assume that any trends will continue without limit. Superintelligent AI (and by extension the singularity) may be the flying cars and jetpacks of the next century, always fifty years away.

I don't think so, because I remain a techno-optimist. But I'm also a skeptic, which means I have to recognize that we just don't know. The future always surprises us.

I am counting on having my brain uploaded into silicon so I can see the future for myself.

44. The Warrens and Ghost Hunting

Evan's Adventure

One cannot explain one mystery by attributing it to another.

—Joe Nickell

Belief in the supernatural seems to be a nearly universal part of the human condition, but the details of specific paranormal belief systems depend on culture and location. In New England we have ghosts—or at least ghost hunters. So it's not surprising that in our younger days as activist skeptics, the investigative team of the New England Skeptical Society (Steve, Bob, Perry, and I [Evan]) cut our skeptical teeth investigating ghost hunters.

Taking on the New England ghostbusting industry led us inevitably to Ed and Lorraine Warren, the patriarch and matriarch of ghost hunting in New England. Ed and Lorraine hunted all types of ghosts (Ed has since passed) and apparitions, demons, and possessed people, places, and things.

Ed and Lorraine's most recent publicity has come in the form of the movies *The Conjuring* and *The Conjuring 2*. The movies are based on two of their most storied cases, the Perron family haunting and the Enfield Poltergeist.

Over the course of the last four decades of Ed's life, he and Lorraine claim to have looked at nearly four thousand cases. They have achieved worldwide fame through books and movies, but as luck would have it, they lived only a couple towns over from us, in Monroe, Connecticut.

We sought to evaluate the phenomenon of ghosts (in the generic sense, referring to all manner of spiritual manifestations) and see if there was any evidence to support the hypothesis that the phenomenon exists. The Warrens claimed to have scientific evidence, which would indeed prove the existence of ghosts—a testable claim that we sunk our investigative teeth into.

We were actually a little intimidated when we started this investigation. The Warrens were famous—they were huge in the ghost-hunting world—and we were just getting started. We were nobody. We knew we had to be careful and extremely thorough if we were going to take them on.

What we found was a very nice couple, some genuinely sincere people, but absolutely no compelling evidence. More precisely, there was a ton of "evidence," but none of it stands up to rigorous scientific testing. None of it. Most of it can't even withstand cursory testing.

Like all pseudosciences, the field of ghost hunting makes bold pretense to being legitimate science. The Warrens called their organization the New England Society for Psychic Research (NESPR), but they were a "research" organization in name only. They still have a presence on the web, and Lorraine still gives ghost lectures. Their original website proudly proclaimed that: "Our mission is to move the area of psychic phenomena out of the dark ages into the mainstream of rigorous scientific thought and inquiry."

But upon inspection, their methods lack the components of genuine scientific inquiry or even the most fundamental attempt at scientific rigor. Rather than an earnest search for the truth, regardless of what that may be, their society seeks only to support their a priori assumption that the phenomenon is real.

Our investigation began with a tour through the Warrens' unique museum, housed in their basement and alleged to be the most haunted place in Connecticut. The museum turned out to be a good representation of the quality of their evidence—all show and no substance. On the museum tour, Ed warned us not to touch anything in the main room, as we would open ourselves up to possible possession. If we did accidentally rub against something (which was nearly unavoidable in that cramped space), we were to report it, so that he could "purify our auras" before we left. The room was a clutter of collected stuff garnered over the Warrens' forty-year career. It included paintings, masks, statuettes, and many books.

Ed claimed that the most dangerous item in the house, however, was a Raggedy Ann doll that was still said to be possessed by a demonic entity. (This became the subject of its own movie in 2017, *Annabelle: Creation*.) They kept it enclosed in a glass case for safety, and they chillingly related the tale of the man who ignored Ed's warnings and taunted the doll, only to die hours later in a tragic motorcycle accident. Bob tempted fate by taunting the doll and even touching the case, and somehow he survived the experience. Twenty years later, however, he is starting to go gray, so there you go.

Ed liked to refer to NESPR as a "theological institute" and stated that his investigations were intimately associated with his religious convictions. In fact, one of his first questions to us, just as with other skeptics he has confronted in the past, was whether or not we believe in God, for without faith we could not understand his research.

Lorraine, who claims to be a "sensitive," or clairvoyant, also found our lack of faith disturbing. In perhaps the funniest moment of our investigation she asked us, "What happened with you boys (referring to our lack of faith)? Was it the science thing?"

Uh, yeah. Pretty much.

By this time we were already well familiar with the "ten-foot stack" phenomenon. Ask a true believer about the evidence for their claimed phenomenon and they will usually point you to a large amount of extremely low-grade evidence (the ten-foot stack). You must pore through all of it, they insist, before you can deny the power of the evidence.

Of course, we have our own saying: No matter how high you pile up crap, it won't turn into gold. Rather than spend our lives wading through the crap, we usually ask for their best evidence. If that's impressive, then we can go from there. We don't ask them to prove their phenomenon with one thing—just show us something compelling.

Our repeated requests for some of the claimed evidence for ghostly phenomena were usually met by changing the subject. Ed was hoping to just give us a lecture instead. The "psychic" hours, Ed told us, are from nine p.m. to six a.m., and the most vicious hauntings occur around three a.m. Why? Because that's an insult to the Holy Trinity. He continued this oration by explaining how a "ghost" is a luminescence without definable form, but on the other hand, an "apparition" has form and features. The countless photos we have seen of balls of light are known as "ghost globules," and the elongated patches as "light rods." He said there are human spirits and then there are the real bad guys—inhuman spirits. These are, of course, the essences of things never alive, or demonic entities. Ed also gave us some tips: Always keep a vial of blessed water on your person to repel entities; if a possessed person meets your gaze, never be the first to break it, as that demonstrates weakness. And on it went, rules and jargon of the trade.

The Photographic Evidence

We did eventually get to some evidence, but alas found ourselves wading through the (ten-foot) muck. The vast majority of the Warrens' physical evidence consists of photographs. They have hundreds of ghost shots, taken by them and those who work with them. The bulk of these photos are simply blobs of light on a piece of film (for those too young to remember, film is an archaic technology for capturing images prior to digital cameras). There are dozens of ways to get such light artifacts onto film, but most fit into one of three categories: flashback, light diffraction, or camera cords. Rare double or multiple exposures create more interesting but still artifactual photographs. It's significant to note that in almost every occurrence of a ghost photograph, the ghost is not seen at the time the photo is taken. It is not until the picture is developed that the ghost or glob or rod is seen, a strong indication that the picture is a result of photographic artifact.

Flashback is simply light from the camera flash reflected back at the lens, causing a hazy overexposed region on the film. The result is often a wispy, blurry light image on the film. It's easy to tell when a flash was used, because of the sharp shadows that are created and because objects in the foreground are brightly lit. The Warrens' website even suggested at one time that using a flash will help create ghost photographs, and "the brighter the flash the better." It also recommended using a foreground object—something to reflect the flash—although they admit that this is paradoxical, especially since they claim that such photographs are the result of psychically created images. However, there is no discussion or any recognition at all that the light images might be the result of photographic artifact created by the flash.

So-called "ghost globules" are spheres of light rather than wispy forms. The images, however, are curiously reminiscent of light diffracting around a point source. A small amount of condensation on the camera lens is enough to mass-produce such ghost globules. Under the right conditions, any discrete source of light can produce this effect.

Before the age of cell phone photography, cameras almost always had a strap, or cord, attached to them. This led to what psychic investigator Joe Nickell deemed "the camera cord effect." The cord or strap of an Instamatic camera can easily fall in front of the lens, and it would go unnoticed with cameras that don't view through the lens but through a separate aperture. Even black cords will look like white blobs or streaks of light when they reflect the light of a flash. We were able to reproduce this effect on our first try, creating a "ghost" photograph as good as any we have seen.

Of course, you don't need a camera cord for the camera cord effect; any close obstruction will do, including a finger.

The age of digital photography has created some new sources of photographic artifact. We were asked to investigate a curious photo with several colored streaks across an otherwise still and focused picture. After some digging (the advantage of digital cameras is that the image files contain all sorts of technical information about the picture—exposure time, etc.) we figured out that the camera was accidentally set to "twilight mode." This mode will use the flash but then keep the shutter open for a second or two to expose a dim background.

Copious examples of all these common artifacts can be seen on the websites of the Warrens and other ghost hunters. What's lacking in all of them, however, is any consideration of alternate explanations for the photographs other than genuine ghosts. There is no investigation into natural sources, no skepticism...no discussion at all, in fact. There's only the simple and unquestioned pronouncement that such blobs of light are evidence of the paranormal.

Video Evidence

The other evidence that the Warrens chose to share with us was a video. Their pièce de résistance is Ed's video of the famous White Lady of Union Cemetery in Easton, Connecticut. We have only been able to view this tape in the Warrens' home, because Ed refused to give us a copy for analysis, a common theme in our investigation. The tape shows an apparent white human figure moving behind some tombstones. Like videos of UFOs, Bigfoot, and the Loch Ness monster, however, the figure is at that perfect distance and resolution so that a provocative shape can be seen but no details that would aid definitive identification. Ed Warren has not investigated the video with any scientific rigor, and he refuses to allow others to do so. Despite Ed's insistence that he was engaged in scientific research, he jealously hoarded his alleged evidence, rather than allowing it to be critically analyzed.

Another investigator working with the Warrens did, however, give us one of their pieces of video evidence. It showed a man "dematerializing" and was taken by a mounted camera in a dining room in the middle of the night during one of their investigations. On the tape, a young man walks into the room, scratches his head, and— *poof!*—disappears. This extraordinary occurrence is quickly followed by a "ghost light" appearing momentarily on the window behind the scene.

We gladly accepted the tape and had it professionally analyzed. The analysis states:

> We are witnessing a wipe in this segment of videotape. Although there are several different ways in video editing to achieve a wiping effect, the most simple of ways has been employed here. Deliberately or accidentally, the camcorder stopped recording on the final frame of the person in the room and resumed recording just a few seconds after the person had moved outside of the view of the camera.
>
> On a related observation, the properties of light alone could dictate a hundred different explanations for the mysterious "dot" of light that appears a few seconds after the man "vanishes." However, I believe that this dot of light was caused by the reflection through the dining room window of the headlights of a passing car. The passing headlights can be seen if you watch the right-hand side of the screen just after the "dot" of light fades out.

So the only piece of evidence we were given turned out to be less than compelling. It was in fact likely just a simple malfunction. Even cursory analysis of this piece of tape would have revealed the same findings to the Warrens, yet no one in their investigatory network bothered to check it out. Rather than take this obvious first step, one of their investigators simply declared that the "ghost light" was "unexplainable." Further, none of the people in the tape were aware that anything had even occurred until the following day when the tape was viewed (again, the fingerprint of artifact)—including the young man who allegedly dematerialized! Ed put his credibility in serious jeopardy when he looked at that tape and, without any verification, stated that experts "can only come to one conclusion: That kid disappeared."

Despite numerous attempts to examine other physical evidence the Warrens claim to possess, we were given only excuses such as "The film was erased," "The people in the film want privacy," "We had just turned off the recording equipment, when…" After forty years of "research" into a phenomenon, they had precious little to show for it.

Eyewitness Testimony

Vastly outnumbering the Warrens' array of low-grade physical evidence is their copious anecdotal evidence. They're great tellers of ghost stories, which led, in no small measure, to their popularity on the lecture circuit. They didn't seem to understand, however, that the case for the reality of ghosts will never be made by stories alone.

In this respect the Warrens are typical of the majority of people, who are compelled by a gripping story and lack a deep understanding of the unreliability of human memory and perception. As we've already discussed at length, perception and memory are fallible. There is no such thing as a "reliable witness."

Further, many sightings or interactions with an alleged entity (whether ghost or alien) take place in the bedroom, late at night or very early in the morning—times and places connected with sleep or, more accurately, the near-sleep state. A classic example is Jack Smirle, investigated by the Warrens themselves, who related the tale of awakening in the early morning, being paralyzed, sensing an entity in the room, being overcome with terror, then being raped by a ghost.

You may recognize these signs from the chapter on hypnagogia.

When we offered hypnagogia to Ed as a possible alternate explanation, he seemed intrigued. "But," he continued confidently, "what about the pressure on the victim's chest when the entity is trying to get into them?" Well, we were sorry to tell Ed that pressure on the chest and shortness of breath are a well-described aspect of hypnagogia.

"Oh," said Ed.

Many investigations of haunted houses take place in the wee hours of the night. Investigators are often called upon to stay up all night, which creates sleep deprivation. In the sleep-deprived state, our brains are highly susceptible to hallucinations, a fertile source for ghostly experiences.

Another prolific cause is the human imagination. Different people have different capacities for imagination. At the far end of the spectrum are individuals particularly prone to fantasy. Coupled with a desire to believe and immersion into a belief system with group support, such fantasy-prone people can generate a tremendous amount of alleged paranormal experiences.

There's good reason to believe that groups such as NESPR would attract such individuals. With their widespread exposure, there is ample opportunity to inadvertently "screen" many individuals. Hundreds or thousands will see one of their lectures in a year. Out of those, dozens will make the effort to go to one of their weekly classes. The ones that stay on for the long haul are invited on investigations. And among those, a few are deemed to be "sensitive," which means they can see things that other people cannot.

Now, we don't expect everybody to be versed in hypnagogia, the effects of sleep deprivation, and the vagaries of the human imagination, but we do expect it from someone who claims to be conducting scientific research in a field where such phenomena play an important role. Although he claims that his critics are closed-minded, he himself dismissed out of hand any alternative explanations of his evidence, without doing investigations that could potentially refute those alternatives. What passes for research in NESPR, and in the field of ghost hunting in general, is passive documentation of anecdote and summary paranormal interpretation.

What Did We Learn?

Our investigation of the Warrens early in our skeptical careers was an eye-opener. The first lesson we learned, one that has been reinforced many times since, is that the purveyors of paranormal and pseudo-scientific claims are ultimately unimpressive. It turns out we had no reason to be intimidated by the Warrens. Instead of the formidable academicians portrayed in the movies, we found an old clueless couple who had managed to parlay telling ghost stories into a living.

We also learned that the specifics of any one belief don't really matter. People make the same mental errors over and over. Ghosts are now a bit passé, although they did have somewhat of a comeback with reality TV. Ever watched one of those ghost-hunting shows? They're about as convincing as the kinds of evidence you will find at a ten-year-olds' sleepover.

The same is true of psychics, Bigfoot hunters, UFO enthusiasts, believers in near-death experiences, vaccine deniers, or that time-cube guy. Same fallacies, different details.

The popularity of various beliefs changes over time and in different regions, but the old classics will come back again and again.

And we'll be ready for them.

45. Loose Thinking About Loose Change

Jay's Adventure

> The truth of the world is that it is actually chaotic. The truth is that it is not The Illuminati, or The Jewish Banking Conspiracy, or the Gray Alien Theory. The truth is far more frightening—Nobody is in control. The world is rudderless.
>
> —Alan Moore

———

Do you remember where you were on 9/11, the moment you heard about jets crashing into the World Trade Center? Well, if you think you do, reread the chapter on memory.

I do have a memory that in 2006 I watched a documentary film called *Loose Change*, one of a series of documentaries covering the events that took place on 9/11. *Loose Change* claims that the attacks on the World Trade Center, the Pentagon, and United flight 93 were actually a US government–run false flag operation. It tells a long tale of incidents that, at every step, reveal a grand conspiracy that would take the coordinated effort of thousands of people not only to pull off but then to keep quiet.

I've lived in New York City twice, first for school and then for work. When I was back living in Connecticut, I was only seventy miles away and would go back and forth all the time to see friends. It was and still is my city. So, five years after 9/11, I was still very emotional about what happened. Watching *Loose Change* reminded me of just how serious and horrible it all was, and how personal. One of my closest friends lost a brother, a firefighter, who went down in the second tower, and I have several other friends who escaped the towers before they came down. I know someone who was in the Pentagon when American Airlines flight 77 hit the building. He told a grim story of hearing the crash, feeling the building moan from the impact, and eventually going outside to see for himself. Other friends were living in the city and reported details of the aftermath. To this day, some details pop into my head like a nightmare. As soon as it was possible to get near ground zero, I did. I had to see it.

Even with all of these personal connections, and even though I was a critical thinker and had already trained myself to have an actively running baloney detector, it wasn't clear to me if the "facts" presented in *Loose Change*—or more importantly, the filmmakers' interpretation of those facts—were legitimate or not. On first viewing, it certainly seems that they have compiled a lot of curious facts about the events on 9/11. There was a moment while I was watching where I thought to myself, "Holy shit, how could all of this be made up?" Some of it had to be real. There were too many pieces of the puzzle that I didn't have answers for. And I remember getting furious during the film, because I was still emotional over the attack. It felt like the US government had to have screwed up somewhere to have let this tragedy happen. At the very least I was left with a lot of questions (which I guess was the point).

Not too long after watching *Loose Change*, we were recording the podcast. I'd been waiting to talk to everyone about it, excited to get into a discussion. A part of me liked the intrigue of it all. Even though I hated rehashing all the pain and death that 9/11 brought, the idea that there was a massive cover-up was alluring. When we were finally on our headsets I asked everyone what they thought, brimming with anticipation. This is close to what was said:

JAY: Guys, any of you ever see *Loose Change*?
PERRY: *What?!*
JAY: Yeah, the documentary about 9/11.

PERRY: *I know what* Loose Change *is, but it looks like you have no goddamn idea!*

STEVE: *Jay, that is all bullshit.*

JAY: No wait. I watched the whole thing and there was a lot of stuff in there that seems like it was way too coincidental for it to all be bullshit.

PERRY: *Steve, can you believe this? Our own friend and brother is taken in by one of the dumbest things I've ever seen in my entire life.*

JAY: All I'm saying, guys, is that it was really interesting. I mean, some of the stuff that they were saying seems pretty damn convincing.

STEVE: *Jesus, Jay, listen. The entire documentary starts with a conclusion and they're just moving chess pieces around to prove their original premise. It's all manufactured, and when you take a closer look into any of their claims, it's all crap.*

PERRY: *I mean, he's on a skeptical podcast . . .*

STEVE: *What point do you think was the most compelling?*

PERRY: *He shares your genetics, Steve.*

JAY: That one part that discusses how the steel . . .

PERRY: *He can't think for himself . . .*

STEVE: *The steel didn't have to melt, it just had to be weakened.*

JAY: Yeah, but would the building fall the way it did if . . .

PERRY: *NONSENSE! Jay, that entire thing is cooked up to fool idiots like you who just sit there and drool while they watch the blinking lights on their computer. Steve, I vote to kick Jay off the podcast until he gets a clue . . . a clue, I say.*

I didn't end up getting kicked off the podcast.

I did, however, have Perry relentlessly make fun of me for at least a year. He could cut a person down with one sentence. Even joking, he could wield words like a lightsaber. I totally deserved it. He was right. I was on a skeptical podcast and had been a skeptical activist for years. In retrospect, I see that I allowed emotion and a slick presentation of one side of a story to throw me off my game, at least temporarily. Similar to learning how an impressive magic trick is done, once I learned why the "facts" in the documentary were wrong, it all of a sudden seemed so obvious and ridiculous.

So why did I fall for it? I now think that in that moment it was lingering emotions and a need to blame someone tangible. This allowed me to make basic logical errors that were happening unconsciously while my mind was lazily being entertained by the documentary.

Don't be like me. There are basic things you can keep in mind so you're not swayed by misinterpreted information, constructed scenarios, and sometimes flat-out lies, and so your friends don't make fun of you for a year.

Watch out when you are only hearing one side of the story. Without other opinions or interpretations, it's easy to be sucked into agreeing with one side of an argument. We all know that most documentaries don't tell both sides of a story, and there's a reason that skeptics look to both sides. Imagine a court case where only one side presents their evidence. They're deliberately driving their audience down a specific path to help them draw a specific conclusion. It was up to me to find the other side of that story.

The volume of information becomes onerous. Large amounts of circumstantial evidence can be very convincing. This is the "where there's smoke, there must be fire" fallacy. In this case there was just a smoke machine. It's easy to buy into a distorted view when you don't already have some reliable information to compare it against. A documentary like *Loose Change* delivers a shotgun of facts wrapped in a compelling narrative. If you ever sit in on a multilevel-marketing or time-share pitch, you'll experience this as well. The speaker will bombard you with information that you can't fact-check on the spot. You have no time to process everything you've heard, and by the end you can be overwhelmed and convinced.

It's easy to underestimate how cherry-picking information can create a false reality. By sheer volume alone, *Loose Change* is an extraordinary example of this. Almost every point of evidence in the film is handpicked to fit into the story they chose to tell. What we, the audience, are unaware of are all the other pieces of information that were left out.

For example, the filmmakers claim that Danielle O'Brien, an air traffic controller at Dulles International Airport, told ABC News that everybody in the control room thought that flight 77 was a military plane. That, of course, would cast doubt on the official story that they were all commercial passenger planes. Here, however, is the full quote from ABC:

> "The speed, the maneuverability, the way that he turned, we all thought in the radar room, all of us experienced air traffic controllers, that that was a military plane," says O'Brien. "You don't fly a 757 in that manner. It's unsafe."

The 757 wasn't flying like a commercial jet should, so their first thought was that it was military. Of course, terrorists having hijacked the plane would also explain the unsafe manner of flight.

The filmmakers also make this claim: "Employees at the Pentagon were seen carrying away a large box shrouded in a blue tarp. Why the mystery?"

Yes, it does look strange if you're unfamiliar with disaster response. The large object under the blue tarp was a tent. Conspiracy theorists hear that and say, "It doesn't look like any tent I've seen." That's because they are not camping: It's a tent used to set up emergency services, like first aid. It's not a pup tent, more like a portable building. This is a good example of how conspiracy theorists can exploit the unfamiliar to make events seem unusual and therefore sinister.

Loose Change takes the events of 9/11 out of context and makes a presumption that they're all related in a way that they are not. It's also a good illustration of how conspiracy theorists can create the impression of something sinister going on just by asking questions. They don't have to prove anything, just make it seem like something is amiss.

Why did the towers fall like they did? Well...how would you know how such buildings should fall under these circumstances? Why did the debris field look like it did when the plane hit the Pentagon? How should it look? It's easy to make naive assumptions about unique and very unusual events and then claim that things don't look like they're "supposed" to.

Finally, your own bias distorts your perception. Confirmation bias is happening to you at all times and with everything that hits your brain. If you actively look for things to confirm your existing beliefs, you will find them. Remember to try to disprove your own hypothesis. It's what all good scientists do.

If you pile enough of these biases and claims up, it can seem like something must be going on. The lesson I learned the hard way is that this something is confirmation bias, sloppy thinking, and conspiracy mongering. AND THAT YOUR BROTHERS ARE RUBES

SECTION 3

Skepticism and the Media

Our modern era has often been referred to as the Age of Information. It didn't take long, however, for observers to notice that it is just as much, if not more, the Age of Misinformation.

Of course, bad, misleading, biased, or distorted information has always been part of human culture. But technology has made the communication of information easier, and a rising tide raises all ships—good, bad, and indifferent.

Science is particularly difficult to communicate well, because it requires a lot of background information, and a lot of work to get it right. In our years of trying to spread science and critical thinking, the press has been both a blessing and a curse. The media is a great way to reach the public, but the forces naturally at work in the media seem to be working against us. We have had to deal with sensationalism, simple journalistic laziness, ideologically motivated misinformation, and even outright scams.

In order to be a skeptical consumer of news, you need to be able to assess the quality and credibility of any news source you rely on for information. This includes recognizing common patterns of journalistic behavior, pitfalls, and signs that a piece of information may not be trustworthy.

In this section we'll discuss some of the common problems we face with journalism in the age of the internet, the web, and social media, and give some dramatic examples of science journalism gone wrong.

46. Fake News

To your request of my opinion of the manner in which a newspaper should be conducted, so as to be most useful, I should answer, "by restraining it to true facts & sound principles only." Yet I fear such a paper would find few subscribers. It is a melancholy truth, that a suppression of the press could not more compleatly deprive the nation of its benefits, than is done by its abandoned prostitution to falsehood.

—Thomas Jefferson

———————

At the SGU we often speak of pseudoscience in our battle against ignorance, but pseudojournalism is just as important to have on your skeptical radar. Pseudojournalism can take many forms, from opinion masquerading as fact to fabricated sources, to "advertorial" sponsored content, to outright plagiarism. In essence, pseudojournalism is any written content that flagrantly ignores media ethics but attempts to pass itself off as legitimate news.

The most pernicious forms of this so-called "fake news" are easy to spot. They push a heavy agenda, come from disreputable outlets, or try to sell you something. They may tell you who to vote for. However, more subtle forms of misinformation are just as common.

The idea that we're living in a "post-truth" era took off in 2016, making it the top pick for the Oxford Dictionaries' word of the year. Right on its heels, though, and perhaps a contender for phrase of the year, is "fake news." Fake news is clearly a thing. There are obviously websites that make up fake news stories for various purposes.

It is simultaneously a bug and a feature of the internet that it's a venue for a war of ideas. There is an obvious benefit to this in an open society—the free exchange of information in the marketplace of ideas. Let's hash it all out with a true democracy of expression and access (well, for those with access).

The bug is that the internet is also a venue for fraud, lying, misinformation, and manipulation. Not everyone is a fair player, and they ruin it for everyone else. And it's not black-and-white. Rather, there is a spectrum of behavior, and most people are at various points along that spectrum on different issues. At the same time, there are extremes: some sites that aspire to a high level of journalism or scholarship and, at the other end, sites that are pure fraud, propaganda, or clickbait.

We are collectively still trying to figure out how to deal with the resulting mess. It seems to me that part of the problem is that we're using the internet to address the problems of the internet. Bad actors can therefore hijack or duplicate the mechanisms of quality control and subvert them. The concept of "fake news" was introduced as a method for sorting through news outlets. It was instantly subverted for nefarious purposes and is already waning in usefulness. Essentially a fake news site can just say, "Oh yeah, well, you're a fake news site." Ideologues can make their own list of "fake" news sites, which are just their ideological opponents. The term becomes a weapon rather than a tool.

We have seen this before in other contexts. The term "skeptic," and in fact the format of skeptical analysis and debunking, has been hijacked by science deniers pretending to be skeptics, tarnishing the brand and sowing confusion.

Characterizing News Outlets

Messy reality doesn't comport well with our desires for easy categorization. Anything we try to categorize will have features that exist along a spectrum and may occur along various spectra simultaneously.

This creates three intellectual pitfalls—the false dichotomy, the false continuum, and the demarcation problem (see these entries in chapters 10 and 22 on logical fallacies and pseudoscience).

Where do we draw the line between "real" news and "fake" news? We must often be content with a fuzzy line and a fractional definition. There are a number of features of fake news, such as lack of vetting of sources and failure to check facts. The more of those features you have, the more fake you are, and beyond some fuzzy hard-to-define point (the demarcation problem again) it is legitimate to call something "fake news."

Some individual features have such a dramatic impact that you can make a clear designation. Let's take scientific journals as an example. They exist along a spectrum of quality, but some journals do not carry out any real peer review, and some don't seem to have any editorial filter. Some journals, in essence, are fake—or broken. Their primary purpose is to make money, not to disseminate quality science. In other words, there may be some black-and-white qualitative differences among outlets that allow for clear categorization as fake.

With all this in mind, let's take a look at the fake-news controversy. Clearly, a simple dichotomy—fake news versus real news—isn't sufficient to cover the range of what's out there. To start, you need an understanding of the various types of news outlets.

Traditional Journalism—These sources consider themselves real journalists and hold themselves ostensibly to some journalistic standard (the *New York Times*, the *Chicago Tribune*, etc.). Their purpose is to report news and they at least try to follow journalistic traditions. Within this category there is a spectrum from great journalism to horrible journalism—essentially a spectrum of quality.

Biased or Ideological Journalism—These are sites that have the format of traditional journalism but have a clear ideological bias or agenda. This is a spectrum orthogonal to quality—the degree of ideological bias. It's probably best to consider all news outlets as biased to some degree: All people have biases, and news organizations are made of people.

Along this continuum are news outlets that make a sincere effort to be ideologically neutral or at least balanced. They may lean toward one political perspective but at least they try to keep their biases in check.

Further along the continuum are news outlets that have a clear and open political bias but make a sincere effort to work within a framework of reasonable journalistic quality. Even if they're liberal or conservative outlets, they are trying to be journalists first.

This category slides seamlessly into outlets that prioritize their political agenda above journalistic quality. They blatantly select the news they choose to report and the spin they put on that news based on the political narrative to which they are catering. At some fuzzy point they cease to be real journalists. They're no longer selling news but selling a narrative they call the news.

Opinion Outlets—There are websites, journals, and newsletters that are not news at all (and don't pretend to be), but a venue for commentaries and opinion pieces with a clear ideological, political, religious, philosophical, or social agenda. They are generally up front about their agenda, as the point is to promote said agenda.

Satirical News—The *Onion* is the best example of a satirical news outlet. This isn't news. This is pure entertainment. They make no attempt to fool the public about their satirical nature. Their articles are obvious satire and humor. Despite that, sometimes the humor is subtle enough that *Onion* articles can be shared as if they're real. There are a number of hilarious instances of world leaders and governments repeating *Onion* articles as if they were news. In 2012, for example, the Chinese government media took seriously an *Onion* article proclaiming North Korean leader Kim Jong-un the sexiest man alive.

There is a range of quality here as well. Some satirical sites just aren't that funny, and without the humor they become increasingly difficult to distinguish from real news outlets or those pretending to be. In other words, there is a demarcation problem between bad satirical news and fake news.

Fake News—Fake news outlets simply make up their news stories.

They have no genuine journalistic process or mechanisms of quality control. Their stories are made-up fictions in the format of real news, optimized for clickbait. They're meant only to push emotional buttons in order to motivate clicks. Sometimes those emotional buttons are political. They may call themselves satire as a whitewash, but the articles are not meant to humorously expose some aspect of society or human nature.

Social Media

Through social media, articles are often detached from their source, and therefore from their context, and are spread as individual isolated units. Articles from all the above categories and ranging over the various spectra of quality, sincerity, and bias are dumped into a single bucket and mixed together. An article comes up on your feed, and it's up to the reader to discover the source and assess its nature and quality. Most people just read the headlines and perhaps the attached blurb, but don't have the time or inclination to do a thorough investigation.

There are sites that curate news for you, but that simply adds another layer that can fall into any of the categories above. However, you only have to do a deep evaluation of the news aggregator once, and you may find a reliable one that can at least filter out the fake, blatantly biased, or low-quality news.

In the end, any news you care about needs to be sourced and further checked against multiple different sources to average out the biases as much as possible. It's also good social media hygiene not to share or spread news that you haven't vetted yourself. If there isn't time to do a reasonable investigation, fine, but it stands to reason that one shouldn't share an article as if it's valid news when you have no idea.

But of course people exist along the same spectra as news outlets. People also have blogs and Facebook pages, so we're all at various times playing different roles in the news cycle—we may be curating news, sharing news, and writing articles. We are all becoming journalists, aggregators, curators, and consumers at the same time.

There is no simple solution to all of this. It's a true, if at times glorious, mess. The only real solution is for each individual to be a savvy and skeptical reader.

Echo Chambers

Added to the concept of fake news is that of the echo chamber. The idea here is simple: We have a tendency to seek out opinions that are in line with our existing worldview and beliefs. This is like confirmation bias on steroids, not only seeking out facts that support our beliefs, but sources of information that are more likely, or even certain, to confirm those beliefs.

People have perhaps always done this, but the internet and social media make it much easier, even the default. There are the aforementioned websites that aggregate or curate news for you. While this is convenient, it also puts a lot of power in someone else's hands. There's a lot of news out there, and a lot of different takes on the news. If from this massive pile only the news and opinions that support a particular worldview are chosen and fed to you, that can have a powerful effect on your view of reality.

Some social media outlets do this automatically, with the intention of giving you what you want. So, if you consume a certain type of news, it will feed you articles and videos that match those preferences. You may not even realize you're looking at the world through this one perspective.

There are also sites dedicated to one topic, such as belief in a particular conspiracy or an anti-vaccine agenda. Such advocacy sites say they're there so that people with similar views can exchange thoughts and ideas without worrying that they will be ridiculed or attacked. While this sounds innocent enough, in practice it means that members are shielded from any disconfirming facts or ideas. Anyone challenging the basic assumptions of the group is branded a "troll" and will likely be banned. The tools exist to easily maintain the intellectual purity of the group.

Astroturfing

In a 2015 TEDx talk, Sharyl Attkisson spoke about the problem of astroturfing, but she wasn't aware that her talk had fallen victim to the very problem she was discussing.

Astroturfing is essentially fake grassroots activism. Companies and special interests create nonprofits, Facebook pages, and social media personas, write letters to the editor, and essentially exploit social and traditional media to create the false impression that there is a grassroots movement supporting some issue. The key to astroturfing is to conceal who is truly behind these fronts.

Attkisson, a journalist for CBS News, cites several examples in which pharmaceutical companies secretly promoted their drug and marginalized criticism. She correctly points out how campaigns of doubt and confusion can work by generating so much controversy that the public loses confidence in the science (and in fact science itself) and throws the baby out with the bathwater.

Attkisson, however, gets some things profoundly wrong, herself falling victim to astroturfing without realizing it. At one point, she uses as an example of astroturfing what she considered pharmaceutical companies' burying of evidence that there is a link between vaccines and autism with fake studies and fake opinions. This may seem surprising for a CBS reporter, but it makes more sense when you know that Attkisson has had anti-vaccine views for many years. She concealed her own agenda, making her reporting on the issue seem like genuine journalism when it was in fact advocacy.

What this shows is that once information itself is called into question because of bias, astroturfing, denialism, or pseudoscience, it then becomes easier to see reality through the lens of your preferred narratives. One person's astroturfing is another person's skepticism.

It's like the movie *Inception*—once you can't trust reality, there is no way to confidently know how many layers deep you are in the dream.

This is why scientists and skeptics who are trying to communicate controversial science to the public are so frequently met with the shill card (using the knee-jerk accusation that they are shilling for industry as a way to dismiss their scientific opinions). It's now a common game on Facebook, blogs, and other social media outlets to count how many comments go by on a post about a controversial topic before someone plays this card. The answer is always—not many.

If you are, for example, an anti-vaxxer, then it's likely part of your narrative that any scientist or opinion writer who expresses a pro-vaccine opinion or counters any of the misinformation generated by the anti-vaccine movement is by definition a shill and part of an astro-turf campaign. The existence of real astroturf campaigns just makes the accusation seem more plausible. Similarly, the existence of real but small conspiracies lends fuel to the conspiracy theorists who weave outlandish grand conspiracies.

Attkisson ends with her advice for the public on how to recognize an astroturf campaign. To paraphrase, she says that the use of words such as "quack," "crank," "pseudoscientist," or "conspiracy" are all red flags for an astroturf campaign, as is voicing skepticism toward those who criticize authority. This advice, however, paints with such an absurdly broad brush that she essentially covers the entire skeptical community. Perhaps this is intentional on her part. She's been criti-cized for her anti-vaccine views, and now she has a plausible-sounding reason to dismiss all of this criticism—she is just a victim of alleged astroturfing.

This is especially disappointing for a journalist. Investigative journalists should know that stories are often not what they appear to be and that this can cut both ways.

There is no easy way, as an individual, to deal with astroturfing. It takes digging. But first you have to realize that Attkisson framed the question as a false dichotomy: Is it real or astroturfing? The actual choice is between real science, astroturfing, and pseudoscience. When you add the third possibility you have a better chance of digging deep enough to know which side of a controversy has the solid evidence and arguments.

While it's not safe to assume that every grassroots movement is exactly as it appears, you also shouldn't assume that accusations of being an astroturfer or a shill are legitimate. For example, the modern skeptical movement as a whole has a history that goes back at least to the 1960s. Many of the current organizations have been in existence for years or even decades. They are led by people who have very public lives. It should be a trivial matter for anyone with investigative skills to find out that I am a real person with an academic appointment who has been a science and critical thinking advocate for decades. Further, there isn't the tiniest shred of evidence that I am a shill for anyone. Anyone accusing me of this is doing it thoughtlessly, as a knee jerk, because it serves their narrative (or deliberately, because they care more about their ideology than the truth).

The best solution for all of us is to apply a high degree of critical thinking and skepticism to any issue. This isn't easy, but it is necessary. Of course, Attkisson and others can just argue that this book is an astroturfing attempt to disguise astroturfing and criticize those who are trying to expose astroturfing. You can always argue that the dream goes one layer deeper.

47. False Balance

In science, "fact" can only mean "confirmed to such a degree that it would be perverse to withhold provisional assent." I suppose that apples might start to rise tomorrow, but the possibility does not merit equal time in physics classrooms.   —Stephen Jay Gould

———————

I have been interviewed as the "skeptical" voice for numerous TV documentaries. For one program about demonic possession, I was concerned that the producer was giving a lot of time to people I felt were dangerous cranks. They were claiming that mental illness is often due to possession and that the treatment should be exorcism. I patiently explained why this was nonsense and harmful and encouraged them to focus on the science. The producer's response was that they were going to portray both sides and let the audience decide. They also stated that if anyone believed the proponents of demonic possession they were stupid and deserved what they got. This was a dramatic example (among other things) of what we call false balance.

The premise of false balance is innocent enough—in fact it carries with it a good measure of utility and virtue. Before the phrase "fair and balanced" became an ironic media slogan, it meant something. Just like science, journalism carries with it a great deal of ethical responsibility. In journalism, that responsibility is, of course, to the reader and not the source. The purpose of a free press is to inform the public, to keep them abreast of the goings-on in their community, their government, and beyond. In an effort to minimize editorializing (allowing a journalist's own opinions and interests to find their way into his or her writing), there is a constant attempt at balance and fairness.

If a fight breaks out on the street between two men, and the reporter tasked with covering it presents only one man's take on what happened, that is hardly a balanced story. And if that same story presents hearsay or unsubstantiated claims as truth, it's neither balanced nor fair. It is generally required that all sides of an argument receive some mention in a news article, and that only substantiated, corroborated facts are reported as such. This approach is especially necessary when covering political news, where there are usually two or more views on any topic. That's because politics is full of opinion and value judgments.

Science, however, is mainly concerned with verifiable facts. There may be a range of scientific opinions on a topic, but those opinions are rarely balanced. For most scientific topics the evidence, and the consensus of scientific opinion, is asymmetrical, favoring one school of thought over others.

A good example is climate change coverage. Say a new study is published in a peer-reviewed journal showing that last year was the hottest year on record. A television news show wants to cover the story, but the editorial staff is faced with a problem: Do they invite only the author of the study onto the show to discuss why climate change is such a problem and what steps could be taken to mitigate its effects? This would certainly be a fair approach, as it would provide unbiased, evidence-based content for the show's audience. Or should the producers invite the author of the study to sit across from a climate change skeptic (what we might fairly call a climate change "denier") to fill the airtime with a back-and-forth about whether or not global warming even exists. This may be seen as a more balanced approach, but it's not a fair or scientifically accurate one.

Nonetheless, this is the approach selected time and time again by professional news organizations. They've fallen into a trap known as false balance, in which they give equal coverage to sides of a story that clearly don't have equal credibility. It happens when Jenny McCarthy sits across the table from a CDC scientist. And when Ken Ham argues with Bill Nye about evolution.

It's one thing to report the news—if an important public figure makes a statement that's in opposition to reality, it should be printed. But not without context. And not without the necessary fact-checking and presentation of opposing evidence to show just why such a statement is untrue. This is when balance and fairness work for the good of the people. But when Deepak Chopra has a seat at the table with Sean Carroll or Brian Greene, that's false balance run amok.

As mentioned previously, anthropogenic climate change is an established fact. Surveys show that the vast majority of scientists agree that it is happening, and the probable outcomes are mostly bad. So on an episode of his weekly comedy program *Last Week Tonight*, John Oliver approached this problem by welcoming ninety-seven climate scientists to come on the show to debate one very lonely climate denier. As expected, hilarity (and chaos) ensued.

There is more advancement on the horizon. In 2014 the BBC published a progress report on their science reporting, instructing their journalists not to interview cranks or science deniers just to provide false balance in their science reporting. Let's hope that approach catches on.

48. Science Journalism

For it is the natural tendency of the ignorant to believe what is not true. In order to overcome that tendency it is not sufficient to exhibit the true; it is also necessary to expose and denounce the false.

—H. L. Mencken

———

Good science journalism is extremely challenging. Part of this comes from an essential tension between the very things it is named after: science and journalism. Science proceeds slowly and cautiously, is very conservative in its claims, and is skeptical toward any new finding (or at least it should be). Journalists, however, want an exciting new and simple story. In fact bad science journalism sounds much closer to self-help gurus than to actual science—"Do X and you will be happy." I know that journalists often don't write their own headlines and that there is a circle in skeptical hell dedicated to headline writers where they are tormented by the twin demons Hype and Sensationalism.

To be fair, I also understand that we live in the real world where ratings and clicks matter to the bottom line. This has been made very clear to me since I have been running my own science Facebook page (The Skeptics' Guide to the Universe). I've posted six to eight science and skeptical news items per day, adjusting variables and seeing directly how much reach each post gets.

What is absolutely clear is that there's a direct relationship between the number of clicks and likes an item gets and several specific variables: how controversial or emotional the subject matter is, how exciting the news item is, how "clickbaity" the headline is, how simplistic the message is, and the presence of cats in any form.

Memes do the best—an interesting picture with a punchy sentiment. Infographics are also popular, as are lists. These are all ways to make information bite-sized and easily consumable in a fast-paced world and in a medium with a lot of competition for eyeballs.

Don't worry—I haven't succumbed to the dark side. My lightsaber glows a reassuring hue of arctic blue. The purpose of my Facebook page is not just to be popular, but to promote science and critical thinking. I have also discovered there is another way to be successful, but it takes a lot of thoughtful hard work.

It's possible, but very challenging, to present interesting science news items in a compelling way that sparks interest without compromising scientific or skeptical integrity. The fact is, science is damn exciting. You can convey this excitement while putting a story into a proper scientific perspective.

Take the 2016 news item of the discovery of a possible ninth planet. What is more exciting than that—another planet in our solar system, ten times the mass of Earth? It hasn't yet been confirmed. The claim is based on a computer model of the gravitational effects of known objects in the Kuiper Belt. Despite the complexity, saying we may have discovered a "possible ninth planet in our solar system" is an interesting and accurate headline and was widely shared on social media. The real challenge for news outlets wanting to attract clicks comes on slow news days or when reporting wonky articles that are fairly dense and esoteric.

We know that there are exciting scientific breakthroughs happening every day, so why is the standard for science journalism so low on average? I think this is partly due to the fact that most journalistic coverage is mediocre, almost by definition. Still, a profession can maintain high general levels of quality, and it would be nice to see higher standards for science journalism.

Part of the problem is that journalists are trained to ask certain questions, like "Why is this important, and why should people care?" When they are reporting science, this leads them to common tropes. Every basic science advance dealing with cells apparently might one day lead to a cure for cancer. Every tiny bit of knowledge about viruses might cure the common cold. And now every metamaterial discovery will result in an invisibility cloak. When in doubt, make the closest possible analogy to some iconic science-fiction technology or, in a pinch, a superhero power. Physicists, it would appear, are all working on tractor beams, transporters, and warp drive.

Sometimes a scientific discovery is interesting because it's a new discovery, not because it will lead directly to some tangential technology.

The recent disruption of traditional journalism by social media has created real challenges for primary journalism. It's less common for news outlets to have dedicated specialist reporters, like trained science journalists. They are more likely to have general reporters covering the science beat and reporting to generalist editors. This is a recipe for disaster, because their background probably doesn't give them the tools to handle complex science news stories.

Is science reporting really this bad, or is it just confirmation bias? Well, in a 2008 study, Gary Schwitzer from the University of Minnesota School of Journalism and Mass Communication published an analysis of five hundred medical news stories over the previous two years. He found that:

In our evaluation of 500 US health news stories over 22 months, between 62%–77% of stories failed to adequately address costs, harms, benefits, the quality of the evidence, and the existence of other options when covering health care products and procedures. This high rate of inadequate reporting raises important questions about the quality of the information US consumers receive from the news media on these health news topics.

About two-thirds of all medical news stories (and these are from the top news outlets) had major flaws. In my opinion this establishes that there is a systemic problem with the quality of medical news journalism in the US. Other countries have similar problems, and I suspect that poor quality is an issue across science news reporting and not isolated to medical news.

It's tempting to blame the reporter for bad science journalism, but reporters are not the only source of the hype and sensationalism distorting the science. A 2012 study of medical science news reporting found that the article abstract, written by the scientists themselves, had positive spin 41 percent of the time. Scientific articles with such spin were far more likely to produce press releases with misleading spin and, therefore, hyped science journalism. So, much of the time, the buildup originates with the scientists or the press release.

I have been involved at every step of this process and seen it happen. In order to improve the quality of science reporting, everyone in the chain needs to be engaged, and they need to communicate well. Scientists, for example, could use more training in how to communicate with reporters and the public.

The worst experiences I have had were when reporters or TV producers had essentially already written their piece. They decided what the story was (the one that sells) and were now just interviewing experts in order to mine them for quotes they could backfill into their narrative. Sometimes the challenge for me was to find out what that narrative was and to see if I could get them to question it and change it if necessary. Otherwise I was just going to be quoted out of context to serve a sensationalist account.

At other times journalists may ask extremely leading questions, putting words in the mouths of their experts. Scientists have to learn not to fall for this. Even if it sounds almost right, the reporter is likely just looking to justify a sensational angle. You need to say, "No, that's not accurate. I would put it this way..." Then give them a bite-sized quote they can use. Give them a long lecture, and they will just ignore it.

How to Vet a Science News Story

In addition to a generally skeptical approach, there are some specific things to keep in mind when reading science news stories.

Who is the reporter and what is the outlet? Specifically, is this someone with a history of good science journalism, or were they likely covering the dog show last week? Is the outlet known for high editorial standards or clickbait? Is the news article just a press release? It's easy, now, to simply aggregate science press releases and present them as news, without any editorial process or investigation at all.

Does the article focus on the tiny bit of speculation at the end of a study and run with that? Researchers will often round off a study with comments about directions for future research or possible implications of their research, and it is not uncommon for journalists to present this speculation as if it's the result of the study. Ask the question, "What do the data in this study actually say?" If you don't fully understand that, then there is likely something missing from the reporting.

Did the journalist put the science news stories into context? Is this a mainstream or minority opinion? What is the power of this one study? Are there other studies that come to different results? Does this actually change our thinking in this area? What are critics saying about the results? Did they talk to experts who didn't publish the study to see how the scientific community looks at it, or did they just let the researchers toot their own horns? Every event or study in science is part of an unfolding messy process. Journalists should report the process, not the study in isolation as if all by itself it fully establishes its conclusions.

Is the journalist exaggerating our prior ignorance or the impact of the study, as in "We knew nothing, and now scientists have finally solved this enduring mystery"? Or are they exaggerating how new or mysterious the findings are? "Scientists are baffled!" And, as discussed above, is the reporter presenting false balance, or are they fairly representing the consensus of opinion among scientists?

It takes a little time and effort to properly vet a science news item.

It helps to have sources you can trust. But it's also a good idea, if you care about the science, to track your way back to the original sources and do a little digging. At the very least, don't take the reporting at face value.

While there are excellent science journalists out there, there is also a system of poor quality control and perverse incentives. The system rewards (at least in the immediate term) sensational journalism and clickbaity headlines. It takes true dedication to quality to eschew these quick tactics for a loftier long-term goal.

There may never be a real solution to this inherent problem. However, the more consumers demand quality, the better that balance of incentives will be. In turn, the demand for quality reporting is driven by quality journalism itself. It's also my hope that scientist bloggers using social media to communicate quality science reporting will move the entire system in the direction of a higher standard.

49. Enter the Matrix

There are many ways in which journalists can mislead a reader with science: they can cherry-pick the evidence, or massage the statistics; they can pit hysteria and emotion against cold, bland statements from authority figures.
                                                    —Ben Goldacre

The *Sunday Mirror* declares, "Scientists Develop Matrix-Style Technique of 'Feeding' Information Directly into Your Brain." And one of our favorite examples of terrible science news reporting had begun. *Discovery News* went with, "Novices 'Download' Pilots' Brainwaves, Learn to Fly." Most other outlets spoke of "uploading" information to the brain and learning in seconds.

The one thing I was certain of as I read these headlines was that none of this was happening. Brain-machine interface technology is progressing rapidly, but we are a long way away from downloading information from or uploading information to the human brain. The news outlets reported that electrical activity was recorded from pilots and then used to stimulate the subject. They also reported that this enhanced learning by 33 percent.

What Actually Happened

The study was titled "Transcranial Direct Current Stimulation Modulates Neuronal Activity and Learning in Pilot Training."

There were thirty-one subjects in the study, divided into four groups. One group had transcranial direct current stimulation (tDCS) to the dorsolateral prefrontal cortex (DLPFC), while a second had sham stimulation of this area. Sham stimulation creates the sensation of being stimulated but does not reach the brain. A third group had tDCS to the primary motor cortex (M1), and a fourth group had sham stimulation of M1.

In prior studies, stimulation of DLPFC has been associated with increasing working memory. While this finding is still somewhat controversial, the studies so far are promising. Stimulation of M1 is meant to enhance the acquiring of motor skills, although this finding is less well established with prior research. The thirty-one subjects were trained on a flight simulator and their performance was tracked. The question for the study was whether or not either type of stimulation would aid in learning this new and fairly complex task.

Here is the main result for DLPFC stimulation:

> Neither the initial nor the final behavioral performance were significantly different between DLPFC stimulation and sham groups... The average trial duration to reach 2-/3-back [a measure of accuracy on the task subjects were learning] was not significantly different between stimulation and sham groups. In addition, the average number of trials to reach 2-/3-back and the average 2-/3-back streak durations were not statistically different between groups. Significant differences in online, offline, and combined learning rates were not observed between stimulation and sham groups.

What they did observe is that the variance for the stimulation groups was 33 percent less than for the sham group. So there was less variability in performance, even though overall average performance was no different. The results for the M1 group were the same—no statistically significant difference in overall learning, and this time no difference in variance either. The researchers also took multiple measures of brain activity, looking for differences in the stimulation and sham groups, and there were some differences, but none that directly relate to actual learning.

Journalism Failure #1: Turning a Negative into a Positive

Overall this was a negative study. Stimulation of either brain region didn't improve learning in the subjects. The only effect was to reduce the variability in learning in the DLPFC stimulation but not the M1 stimulation. Given the small size of this study and the number of variables they observed, it isn't a surprise that there would be a few statistically significant outcomes. These could easily be chance outcomes and shouldn't be considered reliable until they are replicated in larger studies.

More interesting than the outcome of the study itself is how absolutely wrong the mainstream reporting was. There was never any "downloading" or in fact any kind of recording made from pilots. Indeed, there are no pilots in this study! Further, the brain stimulation was standard, not based at all on activity recorded from pilots. Finally, the study was essentially negative: There was no improvement in learning in the stimulation group compared to the sham group. The reported 33 percent "improvement" was not an improvement, just a decrease in variability.

Journalism Failure #2: Stopping at the Press Release

This may be one of the most complete journalistic fails reporting a science news item I have ever seen, but in this case it's the press release from HRL Laboratories that is largely to blame (not to suggest that the journalists are off the hook). The press release reports:

> Dr. Matthew Phillips and his team of investigators from HRL's Information & System Sciences Laboratory used transcranial direct current stimulation (tDCS) in order to improve learning and skill retention. "We measured the brain activity patterns of

six commercial and military pilots, and then transmitted these patterns into novice subjects as they learned to pilot an airplane in a realistic flight simulator," he says.

I found this nowhere in the paper itself. The quote is likely taken out of context, but it's hard to imagine the context since it seems to have no relationship to the study.

The press release also states:

The study, published in the February 2016 issue of the journal *Frontiers in Human Neuroscience*, found that subjects who received brain stimulation via electrode-embedded head caps improved their piloting abilities.

This is true—but then so did the controls. Elsewhere the release says that stimulation "modulates" learning (not "improves"). It modulated learning only in the decrease in variance. However, then the press release states it outright:

While previous research has demonstrated that tDCS can both help patients more quickly recover from a stroke and boost a healthy person's creativity, HRL's study is one of the first to show that tDCS is effective in accelerating practical learning.

But learning was not accelerated.

Conclusion: A press release is not a study.

Using tDCS to alter brain function is an interesting area of research. This type of treatment may eventually play a role in recovery from stroke or other brain injury, but it's still too early to tell.

The current study is small, with thirty-one subjects spread across four groups, and the researchers looked at many variables. This is clearly a preliminary study (one might say a "pilot" study, but not in the way the news reporting indicated), which is useful for guiding further research but lacking the power and specificity to draw any firm conclusions.

What the reporting of this study shows is the massive problem that exists in reporting science news items to the public. The press release and the reporting that flowed from that press release appear to have almost no correlation with the findings of the study itself. What seems to have happened is that the reporters hit upon a great sci-fi movie analogy, *The Matrix*. That was it, and that drove all the details of the reporting. They had no problem altering or just making up details as necessary to fit the sci-fi narrative.

50. Microbiomania

The passengers in our microbiome contain at least four million genes, and they work constantly on our behalf: they manufacture vitamins and patrol our guts to prevent infections; they help to form and bolster our immune systems, and digest food.   —Michael Specter

---

Probiotics cure asthma! Fecal transplants treat schizophrenia! Improve your gut flora to prevent Alzheimer's! A cursory Google search of the word "microbiome" yields a wide variety of articles. Some are solid peer-reviewed studies. Others are popular media coverage of those peer-reviewed studies. And unfortunately, many are overhyped.

"Microbiome" is defined as the aggregate genetic material of all the microorganisms within a particular system. For our purposes, we're discussing the human microbiome—the collective DNA of all the bugs living on us or in us at any given time. Between ten trillion and one hundred trillion microbial cells live symbiotically with our own, the vast majority being bacteria in our digestive tracts. Modern estimates show that for every native human cell, we've got a bacterial cell lurking. It's a rapidly growing area of research, and publication rates are high, especially when compared with the snail's pace of typical academic progress.

Nearly every day, I come across a blog post or magazine article touting the wonders of our microscopic comrades. Some of the coverage is exceptional (I highly recommend Ed Yong's popular science book *I Contain Multitudes* for a thorough examination), but like many areas of scientific inquiry with crossover appeal to the health-and-wellness crowd, microbiome research is a bright red target for pseudoscience, misuse, and reckless reporting.

Credit where credit is due: Evolutionary microbiologist Jonathan Eisen has been policing the worst offenders for nearly a decade on his blog *The Tree of Life*. He even gives out fake awards for the most egregious perpetrators of what he's dubbed "microbiomania." He says, "There are two ends to this. There's ridiculous stuff about why we should kill all microbes. That's the germophobia club, and then there's 'microbiomania,' who think microbes are beneficial and do everything." Of course, the answer is somewhere in between. Discussions about "good" bacteria and "bad" bacteria are woefully simplistic.

Often the hype begins at the source, the very establishment undertaking the research (or a public relations firm working for the institution). University and industry press releases are notoriously aggrandizing, and it doesn't help that they're written like traditional news coverage. This is, of course, completely by design. If a journalist is tasked with an overwhelming number of assigned stories, they might be compelled to simply repackage preexisting press releases to make their quota. It's incredibly common and, in my opinion, incredibly unethical.

A good example of this type of hype is a 2016 paper published in *Scientific Reports* titled "Antibiotic-Induced Perturbations in Gut Microbial Diversity **Influences** Neuro-Inflammation and Amyloidosis in a Murine Model of Alzheimer's Disease" [emphasis added by this skeptic]. As pointed out by Dr. Eisen, the study shows nothing of the sort. Only a correlation can be drawn between changes in gut flora and neuro-inflammation/amyloidosis in the laboratory mice. Antibiotic usage may be influencing both. Or perhaps an intermediary is at play—microbiota of the skin, nose, or ears, or components of the blood? There may be a variable of which we are currently unaware. And then the press release is published: "Antibiotics

Weaken Alzheimer's Disease Progression Through Changes in the Gut Microbiome," a grossly misleading title from the University of Chicago Medical Center, the very institution that undertook the research.

Although the original paper and the associated press release make an attempt to caveat some of the claims made by the authors, you have to dig to find them. And these disclaimers are completely lost in translation as journalists do their own write-arounds. By the end of the game of science-reporting telephone, we're left with articles like, "Antibiotics Could Be Used to PREVENT Alzheimer's Disease by Changing GUT Bacteria," "Antibiotics Might Fight Alzheimer's Plaques," and "Is Popping Antibiotics a Way to Beat Alzheimer's?"

So some correlation between gut bacteria and inflammation turns into a prevention or cure for Alzheimer's disease. Then people wonder why all these promised breakthroughs never materialize.

Poor reporting is one thing, but even more pernicious is the choice of some individuals to pass off agenda-driven, often fabricated rhetoric as true journalism. It's called pseudojournalism, and it belies the basic ethical principles held near and dear by professional news workers. Such fake news is ubiquitous online, and microbiomania is a favorite topic of its most egregious offenders.

Ever the popular wellness crank, Dr. Joseph Mercola is a great example of pseudojournalism. He loves to tout the magical control the microbiome has over every aspect of human health. A cynical person might note that this microbiome hype plays nicely into the products he sells.

According to articles on his website, your gut bacteria are responsible for obesity, diabetes, depression, autism, anxiety, schizophrenia, cancer, asthma, and arthritis, among other random ailments. The solution? Buy his book! (He ends every article on human microbiota with a sales pitch for *Fat for Fuel*.) It apparently contains secrets doctors don't want you to know about. With enough leafy greens, fermented vegetables, and probiotics, you too can prevent disease! Only $16.99 and it comes with six free gifts!

And, no lie, Dr. Mercola recommends you stop showering for optimal microbial health. According to him, our overzealous hygiene is throwing our skin microbiota out of whack. He also claims that the only reason we stink when we don't wash our skin or hair (or wear deodorant) is because we currently keep too clean. And if we just stop altogether, eventually the smell goes away. (I'm sure our coworkers would beg to differ.) And if that weren't enough, he adds insult to injury by recommending bacterial sprays and probiotic soaps. (The creator of one such spray, which contains "friendly ammonia-oxidizing bacteria," hasn't showered in over a decade.)

Many of the red flags of pseudojournalism in medical reporting are here. These red flags include hyped claims that sound fantastical. They typically offer simple solutions to either complex problems or many different problems simultaneously. They generally trade in conspiracy theories and distrust of mainstream sources and experts. They promise secret or suppressed information. And perhaps the biggest red flag—at the end of the day, there is always a sales pitch.

It's difficult to top Mercola's kind of crazy, but if he has met his match in anyone, it's Deepak Chopra. His take on the human microbiome is nothing short of laughable. Skeptic and journalist Kavin Senapathy says it best: "What we know about microbiome is dwarfed by what we don't know, and Deepak Chopra takes liberties to fill in the gaps." And in January 2016, he did so in spectacular fashion. He actually claimed—in public no less—that our bacteria listen to our thoughts.

At an online conference called the Fat Summit (a sort of who's who in health quackery), event host Mark Hyman interviewed Chopra. Hyman said, "I love yoga, and I do it, and I always feel transformed, and it's amazing that not only your genes are listening to your thoughts, but your microbiome, the bacteria are listening to your thoughts." Chopra agreed: "Yeah, the bacterial genes are listening to your thoughts."

I'll just leave that right there.

51. Reporting Epigenetics

One of the things about genetics that has become clearer as we've done genomes—as we've worked our way through the evolutionary tree, including humans—is that we're probably much more genetic animals than we want to confess we are.   —Craig Venter

———————

"Epigenetics" is a recent science buzzword, which should make you very cautious when reading any popular reporting about it. The newer a scientific concept is, the easier it is to exploit for hyped gee-whiz reporting. Even sincere science journalists can be taken in, if they aren't careful to put the new concept into context and really check their understanding.

For example, reporting in the *Advertiser* (an Australian news outlet), Brad Crouch gives a fairly dramatic example of some common tropes of bad science reporting. Crouch's byline says he is a "medical reporter," so he really should know better.

Crouch is reporting in 2014 on a review article on epigenetics recently published in *Science*. It's honestly difficult to see how Crouch got from the review article on which he was ostensibly reporting to his news article. He begins:

LANDMARK Adelaide research showing that sperm and eggs appear to carry genetic memories of events well before conception, may force a rethink of the evolutionary theories of Charles Darwin, scientists say.

Reporters love to suggest that new findings threaten to overturn established science, especially if they can invoke a recognizable name, like Darwin. Crouch also goes on to say that perhaps this research supports Lamarck's theory of evolution rather than Darwinian natural selection. First, the paper is not research, let alone "landmark" research. It is a review article. It's not even a systematic review or a meta-analysis, which a reporter might be forgiven for calling a "study"—it's just a discussion of the topic of epigenetics. That's journalism fail #1: failing to properly identify what kind of paper is being reported on—new research, a review, or opinion.

What is especially interesting is that the words "evolutionary theory," "Darwin," and "Lamarck" appear nowhere in the paper, despite being the focus of Crouch's review article. Here is what the review actually says:

Transgenerational epigenetic effects interact with conditions at conception to program the developmental trajectory of the embryo and fetus, ultimately affecting the lifetime health of the child. These insights compel us to revise generally held notions to accommodate the prospect that biological parenting commences well before birth, even prior to conception.

Essentially, epigenetics are tweaks to the expression of genes in response to environmental factors. If a mother is, for example, living in times when there is good access to food, epigenetic factors will adapt her children to the current abundance. If she is living in lean times, they will be more adapted to food scarcity.

One biochemical mechanism of epigenetic factors that has been discovered is methylation of base pairs (the fundamental units of DNA). This doesn't affect the sequence of genes, but it can affect their expression.

While epigenetics is still fairly new, and the details are yet to be worked out, the experiments so far show that it's a short-term (one or a few generations) tweak to adapt to immediate conditions, and it doesn't have any real impact on Darwinian evolution. It certainly does not indicate that the inheritance of acquired characteristics (often referred to as Lamarckism, even though this notion did not originate with Lamarck, was not unique to Lamarck, and was actually a minor aspect of Lamarck's evolutionary thinking) will now replace Darwinian evolution.

I am not even sure I buy the much more limited conclusion of the review that healthful lifestyles are transmitted to children. This seems like an awful lot of extrapolation from limited research. It is plausible and consistent with existing research that being obese or overeating during pregnancy may cause an epigenetic signal of abundance, leading to children with a greater than average tendency to put on weight. What isn't clear is if this is clinically relevant in humans. I also wouldn't generalize this to "healthful lifestyle."

As the article continues, Crouch goes way beyond overcalling epigenetics and takes a left turn into bizarro world:

> It paves the way for a review of the work of French biologist Jean-Baptiste Lamarck, whose theory that an organism can pass to its offspring characteristics acquired during its lifetime was largely ignored after Darwin's publication of *On the Origin of Species* in the mid-1800s, that work defining evolution as a process of incidental, random mutation between generations.

Wrong. This wasn't Lamarck's theory, but worse still, epigenetics has absolutely nothing to do with the mechanisms of evolutionary change. Crouch does then pivot to the actual subject of the review, the effect of parental habits on the child. He quotes one of the authors, Sarah Robertson:

> "People used to think that it didn't matter because a child represented a new beginning, with a fresh start," she said.
>
> "The reality is, we can now say with great certainty the child doesn't quite start from scratch. They already carry over a legacy of factors from their parents' experiences that can shape development in the foetus and after birth.

"There is now biological evidence that memories of experiences in adults can be transferred through egg and sperm for the lifetime prospects of the child.

"If evolution has developed something like this it can give a child an edge to survive. This will rewrite long held views, that experiences can actually be transferred to offspring."

You will notice that Robertson never says anything to support either the notion that epigenetics calls Darwinian evolution into question or the hyped interpretation that lifestyle transfers to children. Everything she says is compatible with a conservative interpretation of epigenetics.

Here we have journalism failure #2: relying on "common knowledge" about complex scientific topics. In this case Crouch is discussing the review article in the context of what is essentially a myth about Lamarck. He didn't check his background assumptions.

Journalism failure #3 is falling for a common science-reporting trope, hyping how much a new idea or finding contradicts an established pillar of science. Reporters love to ask, "Was Einstein wrong?" or "Do we need to rethink Darwin?"

Journalism failure #4 is reporting on a scientific concept that Crouch clearly doesn't understand. This is where science journalists really need to interview experts, and not just to mine them for quotes but also to check their own understanding of a topic. Further, they should interview experts not involved in the paper on which they are reporting, to see if the scientist is a lone crank or in the mainstream.

Crouch continues to push his clickbait narrative:

Prof Robertson stressed that genes remain the blueprint for a new baby, but said the work of both Darwin and Lamarck may need to be reconsidered.

"The genes are the blueprint and that won't change," Prof Robertson said. "But this is at another level, it is the decoration of the gene, the icing on the cake if you like, a gift to offspring that gives them another layer of information about survival."

She says nothing about Darwin and Lamarck, just like the review says nothing about Darwin and Lamarck. The statements that the author uses to support his interpretation in fact contradict it. Epigenetics is another level of information, which doesn't change the basic fact that genes are the "blueprint."

As an aside, I don't like the "blueprint" metaphor for genes. It's misleading. Genes are not a blueprint. There is no representation of the final product in the genes. Genes, rather, are more of a recipe, a set of instructions that, if followed, result in the final organism. Yes, this is nitpicking, but metaphors in science communication should strive to be conceptually accurate.

Epigenetics is not a new concept, as one might think from reading Crouch's article, but it is still relatively young, scientifically speaking, and there is likely scant public awareness of the field. Reporting on this review was a good—and missed—opportunity to introduce readers to the concept of epigenetics and put it into some historical and scientific perspective.

Anyone reading Crouch's piece who was not already familiar with evolutionary theory and epigenetics wouldn't have come away with any real insight, but rather completely misinformed. That's a good overall marker for the quality of a piece of science journalism. Will the average reader be better educated about the relevant issues or confused and misled? Good science journalism would even go the extra step of anticipating common myths, tropes, or misconceptions and correcting them.

Here's Crouch's headline: "Darwin's Theory of Evolution Challenged by University of Adelaide Genetic Memory Research, Published in Journal Science."

Here's the real headline: "Local Scientists Write a Review Article on the Decades-Old Concept of Epigenetics." Okay, not so exciting, but you get my point.

SECTION 4

Death by Pseudoscience

Among the most difficult types of e-mails we receive from listeners of the show are those that are essentially pleas for help. They are often from people who have a loved one who is ill and, in desperation, is falling prey to dangerous quackery. We hear from people who are trying to deal with a spouse who does not want to vaccinate their children. Or perhaps their family member is just obsessed with some pseudoscience and is wasting their time and money chasing a fantasy.

We frequently hear direct stories of the harm that a belief in nonsense can bring. So we take it very seriously when we are also frequently asked, about belief in magic, "What's the harm?" The implication is (and sometimes the listener states this explicitly): So what if people believe in ghosts or aliens? Let people have their fun or comforting beliefs. You're just going around popping people's bubbles. Do you tell young children that Santa Claus doesn't exist?

Um…No.

If you're reading this guide, you likely want to think for yourself and be purged of misinformation, sloppy thinking, and false beliefs. But if you also want to believe in magic and fairy tales, go right ahead. Personally, I like separating my entertaining fantasy from reality.

As skeptics, we don't get too involved in what people believe. Seriously, what you believe is a personal matter that happens inside your head. Where scientific skepticism gets involved is when people make claims, have a pretense to knowledge, or state something as a fact or as if it's a valid conclusion of logic or science. If you make any such claims, you are stepping into the arena of science, so don't complain if you are held to its clearly posted rules. (Okay, they may not be clearly posted, but they are there, and we'll happily point them out.)

We also get involved when people try to impose their beliefs on others or say that we should run society according to their beliefs. If you think your beliefs should hold sway, then questioning the basis for those beliefs is fair. Also, part of our mission is consumer protection. There is a lot of fraud and potential harm out there, and for those who want to protect themselves we offer our critical analysis of specific claims and guidelines for how to avoid being scammed.

One of those guidelines is not to fall for the notion that there is no potential harm to believing that fantasy or nonsense is real. We live in a complex world and have to make a lot of complex decisions, individually as a family or small group and collectively as a society. We're better off if those decisions are based on reality.

The "What's the harm?" question is itself potentially harmful, because it trivializes the real challenge people face trying to understand complex questions that inform important decisions. It also whitewashes, and even excuses, the direct harm done by many con artists and charlatans.

There are countless dramatic examples of harm, even fatal harm, coming from believing that fantasy is real. That's why one of our informal segments on the SGU is called "Death by Pseudoscience," to occasionally highlight some of those examples. There may be no harm in believing in magic, but there can be from acting on it. The more exposure these cautionary tales get, perhaps the fewer e-mails we'll receive from desperate listeners.

52. Death by Naturopathy

Naturopathy is nothing more than a hodge-podge of mostly unscientific
treatment modalities based on vitalism and other prescientific notions
of disease.   —David Gorski

You don't often hear of heart attacks in thirty-year-old women, especially those whose primary medical complaint is eczema, an itchy yet
otherwise benign skin rash. So when Jade Erick lost her life in March
2017, the San Diego medical examiner went to work. The official
cause of death was found to be "anoxic encephalopathy due to prolonged resuscitated cardiopulmonary arrest due to adverse reaction to
infused turmeric solution."

Yes, you read that correctly. An intravenous infusion of turmeric
caused a young, otherwise-healthy woman to have a heart attack, and
while she was down, her brain was starved of oxygen and she died. For
those of you unfamiliar with the aromas and flavors of Indian cuisine,
turmeric is a spice often used in making curry, and it can be found in
dishes all over the Middle and Far East. It has a yellow-orange tint
and a wonderfully aromatic, slightly bitter taste. It is frequently used
in place of pricier saffron or blended with other spices.

But turmeric is a food, not a drug.

The spice is made from the plant *Curcuma longa*, and although it's
been used in traditional Ayurvedic (a system of treatment common

in India) practices for thousands of years, modern science has shown little promise for its efficacy. A recent review in the *Journal of Medicinal Chemistry* looked at over 120 clinical trials of curcumin, the major chemical component of turmeric. The authors conclude: "No double-blinded, placebo controlled clinical trial of curcumin has been successful... Curcumin is an unstable, reactive, nonbioavailable compound and, therefore, a highly improbable lead."

Curcumin is an experimental agent—it isn't FDA approved to prevent or treat any disease. In fact, the FDA listed curcumin on its website as one of "187 Fake Cancer 'Cures' Consumers Should Avoid."

Curcumin is a good example of why you should be generally skeptical of herbal remedies. Curcumin has almost no bioavailability, which means no active ingredients get into the blood, where they can have an effect. Further, curcumin has properties that cause it to react to many experimental assays. So if you're testing it in a petri dish, it seems to do everything. When you give it to animals or people, it does nothing.

Yet in the unregulated, pseudoscientific world of naturopathy, turmeric is a magical panacea. Most notably, it's touted as a powerful anti-inflammatory agent, with properties claimed to treat not only pain but also stomach, skin, liver, and gallbladder problems, arthritis, fatigue, cancer, colitis, diabetes, and even Alzheimer's. But this type of cure-all rhetoric is hardly specific to turmeric. Naturopaths have a long and storied history of promoting quack cures with no scientific backing.

The American Association of Naturopathic Physicians defines naturopathy as a "distinct primary health care profession, emphasizing prevention, treatment, and optimal health through the use of therapeutic methods and substances that encourage individuals' inherent self-healing process." Quackwatch defines naturopathy as "a largely pseudoscientific approach... riddled with quackery."

Modern naturopathy reared its head in America around the turn of the twentieth century, and as of this writing, eighteen states offer legal licensure to "certified" practitioners. Most states deny prescriptive and surgical authority to naturopathic practitioners, and they disqualify them from using the term "physician." This doesn't, however, prevent them from calling themselves "doctors"—a distinction many prospective patients are unlikely to make. In Colorado, naturopaths can suture wounds. In Oregon, Montana, Utah, and Washington they may use local anesthesia. In Vermont they can perform episiotomy and perineal repair associated with "naturopathic childbirth." And in California, "a naturopathic doctor may utilize routes of administration that include oral, nasal, auricular, ocular, rectal, vaginal, transdermal, intradermal, subcutaneous, intravenous, and intramuscular."

This brings us back to Jade Erick, dead at thirty by a turmeric infusion. The naturopath responsible has been named by the San Diego medical examiner as Kim Kelly, whose archived website includes the following boast:

> I am excited to announce that I've started administering intravenous curcumin…People have used it to help with pain, inflammation, immune system, arthritis, liver conditions and cancer. It's also been found that intravenous curcumin in combination with vitamin C and glutathione (I call it the "Mother of All Antioxidants") has a potentiating effecting [*sic*] in helping people with chronic health conditions e.g. hepatitis C, liver fibrosis…If you are suffering from any type of inflammatory condition, whether it be arthritis, autoimmune condition (e.g. scleroderma, lupus or rheumatoid arthritis), Alzheimer's or dementia, this may be a great modality of treatment to try…The safety, tolerability, and nontoxicity of curcumin at high doses have been well established by human clinical trials. Promising effects have been observed in patients with various pro-inflammatory diseases.

No evidence is provided to back up his claims. Even the website for the National Institutes of Health's National Center for Complementary and Integrative Health, historically sympathetic toward naturopathy and its practices, says, "Claims that curcuminoids found in turmeric help to reduce inflammation aren't supported by strong studies."

Naturopath-turned-skeptic and journalist Britt Marie Hermes has written at length about both Ms. Erick's particular case and the harms perpetrated by naturopathy at large, stating:

> Naturopathic doctors frequently offer treatments that have not been fully vetted for safety or effectiveness, and many therapies used in naturopathic practice have been disproved by rigorous trials. This is likely the result of naturopathic education blurring the line between treatments backed by good evidence and practices using "natural" substances that turn profits.

Common pseudoscientific practices promoted by naturopaths include homeopathy, detoxification, biotherapeutic drainage, chiropractic, supplements and nutraceuticals, colon hydrotherapy, iridology, chelation, acupuncture, and rolfing (among others). Most alarming, Ms. Hermes points out, is another form of intravenous therapy, "oxygenation" therapy using ozone gas or hydrogen peroxide, both of which can and have killed. They are not backed by sound science, nor are they FDA approved (or legal, for that matter).

Oxygenation therapy is a good illustration of how incoherent naturopathic philosophy can be. Ozone is described by the FDA as "a toxic gas with no known useful medical application in specific, adjunctive, or preventive therapy." Simply put, ozone is O_3, three oxygen atoms bound together. It's a highly reactive oxygen molecule with two potential effects: providing more oxygen to tissues and using oxidative reactions to kill bacteria, viruses, and cancer cells. Therapies involving hydrogen peroxide (H_2O_2) follow a similar reasoning.

This is ironic because it flies in the face of the whole antioxidant craze, and it brings up a major weakness of that fad. Oxygen and oxygen free radicals are a two-edged sword. Tissues need oxygen, and they use oxygen free radicals as part of the immune system to kill invaders, but the free radicals also damage host tissue. So there needs to be a delicate balance. Pushing this balance in one direction or the other with antioxidants, ozone, or hydrogen peroxide is unlikely to provide side-effect-free benefits and may cause more harm than good.

That said, ozone and hydrogen peroxide do have legitimate uses as a surface sterilizer and external disinfectant (mainly topical, dental, or on blood products). In the laboratory, ozone has been shown to inactivate HIV, but so do many other things that aren't useful in therapy. These so-called "treatments" are ineffective against cancer, HIV, multiple sclerosis, or the myriad other disease states their naturopathic practitioners claim they cure. The further claim that they detoxify the body is pure pseudoscience. Nonspecific detoxification itself is a scam, whether by oxygenation or any other method.

Although it's impossible to tally the overall number of casualties involved, there are multiple documented deaths from ingestion or intravenous infusion of oxygenation products. And although neither hydrogen peroxide nor ozone are legal to administer by naturopaths, the therapies continue. In fact, as recently as December 2015, the American Association of Naturopathic Physicians offered continuing education credits for a webinar called "Clinical Applications of Medical Ozone Therapy."

Ozone therapies are just as lacking in rationale and evidence as the IV turmeric that caused the death of Ms. Erick. Both are examples of the absence of a good science-based standard among naturopaths. They eschew science and evidence for philosophy, the appeal-to-nature fallacy, and their own anecdotal experience and wisdom. But they still want the patina of science, because it sells well. This kind of approach leads them to embrace treatments like curcumin that have misleading basic science studies but negative or lacking clinical trials.

Of course, even legitimate treatments can sometimes cause harm. That's why real medicine is based on a careful and evidence-based risk versus benefit approach. You can't do a risk versus benefit analysis, however, unless you have some statistics about both. This only comes from rigorous clinical studies in the context of scientific plausibility.

Carefully considering the science and evidence can be hard work. It's much easier to just go with instincts. Naturopaths essentially shoot from the hip, and sometimes they shoot to kill.

53. Exorcism: Medieval Beliefs Yield Medieval Results

Nothing in life is to be feared, it is only to be understood. Now is the time to understand more, so that we may fear less.   —Marie Curie

———————

The skeptical, critical thinking, and scientific principles outlined in this book are like the rungs of a long ladder that humanity has used to climb laboriously out of the swamp of superstition, bias, hubris, and magical belief. We tend to look back now at medieval beliefs and congratulate ourselves for being born in a later age. But not every individual or even every institution has followed our best thinkers out of the muck. We all need to climb the ladder for ourselves.

It does seem amazing that in the twenty-first century there are still those who embrace the worst possible nonsense that infected our thinking in more primitive times. And yet belief in things like demons and possession continues in some quarters, victimizing those who get caught in this web of ignorance, often innocent children.

An exorcism is a religious ceremony, ritual, or practice to rid a person, a place, or an object of a demon, a devil, or some other evil spiritual entity believed to be inhabiting it. A person, place, or object is said to be "possessed" if they're deemed to have been so infected.

Due to the popularity of the book and the movie *The Exorcist*, exorcism tends to be associated with Christianity in the minds of most people from Western cultures. However, exorcism isn't exclusive to the cache of religions under the Christian canopy. Exorcism is practiced as part of Islam, Hinduism, Judaism, and Taoism, to name a few.

The Roman Catholic Church believes in "diabolic possession," and its priests still practice what is called "real exorcism," a twenty-seven-page ritual to drive out evil spirits. The ritual involves the use of holy water, incantations, various prayers, incense, relics, and Christian symbols such as the cross.

Interest in exorcisms has actually been increasing in recent years. There are 500–600 Catholic exorcists worldwide. In the US the number increased from twelve to fifty between 2006 and 2016. There is no central database of how many exorcisms are performed, but one active priest, Father Gary Thomas, the official exorcist of the Diocese of San Jose in Northern California, claims to have performed fifty to sixty exorcisms over that ten-year period.

While there are official rules and regulations for performing the exorcism ritual, and officially recognized exorcists who follow them, there are plenty of unqualified members of any given church who unofficially take it upon themselves to perform exorcisms. In many of these cases it's the parishioners and worshippers—the most fervent believers in diabolic possession—who take it upon themselves to determine that a person might be possessed and that they (being the hard-core devotees of the religion) are capable of exorcising the evil forces from the possessed one. And why not? In the case of Roman Catholicism, the concepts of possession and exorcism emanate from the highest authority, the Pope.

In 1999 the Vatican did update its 1614 guidelines for expelling demons (it only took 385 years to see the necessity of "updating"), urging exorcists to avoid mistaking psychiatric illness for possession. It is debatable if these updates have had the desired effect. As paranormal investigator Joe Nickell pointed out in his February 2001 article in *Skeptical Inquirer* on the subject of exorcisms:

Belief in spirit possession flourishes in times and places where there is ignorance about mental states. Citing biblical examples, the medieval Church taught that demons were able to take control of an individual, and by the sixteenth century demonic behavior had become relatively stereotypical. It manifested itself by convulsions, prodigious strength, insensitivity to pain, temporary blindness or deafness, clairvoyance, and other abnormal characteristics. Some early notions of possession may have been fomented by three brain disorders: epilepsy, migraine, and Tourette's syndrome (Beyerstein 1988). Psychiatric historians have long attributed demonic manifestations to such aberrant mental conditions as schizophrenia and hysteria, noting that—as mental illness began to be recognized as such after the seventeenth century—there was a consequent decline in demonic superstitions.

While the church professes to rule out psychiatric or neurological disorders before declaring someone truly demon-afflicted, this isn't really possible. In the end their evidence is the behavior of the person, who among other things could simply be acting out what they believe a possession should look like. Sometimes belief in possession is based entirely in others, not the person allegedly possessed. These innocent people become the victims of abuse, torture, and death by exorcism.

Here are just a few gruesome examples:

In 2014, Zakieya Avery of Germantown, Maryland, mother of four children aged eight, five, two, and one, stabbed and choked her two youngest children to death in front of the two older children. She then stabbed the two older children. She told investigators that evil spirits were moving between the bodies of the children, and with the help of an adult housemate, she had attempted to exorcise the spirits.

In 2012, four women from South Africa—Fundiswa Faku, Lindela Jalubane, Nokubonga Jalubane, and Nonhlanhla Mdletshe—attempted to exorcise a demon from Faku's fifteen-year-old cousin, Sinethemba Dlamini. In the fervor and struggling of the victim, she succumbed to death only after her intestines were ripped from her body through her vagina.

In 1976, one of the most famous cases of death by exorcism, Anneliese Michel of Germany, was documented. Anneliese's story is dramatized in the films *The Exorcism of Emily Rose* (2005), *Requiem* (2006), and *Anneliese: The Exorcist Tapes* (2011).

Her parents had become convinced that Anneliese was possessed by demons. She was epileptic and previously had been treated for depression. Despite this, the parents called upon the services of two priests to perform the exorcism ritual. She died several months later, the cause of death determined to be dehydration and malnutrition. Her parents, with the guidance of the priests, effectively caused her to stop eating and drinking, and Anneliese slowly died a horrific, prolonged death. A court found the parents and priests guilty of manslaughter.

Even when exorcism doesn't lead to death, there is tremendous potential for harm. Perhaps the worst thing you can do to a delusional psychiatric patient, for example, is to confirm their delusions. Telling them they are actually possessed by demons will simply exacerbate their illness, severing any tenuous attachment they have to reality. It would be like telling a paranoid patient that the CIA really is listening to their thoughts through the fillings in their teeth.

This torture is inflicted because of so-called "evidence" presented for possession, which is always entirely unconvincing. I have seen many exorcisms recorded on tape, and always the victim is either doing nothing extraordinary or doing a very bad job of auditioning for the sequel to *The Exorcist*. Never do we see anything preternatural or even slightly compelling. There never has been a single proven case of possession in recorded history. And yet there are millions, arguably billions, of people living today who have no problem accepting the ideas of "possession" and "exorcism" as being real.

This is what the world is like with some skeptical principles as a guide and caution. While demons and devils may not exist, people's ability to do great harm in the name of nonsense is all too real.

54. Death by Denial

Science is a way to teach how something gets to be known, what is not known, to what extent things *are* known (for nothing is known absolutely), how to handle doubt and uncertainty, what the rules of evidence are, how to think about things so that judgments can be made, how to distinguish truth from fraud, and from show.   —Richard Feynman

Sometimes harm comes from embracing pseudoscience, and sometimes it comes from denying established science. One of the most pernicious and frustrating manifestations of the latter is HIV denial.

On June 5, 1981, the CDC published a report of five cases of a rare pneumonia (PCP pneumonia) in five otherwise-healthy young gay men. In 1982 it was recognized that these and other patients were suffering from an immune deficiency syndrome and opportunistic infections. In 1983 a retrovirus was discovered as the probable cause of this new syndrome (AIDS—acquired immunodeficiency syndrome). The virus was initially called lymphadenopathy associated virus but was later renamed human immunodeficiency virus (HIV). In 1985 the first blood test for HIV was approved, and in 1987 the FDA approved the first drug for the treatment and prevention of HIV—AZT.

Over the next twenty years scientists developed what is called highly active antiretroviral therapy, or HAART, treatment. It is a cocktail of drugs designed to interfere with HIV at multiple steps in its infection and replication. As these drugs continue to be developed, the life expectancy of those with HIV increases. By 2013 data showed that those with HIV who start HAART treatment early have essentially a normal life expectancy. In thirty years HIV went from a short-term death sentence to a manageable chronic illness.

There have been only a few "cures" of HIV, meaning that those with a documented infection are now virus free after treatment. One such cure was effected with a bone marrow transplant, another with high-dose HAART in an infant infected at birth. These approaches will likely not be applicable to most patients with HIV.

Research into an HIV vaccine is ongoing. HIV is particularly challenging for various reasons—it mutates throughout the infection, it evades the immune system, and it can hide in a dormant state. The best we've done so far is a phase III trial showing 31 percent efficacy in preventing initial infection. This is a modest result, but it shows we are making progress.

In short, HIV is one of the stunning success stories of modern medicine. Science rapidly discovered the nature and cause of AIDS, we've had a steady increase in our understanding of this virus, it is now a manageable disease, and we continue to make progress toward a vaccine and even a possible cure.

This obvious success has not prevented people from denying the science.

In the 1980s, HIV denial—denying that HIV was a real virus that was primarily responsible for AIDS—was unjustified pseudoscience, but the deniers could legitimately point to the fact that existing treatments were not very effective. The real test of any scientific theory is how useful it is going forward. Does it make useful predictions? In the case of medicine, does it lead to useful treatments or other interventions?

Essentially we conducted a meta-experiment. If HIV was actually the cause of AIDS, then the research derived from that premise would likely bear fruit. We would expect to deepen our understanding of

AIDS and hopefully develop treatments that were increasingly effective. The deniers predicted that HIV was a scientific dead end, and we would only worsen outcomes by exposing AIDS patients to toxic drugs that were of no use.

In the last three decades it has become clear that the HIV theory is not only viable, it's a remarkable success. Drugs designed specifically to target the mechanisms of HIV have had dramatic clinical benefits for patients.

HIV denial seemed to wane a bit in the late 1990s and early 2000s because the older crop of deniers were (to be blunt) dying off. It was also difficult to recruit new deniers, given how effective the standard treatment was becoming. In 2009 Christine Maggiore, founder of HIV denialist group Alive & Well AIDS Alternatives, died of pneumonia, likely as a complication of HIV. Before her death, Maggiore's dedication to HIV denial had other tragic consequences. She decided to breastfeed both her children, despite the fact that breastfeeding increases the risk of contracting the virus. Her daughter, Eliza, died at the age of three, apparently from pneumonia that was likely an opportunistic infection due to advanced AIDS. Maggiore never had her children tested or treated for HIV. Her pediatrician, interestingly, was anti-vaxxer Dr. Jay Gordon. He claims to support the conclusion that HIV causes AIDS, but his website used to contain some squirrelly comments on HIV that suggested he may have some denialist sympathies.

The autopsy report is unequivocal. There was strong evidence that Eliza had an advanced case of AIDS from HIV, including AIDS encephalitis (brain infection). Her cause of death was pneumocystis carinii pneumonia (PCP pneumonia) caused by AIDS—this is a common opportunistic infection in AIDS but it's extremely rare in an immunocompetent individual. The autopsy left no reasonable doubt that Eliza died from complications of AIDS.

Maggiore never acknowledged that her daughter's death was due to HIV or AIDS. The HIV denial community rallied around her, claiming the death was due to an allergic reaction to amoxicillin. Denial is a powerful thing.

But denial doesn't just affect individuals. When denialism occurs at a government level, the harm can be magnified thousands of times.

There is an AIDS epidemic in Africa, with estimates as high as 10 percent of the population being affected. Efforts to fight it are hampered by the endemic social problems of that continent. Chief among them is the lack of sufficient modern health resources, but this is greatly exacerbated by a history of HIV denial, especially in South Africa, where it is estimated that 18.8 percent of the adult population is infected with HIV.

From 1999 to 2008 Thabo Mbeki was president of South Africa, and during his tenure he maintained doubt about HIV as the cause of AIDS and dramatically harmed efforts to control and treat HIV in his country. Problems began in 2000 when he announced that he was unconvinced as to the true cause of AIDS. He then put together a panel to explore the issue, packing it with HIV deniers. Most prominent among them was Peter Duesberg, a German American molecular biologist who doubts the role of HIV in causing AIDS. They, of course, confirmed his denial.

Acting on this doubt, his government denied HAART medications to patients with HIV. Mbeki delayed and cut funding for testing, medication, and other treatments. He also appointed a health minister, Manto Tshabalala-Msimang, sympathetic to his position.

Tshabalala-Msimang used politics as cover for what were essentially pseudoscientific positions. For example, she warned that, "We cannot use Western models of protocols for research and development," a convenient way to use national and cultural pride to deny science. Instead of proven "Western" therapies, she favored indigenous medicine, such as garlic and beetroot extracts, to treat HIV. These treatments, which were not based on any scientific plausibility, proved worthless.

When challenged about the lack of evidence to support beetroot and other such treatments, she again criticized "Western" medicine for being "bogged down" in clinical trials. Doing research only delayed getting effective treatments to patients, she argued. This is, sadly, not an uncommon position, as if we can rely on anecdotes and tradition to determine what works, and any scientific trials are just a waste of time.

As a clear demonstration of how political the issue had become, in 2004 Mbeki characterized AIDS as a "disease of racism." Writing for the *New York Times*, Roger Cohen summarized Mbeki's position as being that HIV/AIDS was "a fabrication foisted on Africans by whites determined to distract the continent from real problems of racism and poverty, and accepted by blacks afflicted with the slave mentality engendered by apartheid."

The deputy health minister, Nozizwe Madlala-Routledge, tried to mitigate these disastrous policies by finding ways of providing effective treatment to South Africans with HIV. In response, Mbeki fired her.

A 2008 study by Pride Chigwedere and others found that the policies of Mbeki between 2000 and 2005 resulted in more than 330,000 premature deaths, with a loss of 2.2 million person-years (the number of years those who died collectively would have lived). Further, 35,000 babies were born with HIV that could have been avoided with standard treatments.

That is a lot of death by pseudoscience.

There is a somewhat happy ending to the story. While nothing can reverse the unfortunate losses and ruined lives from Mbeki's science denial, he did leave office in 2008. His successor appointed a new health minister, Barbara Hogan, who immediately reversed Mbeki's policies on HIV/AIDS and started providing treatment to patients.

There is no denying the danger of pseudoscience and denial. This extends beyond HIV and AIDS—there are still those who deny the germ theory of disease, that vaccines are safe and effective, or that chemotherapy can be an effective option for certain cancers. The tobacco industry did its best to deny the connection between smoking and lung cancer. The denial of mental illness leaves many vulnerable individuals untreated.

It is especially tragic when we have the science—we did the research, did the hard work, and have real solutions for people who need them. But stubbornness and ideology can't help but get in the way.

55: Suffer the Children

Isn't this enough? Just this world? Just this beautiful, complex, wonderfully unfathomable natural world? How does it so fail to hold our attention that we have to diminish it with the invention of cheap, man-made myths and monsters? .

—Tim Minchin

What could possibly motivate a parent to harm their own child—to withhold from them needed medical care, to watch them suffer and even die when help is available? As a parent myself, I find it unfathomable, and yet it happens all too often. As you might guess by now, it only takes an alluring narrative or a compelling anecdote, coupled with a lack of critical thinking, to create such heartrending tragedies.

What is especially painful about such cases is how wide-ranging the harm can be and how profound the failure of those who should have protected the victims. Parents, of course, are often the focal point of bad decision-making. Because of religious beliefs, sometimes just being enamored of New Age thinking or maybe a little too suspicious of the mainstream, they fail to provide the basic medical care that it is every parent's duty to provide for all their children.

Often there are practitioners involved too—preachers shaming the parents into not losing faith, or practitioners of dubious medical treatments making unfounded promises. There may also be regulators who fail to make the child's interest paramount and effectively allow the parents to neglect them.

Sadly there are many examples we could give, but here is one to show how the power of belief can bring a parent to neglect or harm their own child.

When Candace Newmaker was ten years old, she was wrapped tightly in a flannel sheet. Three adults, including her adoptive mother, Jeane, stood by and taunted her as she struggled to break free from this restricting cocoon. Also present were Connell Watkins and Julie Ponder, self-proclaimed experts in reactive attachment disorder, from which they believed Candace suffered. Over the next forty minutes, all captured on video, Candace begged to be set free, complained that she couldn't breathe, and even vomited and choked. The adults, deaf to her pleas, just kept taunting her. After forty minutes Candace went silent, and yet after that point the adults attending her stood around chatting for another half hour. At that point they decided to see how Candace was doing. When they unwrapped her she was blue and not breathing, and when they checked for a pulse, nothing. Candace had asphyxiated and was later pronounced brain-dead.

You may think the adults in this scene were sadists, but they apparently meant well. Such is the perverse power of pseudoscience.

Candace was having some behavioral problems and her mother, a pediatric nurse, attended a conference promoting so-called "attachment therapy." The idea is that children who were neglected and abused when young may fail to form a bond with their adoptive parents, resulting in behavioral problems. The "therapy" is intended to form such an attachment.

This is, unfortunately, not the only death at the hands of this dubious therapy.

According to Nancy Thomas Parenting and Families by Design (attachment.org):

A child with Reactive Attachment Disorder needs treatment for the condition they have. Children with this disorder exhibit extreme levels of behavior and manipulation. A therapist who is not trained to deal with these extremes can do more harm than good.

Reactive attachment disorder (RAD) is recognized as a psychiatric diagnosis. Like all diagnoses in *DSM* (*Diagnostic and Statistical Manual of Mental Disorders*, the standard manual of mental illnesses) it is based on clinical features. These features can be nonspecific, however, especially in older children, where the diagnosis is more controversial. According to the Mayo Clinic:

> With treatment, children with reactive attachment disorder may develop more stable and healthy relationships with caregivers and others. Treatments for reactive attachment disorder include psychological counseling, parent or caregiver counseling and education, learning positive child and caregiver interactions, and creating a stable, nurturing environment.

You will notice that nowhere in that description of standard treatments is wrapping the child in a sheet so they can't breathe. Self-described attachment therapists employ a controversial theory and derived treatments for RAD that are not supported by published science. But they feel that they're uniquely qualified to treat children with this diagnosis.

Further, they tend to subscribe to a concept of RAD that sees such children as manipulative. While this may be true, it's very tricky to determine if someone is truly being manipulative, and professionals know to be particularly careful and skeptical when making such a determination. The problem is that the conclusion that someone is "manipulative" can be self-fulfilling. Any behavior can be perceived as manipulative. In this way it's like a conspiracy theory—any counter-evidence or lack of evidence is just more evidence of manipulation.

This is exactly what happened to poor Candace. Convinced by the seminar that Candace had RAD and needed attachment therapy, her

428 THE SKEPTICS' GUIDE TO THE UNIVERSE

mother brought her to Evergreen, Colorado, for two weeks of intensive attachment therapy. During treatment sessions, Jeane would lie on top of her much smaller daughter to restrain her while she licked her face. Candace would have her face held while "therapists" screamed at her. These interventions culminated in the "rebirthing" therapy in which Candace was wrapped in a sheet, simulating the birth canal, so that she could be "reborn."

What I think this case most represents is the absolute hubris of many gurus. Those practicing on the fringe think they have a special gift, talent, or insight. They don't need scientific evidence, because they have intuition. They don't need to follow standards of safety, informed consent, evidence-based medicine, or even basic ethics, because they are healers.

Professions have standards for a reason, and that is at the core of what makes them professional in the first place. In the case of Candace, her therapists ignored basic safety standards because they were so cocky about their diagnosis and their methods. They interpreted Candace's pleas as manipulation. Her insistence that she couldn't breathe, and her subsequent silence, were just more of the same.

One clear lesson my decades of medical experience have taught me is the need for humility. Our collective knowledge is imperfect, and my individual knowledge is imperfect. We have to try our best and double-check our understanding, be careful with our interventions, and always put the patient's interests first. Candace was killed largely by the unchecked arrogance of those who promised to treat her.

On April 21, 2001, Connell Watkins and Julie Ponder were convicted of reckless child abuse and sentenced to sixteen years in prison in the death of Candace.

SECTION 5

Changing Yourself and the World

Hopefully this has been a fun and fascinating adventure through the vagaries of human psychology and the challenges this creates for our humble species. While I think these topics are endlessly interesting, the real purpose is to apply them to your life. I know this brings us dangerously close to being a self-help book, but let me make one thing perfectly clear: This is not a f***ing self-help book.

Think of critical thinking more as an applied science, like medicine. I love learning about the body, and especially the brain, but it's all the more interesting when you try to put that knowledge to practical use. It makes it real, not just abstract or hypothetical. So how do we do that? How do we live a skeptical life? Some version of this question is by far the most common question we get from our listeners. They want to know how to apply some aspect of skeptical thinking to a specific problem in their life.

In the following chapters we will put all this critical thinking stuff to personal use as people, as parents, and as members of society. This may be the most challenging section, because it takes critical thinking out of the abstract, out of the many examples we have given throughout the book, and points skepticism directly at you, the reader.

But don't worry—this part will be fun too.

56. Being Skeptical

You look at science (or at least talk of it) as some sort of demoralising invention of man, something apart from real life, and which must be cautiously guarded and kept separate from everyday existence. But science and everyday life cannot and should not be separated.

—Rosalind Franklin

The first rule of skepticism is *not* that we never talk about skepticism. Talk about it as much as you want. The first rule is to apply these principles of critical thinking to yourself foremost. Remember, all the cognitive biases, flaws in memory and perception, heuristics, motivated reasoning, the Dunning-Kruger effect—it all applies to you, not just other people. Really let that sink in. These concepts are not weapons to attack other people and make yourself feel superior, they're the tools you need to minimize the bias, error, and nonsense clogging up your own brain.

Realize that you will never achieve the goal of ridding yourself of bias and error. All you can do is remain vigilant and work hard to keep them to a minimum. And while you're keeping that light focused inward, recognize that you have sacred cows, ideas that are part of your identity and will cause you emotional pain to change. Rather than try to deny your humanity, embrace it and work with it.

For me this meant taking pride in being able to change my mind. I identify with a skeptical and critical process of approaching claims, of letting the facts be supreme, of being honest, fair, and transparent. That is my identity, not any particular belief or conclusion. Therefore, if I'm wrong about something, that just gives me the opportunity to change my mind and show that I can correct error. It's not easy. Being wrong will still hurt the ego. You need to get to a place where refusing to correct an error will hurt more.

When it comes to dealing with other people, I would not presume to tell you what you "should" do. I can only tell you my approach, the SGU approach, and how that has worked out. I remind myself that we're all flawed humans just trying to get by in a complex and often scary world. We are mostly the products of our circumstances.

For example, most people will end up accepting the belief system they are born to. So it hardly seems fair to blame people for the one thing in their life they had zero control over, the situation into which they were born. We're all the product of our opportunities, our mentors, and our peers, in addition to the neurological proclivities with which we were endowed.

This attitude helps me be less judgmental. I prefer to take a nurturing approach. I am happy with where I am in my life and I owe a huge debt of gratitude to everyone who nurtured and taught me along the way. I feel the best thing I can do in return is to mentor and nurture others.

This does not mean I don't hold people responsible for their actions. Con artists never cease to anger me, and I think they deserve to be punished. But on the SGU we address beliefs and arguments, and we try hard not to attack the believers. They're just flawed people like us. On the other hand, we will criticize the purveyors of nonsense who are spreading misinformation, lies, and pseudoscience. They stepped into the public arena to promote their nonsense, and so they're fair game for scrutiny, skeptical analysis, and rebuttal. Even still, we always try to be fair.

Other proponents of science and reason prefer to take a harder-edged approach, and that's fine. Whatever works for your talents and personality. I can't prove that one approach works better than another. I do suggest that if you are trying to change the minds of other people, or spread your worldview, you at least think about your strategy and monitor your results. Your instincts may or may not be serving you well, so apply some of that critical thinking to how you spread critical thinking.

If my first suggestion, therefore, was to be humble, and my second suggestion was to be nurturing, my third is to be courageous. It can be difficult and uncomfortable to challenge other people's beliefs or to push back against pseudoscience. One common question we get is what to do when you are sitting in a doctor's office and they are recommending pure nonsense, like homeopathy? It might be easy just to say thank you and never return. But then you're missing an opportunity to give that professional some valuable feedback. They need to know what you think about their suggestion, that you value science-based medicine and that is what you expect from your physician. You may still decide not to return, but you've dropped a little bit of critical thinking into the world, and it all adds up.

The same goes for your teachers or your kid's school, for your workplace, and for your social circle. You can remain polite if you don't like confrontation, but I would never be afraid to stand up for reason. Of course, use judgment when choosing your time and place, and know when to back off. There's courageous, and then there is a pain in the ass. If your company takes a contract from a charlatan, do you just keep your head down and do your work, or do you say something to your boss? That depends on a lot of variables, but I wouldn't rule out raising the question of where the company's responsibility lies.

The way I see it, every social circle can use its own skeptic, that person who will give the skeptical point of view on any topic. Before long, people will come to you to ask what the skeptic thinks, and they will know what to expect, so they won't necessarily be put off.

When you do decide to engage, then really engage. Don't just give your opinion, find out what they think and why. Address their narrative and their understanding of the topic. If you don't have a good answer to something, that is a great opportunity to say you don't know, and then find out together. The best thing you can do is not just to give them a skeptical answer, but to model for them how to work it out for themselves. We really try to do that in the SGU, not tell people *what* to think but teach them *how* to think.

Show them the process, messy as it is. Science is messy. Thinking for yourself is an endless freak show. But you work through to at least a reasonable tentative answer by following a logical process that respects facts and accuracy.

Skepticism is also fun and empowering. One of the most common bits of feedback we get from listeners is that they've finally shed a belief system, and how liberating it is. They don't have to labor under an oppressive belief anymore. They can think freely and only believe what actually makes sense to them.

I like showing other people how much I enjoy being skeptical. Sure, it can be frustrating to deal with the abject nonsense that plagues our species, but in the end it's a lot of fun to explore interesting questions, exchange ideas with people who disagree with you, and learn more about yourself and the universe. Being skeptical can't be all negative, it needs to be positive too.

I think that's why Carl Sagan is so universally respected among skeptics and science communicators. More than anyone else, he embodied that sweet spot of opposing pseudoscience while embracing the wonder and magic of a scientific view of the universe. Science can give you a perspective that is just as mind-blowing, and far more wondrous than any fantasy.

How to Talk to Friends and Family

There was a time when Sunday dinner at the Novella house wasn't over until our sister ran crying from the room. A simple disagreement over a little thing like the nature of reality and the existence of an afterlife would somehow become personal. At one point I was banned from Seder dinner with Evan's family just because I wrote something about his sister believing in ghosts. Cara even broke it off with a boyfriend because of his anti-vaccine and alternative-medicine views.

We've all been there. No one can push your buttons like family and those close to you. You also have a lot more invested in relationships you are likely to maintain for a lifetime—and you want to keep for a lifetime. The stakes are higher.

We're not just talking about whether the Loch Ness monster likely exists; we're often confronting serious life decisions. We get questions from listeners whose spouses have cancer and are considering not going through with the recommended treatment in favor of vitamins or coffee enemas, parents who want to treat their children with homeopathy, or people whose siblings are investing heavily in multilevel marketing schemes.

These conversations can be truly frustrating with friends and family. During a disagreement, it's rare to get the response you want. Don't expect people to say, "You know, I never thought of it that way," or "I really can't argue against what you're saying." Chances are you will be met with defense, misdirection, and even personal attacks. You may also return to the topic months later and find that nothing has changed, as if your prior discussions hadn't even occurred.

By now at least this shouldn't be a surprise. We reviewed a long list of psychological mechanisms that seem to conspire to maintain our existing beliefs or the beliefs we desire. We tend to resist changing core or valued beliefs with every ounce of our will. When dealing with people close to you, it's especially important to remember the advice above, to keep humble, patient, and focused on educating rather than just winning an argument. And remember that sometimes you may in fact be wrong. Always consider that possibility first.

The Long Game

When people ask us for specific advice about how to deal with a family member or friend, we always say that it depends. It depends first on your goals. That may seem obvious, but it's important to consider exactly what you want to get out of the interaction. Who is your real audience, and are you trying to change opinions or behavior? Second, how much time and effort are you willing to invest?

If you are chatting with someone on social media, you will probably never meet them or have a relationship, and therefore you may be satisfied making a few pithy points and then going to get pizza. When dealing with a spouse, however, you may be willing to invest literally years of effort to slowly change their worldview. This means you need to be incredibly patient. Don't expect the scales to suddenly fall from their eyes.

So how do you do this?

1. As we're fond of saying on the SGU—plant the seed. Just leave the other person with a question or a thought and invite them to think about it or investigate further. Don't ask or expect them to express their opinion at that time. They may have an easier time really questioning their beliefs in the privacy of their own mind.

2. Don't make it into a competition or confrontation. The more personal you make it, the more defensive they'll get. Just explore the question together. In fact, it is helpful not to come out swinging but to begin by asking questions, admitting what you don't know, and then making the topic into a joint quest.

3. Find common ground. As we discussed in the chapter on logical fallacies and how to argue, every argument has premises and logic. If you and someone else disagree, you must be starting with different premises or one or both of you are using invalid logic. First find out what you agree on, and then try to identify where you disagree and see if you can resolve the differences together. You may get down to subjective opinions or value judgments, but then at least you'll know where you disagree.

    Don't fight over facts, resolve them. Agree ahead of time that you'll simply look up any details in dispute, and try to agree on sources you both trust.

4. Don't attack head on, but rather engage and nurture their skepticism. You may find much more success by not addressing the belief that concerns you the most. Rather, find a topic about which you agree. I have a close friend with whom I disagree a lot, but he saw a good documentary on the Roswell incident and understood why claims of a crashed flying saucer were baseless. That was a topic on which we could agree, and it became a great way for me to mentor his critical thinking. On that topic, without any confrontation, we could discuss what kinds of evidence are useful, how misinformation spreads, and how people can come to fool themselves and believe things that are not true. Everyone is skeptical of something. Find that thing.

   Rather than try to just convince someone not to believe in something, or to accept the scientific consensus on a topic, spend the time to teach them critical thinking skills. Eventually they may question the belief themselves.

5. Think about your tone. It's amazing how controversial the simple advice to be nice has become within the skeptical community. Phil Plait (the Bad Astronomer) gave his famous "Don't Be a Dick" talk at an Amazing Meeting (a skeptical conference) and received a surprising amount of pushback from people who thought he was being a dick.

   There is a legitimate controversy here. There is something called "tone trolling" that occurs in social media: Instead of addressing the points that someone is making, tone trolls nitpick their perceived tone. It's a strategy to push someone out of a discussion or get them to shut up. I have often been told I'm mean just for pointing out a factual error or saying that an argument is a logical fallacy. Tone becomes a misdirection to avoid the real issues and to shout down someone whose opinions you don't like.

   Further, some people jealously defend their right to have whatever tone they please and to be direct and even ridiculing when they choose. Some issues, they argue, require a forceful tone. I've encountered this attitude when talking to or about con artists or charlatans. They or their defenders demand respect and will commonly "tone troll" their critics. But whether or not they or their opinions are deserving of respect may be the very question at hand.

6. Try to understand the other person's narrative. We don't just have specific beliefs about facts, we build narratives about how the world works, how to make sense of often overwhelming and confusing information, what motivates other people, and what has happened in the past. If you want someone to change their mind, you cannot simply take away their narrative by telling them they are wrong. That will leave them feeling vulnerable and insecure. You need to replace their narrative with another one that offers better explanatory power, preferably one based on science and reason.

In other words, help them make sense of the facts and the claims with your skeptical viewpoint. In the Roswell incident example above, my friend's narrative was that the government was covering up evidence of a crashed alien flying saucer. My goal was to help them replace that conspiracy narrative with a skeptical one, involving mythology, self-deception, bad media reporting, and some self-promotion by those wanting to sell the conspiracy narrative.

With all that in mind, it's likely that with friends and family you want to maintain an amicable relationship. You can probably make all the points you need to make without getting nasty or insulting. It can sometimes be hard to be negative—essentially saying that someone is wrong—without coming off negatively. But it's possible. Just be thoughtful.

Whatever approach you decide to take, we have found that this works well for us. We frequently, for example, get confrontational and insulting e-mails from listeners who disagreed with a segment or perhaps just don't like the sound of Evan's voice or any of Jay's jokes.

Our policy is to respond (when we respond) in a professional and polite way. It's amazing how disarming this can be. Often the original e-mailer will backpedal, even apologize for their initial tone. You don't have to call someone a child. Just model being an adult and let the contrast stand.

Finally, if I think someone is worth engaging with, doing it nicely is the only way I seem to make any progress. But again, this is context dependent. Sometimes a sharper knife is needed, depending on your goals.

The good news is that people can change, although there is no formula and it's not easy. Psychological research has some encouraging findings—that if you take the time to explore arguments and evidence, people can change their mind and come to a more sophisticated understanding of the question at hand. Being a skeptical influence in the lives of people around you will have an impact in the long run.

Skeptical Parenting

There I was, listening to Gimli—dwarf, warrior, and member of Tolkien's Fellowship of the Ring—give a lecture on breastfeeding to my sister-in-law. No, I wasn't dreaming. Life just takes you to some strange places sometimes.

Let me back up…Most years we go to Dragon Con, a large convention of everything science fiction, fantasy, and nerdy, to talk for the skepticism and science sections of the conference (and eye all the great costumes). One year we went with Jay's wife, Cortney, while she was pregnant with her first child.

We got in line to get a picture autographed by John Rhys-Davies, the actor who played Gimli in the Lord of the Rings trilogy. He noticed that Cortney was pregnant and decided to chat with her about whether or not she was going to breastfeed. She was uncertain at the time, and apparently Rhys-Davies had explored the topic thoroughly for his own children. Make no mistake, he was a super-nice guy, and it was a pleasant conversation. But he was also a bit of a "breast is best" warrior.

Several things occurred to me during that bizarre conversation. The first was that the poor guy had no idea who he was talking to. I wanted to say, "But Gimli, published scientific evidence doesn't support all the typical 'breast is best' claims. Putting pressure on mothers to breastfeed can be counterproductive." But there was a line of annoyed fans behind us who probably weren't interested in hearing a debate about the relative benefits of breast milk and formula. Spider-Man in particular was getting a bit jumpy.

It also drove home something I had previously observed— everyone thinks they are entitled to an opinion on certain topics, such as pregnancy and raising children, even when there are complex scientific issues at stake. Not only do people think their opinions are as good as expert reviews of rigorous scientific data, they believe they have almost a duty to share those opinions dogmatically with others, even while in costume at a comic convention.

Psychologists call this the "illusion of explanatory depth." We tend to think we know how things work, even when we don't. This phenomenon is partly due to the fact that we are used to relying on the expertise of others and confuse it with our own. Most of us drive our cars or use our computers without really knowing how they work. We may overestimate our superficial knowledge of such things and underestimate the technical knowledge of real experts. By a similar mechanism, we feel we are entitled to an opinion about breastfeeding, even when we aren't medical experts.

There is also a lot of moral judgment tied up in what are really questions of either medical science or personal preference. If you don't breastfeed, or if you take pain medication during delivery, or if you feed your child anything other than gluten-free hypoallergenic organic free-range kale, you're a bad parent.

What this means is that as soon as you announce to the world (or it's apparent) that you are going to have a child, you will be subjected to unsolicited advice of every kind.

Encourage Critical Thinking

Being a skeptical parent means knowing not only how to deal with questions of science regarding having and raising children, but also how to encourage your children to grow into critical thinking adults. I wish I had a formula for guaranteeing such an outcome, but of course every child is different and environments vary. As in all things, you can't control your children, you can only influence them. I will also state up front that the general advice of psychologists is to have a loving and nurturing relationship with your child. That comes first.

But along the way, you can gently encourage them to be curious, to love science, and to engage in critical thinking.

Seize Opportunities

When something comes up that you could use as a critical thinking lesson, get into the habit of using it. Show your kids how to question what they see. This is the very beginning of their baloney detector. Essentially, model the good thinking practices you're trying to teach. If they ask a question, don't just give them an authoritative answer, no matter how simple the question. I would give them positive feedback for asking and being curious, then reflect on how interesting the question is and reflect it back.

For example, my daughter asked me what color dinosaurs were. That is a deceptively insightful question, because it assumes that the way dinosaurs are typically depicted may not be accurate. I asked her back, "Well, how could we know what color they were?" That's the more interesting question. We then explored it together.

I also find it helpful to admit when I don't know the answer, or when even scientists don't know. In fact, with most questions you can push to the limits of what is known and point out what science has not yet discovered. This teaches several things: It's okay to question anything; knowledge is not absolute but is a process of discovery; there is no absolute authority, as any claim is only as good as the evidence that supports it.

With this approach I found that my children became naturally skeptical. They particularly delight in proving me wrong on some detail, which is also an opportunity to model how to behave when confronted with error.

Another rule I found helpful: Aim a little higher than your instincts tell you. In other words, we tend to assume that children don't understand things, but they probably know and understand more than you think. Avoid talking down to them or underestimating them. Keep the carrot out in front of them and make them reach for it a little, but obviously not too much. At the very least they'll really appreciate you talking to them as if they were a real person.

Giving a Love for Science

I love science, we all do on the SGU, and we have from a young age. We also have in common the fact that at least one of our parents shared with us their love for science when we were young. This has been the easiest part of being a parent, enthusiastically sharing my love for science with my children.

The consensus opinion is that sharing fun quality time with your kids is perhaps the best thing you can do for them. Experts also agree that it is helpful to regularly read to your children and have as many books around as possible. All you have to do, really, is make sure that some of those books, and some of that quality time, deals with science. I would take my daughters on exploring adventures through the woods in our backyard. If you don't have access to something like that, there are plenty of other things you can do.

I would also take them on imaginary journeys through the solar system. This requires no more than a box, or even a closet—and your imagination. Rather than just tell them about the solar system,

I would "blast off" with them in our imaginary spaceship and we'd travel to each planet in order. I would describe how long it took to get there, what it looked like from orbit, and what it was like to be on the surface, gravity and all.

You can also find something they are interested in and encourage it. Don't force a topic on them: Try out different ones, and when they show interest, run with it. My older daughter became fascinated by birds, for example, while my younger daughter is interested in engineering. The topic doesn't really matter; it could be plants, insects, dinosaurs, the planets—anything. Whatever excites them gives you the opportunity to show how to observe, how categorization works, how scientists know stuff, how to separate reliable claims from common myths, and how to explore a topic through reading. You can also look for ways to show them how the topic they are most interested in relates to other areas of knowledge.

Even if you are not a parent yourself, you may have children in your life through friends or family. There are many opportunities to spread a little critical thinking or love for science to those around you, at any age.

Ultimately, being a skeptical "parent" is really about being a skeptical person who values teaching and nurturing. We are all students and teachers throughout our entire lives. We never have to stop learning, and no matter what your level of understanding, you have something you can teach others. In fact, teaching is the best way to learn—teaching and learning are both a dynamic part of the same process.

People who are more experienced or who have a deeper fund of knowledge than you have a lot to share. But those with less experience or knowledge also have a lot to share—their enthusiasm and their fresh perspective. The most penetrating and insightful questions often come from those with no preconceptions. This is why children may sometimes ask the best questions, because they don't know enough to have hidden assumptions that constrain their thinking.

Such questions are an opportunity for you to confront assumptions you did not know you were making.

This student-teacher approach is also the approach we take on the SGU. We are teaching and learning at the same time as part of the same process. I don't expect my endless process of teaching and learning will end until I do.

Epilogue

Being a member of the skeptical community has been one of the most fulfilling adventures of my life. It has been a personal intellectual journey and a collective journey with those close to me. Through the community, and especially through the *Skeptics' Guide* podcast, I have been able to expand that circle by orders of magnitude.

It has been fun, enlightening, frustrating, and empowering, and I know it is a journey that will never end.

One of the most common observations our regular listeners make is that while listening they feel as if they're the sixth Rogue in the room, just the silent one. The show for them is more than just an educational tool or a means of distraction while performing mindless tasks. For many it is a connection with like-minded friends, and we love to hear that because it's exactly what we are going for. For many it is also an "escape to reality." It can often feel like the world is crazy, and it's nice to know there are other people out there who value science, logic, and reason.

If this book was your introduction to the SGU and to skepticism, then I hope you found it a valuable and entertaining one. It may have even pissed you off in some sections, and that's good. That means we successfully challenged something you care about, perhaps more than you thought.

To close out, we'd like to leave you with one final thought: Don't trust us. This might sound strange after reading the book, but the point is, you shouldn't really trust anyone when it comes to empirical knowledge. Think for yourself. If something is important, verify it as best you can. We do all need to rely on experts and other people as sources of information, but at the same time we should strive to understand how we know what we know. Why do experts have the opinions they do? Is there more to the story?

Everyone makes mistakes, including us. I hope we didn't make any here. We try our best, but error is unavoidable. All we can do is be open to correcting them when they come to our attention.

you did

I think over the years we may have learned as much collectively from our listeners as they have from us. Almost every week someone who is a world expert on a narrow topic we discussed writes in to give us more information. Listeners give us new perspectives, or fill holes in our discussion, or even challenge our facts or arguments. I do think this is the most positive thing about new media—it's interactive. We're not just talking at our listeners. We're having a conversation on the show that we're happy to bring our listeners into when they have something to add.

One of our goals for this book is that it too will become part of the conversation. It's the guide to the *Skeptics' Guide*. We have often been asked for a thorough guide to all of the concepts we reference regularly on the show. Well, here it is. Hopefully this is just one more step on the endless journey of critical inquiry.

Until next time, this is your *Skeptics' Guide to the Universe*.

Acknowledgments

We are so grateful to the many mentors and science communicators who taught and inspired us for decades. Carl Sagan, Isaac Asimov, Stephen J. Gould, Eugenie Scott, Joe Nickell, Paul Kurtz, James Randi, Martin Gardner, Mary Roach, Bill Nye, Neil deGrasse Tyson, Oliver Sacks, and many others (yes, even Leonard Nimoy) are the giants on whose shoulders we stand. In addition, we are part of a vibrant, supportive community of fellow skeptics who encourage us every day in their careers and daily acts to make the world a little more rational.

At the top of this list are our SGU listeners, who have taught us as much over the years as we have taught them. We would like to thank every listener who has ever sent us an e-mail, a question, feedback, pedantic correction, suggestion for the show, or just taken the time to tell us how we have contributed to their lives. And especially we are grateful to our members and patrons who support our work.

Perry DeAngelis, to whom this book is dedicated, was a founding member of the New England Skeptical Society and the SGU and was no doubt responsible for much of its early success. Perry passed away in August 2007, but his influence will always remain. Mike Lacelle started as our self-proclaimed "number one fan" but became a close friend and behind-the-scenes worker bee. We also lost him much too early.

Rebecca Watson was part of the show for nine years, from its early days of obscurity, adding her unique style and wit, and helping to shape the tone and popularity of the SGU. Fellow podcaster, skeptic, and musician George Hrab has been a frequent guest on the SGU and is our "partner in crime" in so many other ways. He developed a stage extravaganza with us, shot videos, helped produce live streaming shows, and other special projects. More important, he is one of our closest friends and a selfless supporter of our shared mission.

We have developed close ties of collaboration and friendship around the world, especially with our many friends down under—Richard Saunders, Eran Segev, Rachie Dunlop, Jo Benhamu, and the New Zealand skeptics.

We have had the privilege of working with many other collaborators and volunteers as well, including Liz Gaston, Phil Hudson, Douglas Sobon, David Young, Jake Willson, Ian Callanan, Joel Bellucci, Reid Gower, Kernan Coleman, all our friends at the Northeast Conference of Science and Skepticism, everyone who helped on our video projects, everyone who has ever made Jay laugh, and that guy who did that thing (you know who you are).

This book is just the latest of our projects that was helped along by many hands. Thanks to Richard Wiseman for his advice and example early in the process, Kiera Wilhelm for suggesting our subtitle, our agent Rob Kirkpatrick for helping make this project a reality, the tremendous work of our editor Maddie Caldwell, and all the people at Grand Central Publishing for their work on this book.

We especially have to thank our family and close friends who have supported us, helped us, and put up with us over the years. We could not have maintained this project without the love, support (and hard work) of Jocelyn Novella, Cortney Novella, Jennifer Bernstein, and Gabriel Lewis. Most important, the Novella brothers owe their lifelong love of science, science fiction, and critical thinking to their father, Joe Novella, who, over pasta and meatballs, inspired the conversations that became the SGU.

Finally, for those of you being introduced to skepticism for the first time through this book, our hope is that you will be inspired—in a small way or a big way—to change the world.

References

SECTION 1: CORE CONCEPTS EVERY SKEPTIC SHOULD KNOW

1—Scientific Skepticism

Richard Feynman, "The Problem of Teaching Physics in Latin America," *Engineering and Science* vol. 7, no. 2, November 1963, pp. 21–30.

John Cook, Stephan Lewandowsky, and Ullrich K. H. Ecker, "Neutralizing Misinformation Through Inoculation: Exposing Misleading Argumentation Techniques Reduces Their Influence." *PLOS One*, May 5, 2017, https://doi.org/10.1371/journal.pone.0175799.

NEUROPSYCHOLOGICAL HUMILITY

Robert E. Bartholomew, "Two Mass Delusions in New England," NESS, April 1998, http://www.theness.com/index.php/two-mass-delusions-in-new-england/.

2—Memory Fallibility and False Memory Syndrome

Steven Novella, "Did Hillary Lie?" NESS, March 30, 2008.

Elizabeth F Loftus and Jacqueline E Pickrell, "The Formation of False Memories," *Psychiatric Annals* vol. 25, 1995, pp. 720–25.

Isabel Lindner et al., "Observation Inflation: Your Actions Become Mine," *Psychological Science* vol. 21, no. 9, August 5, 2010, https://doi.org/10.1177/0956797610379860.

Julia Shaw and Stephen Potter, "Constructing Rich False Memories of Committing Crime," *Psychological Science* vol. 26, no. 3, 2015.

Travis J. Tritten, "NBC's Brian Williams Recants Iraq Story After Soldiers Protest," *Stars and Stripes*, February 4, 2015.

Ian Skurnik et al., "How Warnings about False Claims Become Recommendations," *Journal of Consumer Research*, vol. 31, no. 4, 2005, pp. 713–24.

Ulric Neisser and Nicole Harsch, "Phantom Flashbulbs: False Recollections of Hearing the News about Challenger," in *Emory Symposia in Cognition, 4. Affect and Accuracy in Recall: Studies of "Flashbulb" Memories*, ed. Eugene Winograd and Ulric Neisser (New York: Cambridge University Press, 1992), pp. 9–31.

Takashi Kitamura et al., "Engrams and circuits crucial for systems consolidation of a memory," *Science* vol. 356, no. 6333, pp. 73–78, doi: 10.1126/science.aam6808.

Ellen Bass and Laura Davis, *The Courage to Heal: A Guide for Women Survivors of Child Sexual Abuse*, 20th anniversary edition (New York: HarperCollins, 2008).

J. S. La Fontaine, *Speak of the Devil: Tales of Satanic Abuse in Contemporary England*, (Cambridge, UK: Cambridge University Press, 1998), https://books.google.com/books?id=JBxfvDeQdmoC&printsec=frontcover&hl=en#v=onepage&q&f=false.

R. J. McNally, "Searching for Repressed Memory," Nebraska Symposium on Motivation, vol. 58, 2012, pp. 121–47.

Cara Laney and Elizabeth F. Loftus, "Traumatic Memories Are Not Necessarily Accurate Memories," *Canadian Journal of Psychiatry* vol. 50, no. 13, November 2005, pp. 823–28, https://doi.org/10.1177/070674370505001303.

Kenneth Lanning, "Investigator's Guide to Allegations of Ritual Child Abuse," January 1992, http://www.sacred-texts.com/pag/lanning.htm.

Ed Cara, "The Most Dangerous Idea in Mental Health," *Pacific Standard*, November 3, 2014, https://psmag.com/social-justice/dangerous-idea-mental-health-93325.

Elizabeth F. Loftus, "The Reality of Repressed Memories," *American Psychologist* vol. 48, 1993, pp. 518–37.

3—Fallibility of Perception

Christopher Chabris and Daniel Simons, "The Invisible Gorilla," 1999, http://www.theinvisiblegorilla.com/videos.html.

"Missing the Gorilla: Why We Don't See What's Right in Front of Our Eyes," *Medical Xpress*, April 18, 2011, https://medicalxpress.com/news/2011-04-gorilla-dont-front-eyes.html.

Richard Wiseman, "Colour Changing Card Trick," https://www.youtube.com/watch?v=v3iPrBrGSJM.

Trafton Drew, Melissa L. H. Vo, and Jeremy M. Wolfe, "The Invisible Gorilla Strikes Again: Sustained Inattentional Blindness in Expert Observers," *Psychological Science* vol. 24, no. 9, September 2013, pp. 1848–53, doi: 10.1177/0956797613479386.

Alan D. Castel, Michael Vendetti, and Keith J. Holyoak, "Fire Drill: Inattentional Blindness and Amnesia for the Location of Fire Extinguishers," *Attention, Perception, & Psychophysics* vol. 74, no. 7, October 2012, pp. 1391–96.

Daniel Simons and Daniel Levin, "Failure to Detect Changes to People During a Real-World Interaction," *Psychonomic Bulletin & Review* vol. 5, no. 4, 1998, pp. 644–49, http://psych.unl.edu/mdodd/Psy498/simonslevin.pdf.

Janelle K. Seegmiller, Jason M. Watson, and David L. Strayer, "Individual Differences in Susceptibility to Inattentional Blindness," *Journal of Experimental Psychology: Learning Memory and Cognition*, May 2011, pp. 785–91.

4—Pareidolia

Elizabeth Svoboda, "Facial Recognition—Brain—Faces, Faces Everywhere," *New York Times*, February 13, 2007.

Nancy Kanwisher and Galit Yovel, "The Fusiform Face Area: A Cortical Region Specialized for the Perception of Faces," *Philosophical Transactions of the Royal*

Society of London, B Biological Sciences vol. 361, no. 1476, December 2006, pp. 2109–28, doi: 10.1098/rstb.2006.1934.

Bettina Sorger et al., "Understanding the Functional Neuroanatomy of Acquired Prosopagnosia," *Neuroimage* vol. 35, no. 1, April 2007, pp. 836–52.

Leonardo Da Vinci and Edward McCurdy, *Leonardo da Vinci's Note-Books* (London: Duckworth, 1906), p. 173.

5—Hyperactive Agency Detection

Justin L. Barrett, *Why Would Anyone Believe in God?* (Lanham, MD: AltaMira Press, 2004).

Bruce M. Hood, *SuperSense: Why We Believe in the Unbelievable* (New York: Harper-Collins, 2009).

6—Hypnagogia

Brian A. Sharpless and Jacques P. Barber, "Lifetime Prevalence Rates of Sleep Paralysis: A Systematic Review," *Sleep Medicine Reviews* vol. 15, 2011, pp. 311–15. Accessed May 25, 2017. doi: 10.1016/j.smrv.2011.01.007.

American Academy of Sleep Medicine, "Sleep Paralysis—Overview & Facts." Accessed May 25, 2017. http://www.sleepeducation.org/sleep-disorders-by-cate gory/parasomnias/sleep-paralysis/overview-facts.

"Ghost Attacked Me, Says Spooked Alba," *Sydney Morning Herald*, February 6, 2008. Accessed May 27, 2017. http://www.smh.com.au/news/people/ghost-atta cked-me-says-jessica-alba/2008/02/06/1202233907134.html.

Keith Hillman, "What Is Hypnagogia?" Accessed May 25, 2017. http://www .psychology24.org/what-is-hypnagogia/.

Whitley Strieber, *Communion: A True Story* (Sag Harbor, NY: Beech Tree, 1987).

7—Ideomotor Effect

British Dowsers, https://www.britishdowsers.org/learn/.

Ray Hyman, "How People Are Fooled by Ideomotor Action," Quackwatch, http:// www.quackwatch.org/01QuackeryRelatedTopics/ideomotor.html.

William Benjamin Carpenter, "On the Influence of Suggestion in Modifying and directing Muscular Movement, independently of Volition," proceedings, Royal College of Surgeons of England, Royal Institution of Great Britain, London, March 12, 1852, pp. 147–53.

"Using a Pendulum...to Communicate with Spirit and Your Higher Self," Healing Crystals For You, http://www.healing-crystals-for-you.com/using-a-pendulum .html.

American Psychological Association, "Facilitated Communication: Sifting the Psychological Wheat from the Chaff," November 2003, https://www.apa.org/ research/action/facilitated.aspx.

John Jackson, "What Is the Ideomotor Effect? A Natural Explanation for Many Paranormal Experiences," UK-Skeptics, 2005. Accessed May 25, 2017. https:// www.scribd.com/document/56837545/Ideomotor-Effect.

Mark Tutton, "Trapped 'Coma' Man: How Was He Misdiagnosed?" CNN, November 24, 2009, http://www.cnn.com/2009/HEALTH/11/24/coma.man.belgium/index.html?eref=igoogle_cnn.

METACOGNITION
8—Dunning-Kruger Effect

Justin Kruger and David Dunning, "Unskilled and Unaware of It: How Difficulties in Recognizing One's Own Incompetence Lead to Inflated Self-Assessments," *Journal of Personality and Social Psychology* vol. 77, no. 6, December 1999, pp. 1121–34, http://dx.doi.org/10.1037/0022-3514.77.6.1121.

David Dunning, "We Are All Confident Idiots," *Pacific Standard*, October 27, 2014.

9—Motivated Reasoning

Julia Galef, "Why You Think You're Right—Even If You're Wrong," presented at TED Talks, University Park, Pennsylvania, February 2016, https://www.ted.com/talks/julia_galef_why_you_think_you_re_right_even_if_you_re_wrong.

Troy H. Campbell and Aaron C. Kay, "Solution Aversion: On the Relation Between Ideology and Motivated Disbelief," *Journal of Personality and Social Psychology* vol. 107, no. 5, November 2014, pp. 809–24, http://dx.doi.org/10.1037/a0037963.

Leon Festinger, *A Theory of Cognitive Dissonance* (Evanston, IL: Row, Peterson, 1957).

Toby Bolsen, James N. Druckman, and Fay Lomax Cook, "The Influence of Partisan Motivated Reasoning on Public Opinion Political Behavior," *Political Behavior* vol. 36, no. 2, June 2014, pp. 235–62, doi 10.1007/s11109-013-9238-0.

Brendan Nyhan and Jason Reifler, "When Corrections Fail: The Persistence of Political Misperceptions," *Political Behavior* vol. 32, no. 2, June 2010, pp. 303–30.

Thomas Wood and Ethan Porter, "The Elusive Backfire Effect: Mass Attitudes' Steadfast Factual Adherence," forthcoming *Political Behavior*, written December 2017, https://papers.ssrn.com/sol3/papers.cfm?abstract_id=2819073.

Drew Westen et al., "An fMRI Study of Motivated Reasoning: Partisan Political Reasoning in the U.S. Presidential Election," working paper, https://www.uky.edu/AS/PoliSci/Peffley/pdf/Westen%20The%20neural%20basis%20of%20motivated%20reasoning.pdf.

Jonas T. Kaplan, Sarah I. Gimbel, and Sam Harris, "Neural Correlates of Maintaining One's Political Beliefs in the Face of Counterevidence," *Scientific Reports* vol. 6, article no. 39589, 2016, doi:10.1038/srep39589.

Arms Control Association, "Nuclear Weapons: Who Has What at a Glance," https://www.armscontrol.org/factsheets/Nuclearweaponswhohaswhat.

10—Arguments and Logical Fallacies

Madhucchanda Sen, "Evaluating Arguments: Inferences and Fallacies," in *An Introduction to Critical Thinking*, edited by Madhucchanda Sen et al. (Pearson Education India, 2011), p. 46.

11—Cognitive Biases and Heuristics

Daniel Casasanto and Evangelina G. Chrysikou, "When Left Is 'Right': Motor Fluency Shapes Abstract Concepts," *Psychological Science* vol. 22, no. 4, April 2011, pp. 419–22, doi: 10.1177/0956797611401755.

Rugg, cited in Scott Plous, *The Psychology of Judgment and Decision* (New York: McGraw-Hill, 1993).

Daniel Kahneman and Amos Tversky, "Subjective Probability: A Judgment of Representativeness," *Cognitive Psychology* vol. 3, no. 3, July 1972, pp. 430–54.

Andrew B. Geier and Paul Rozin, "Univariate and Default Standard Unit Biases in Estimation of Body Weight and Caloric Content," *Journal of Experimental Psychology: Applied* vol. 15, no. 2, June 2009, pp. 153-62, doi: 10.1037/a0015955.

12—Confirmation Bias

Jon Ronson, *So You've Been Publicly Shamed* (New York: Riverhead Books, 2015).

Chris Lee, "Confirmation Bias in Science: How to Avoid It," *Ars Technica*, July 13, 2010, https://arstechnica.com/science/2010/07/confirmation-bias-how-to-avoid-it/.

Daniel Klein, "I Was Wrong, and So Are You," *The Atlantic*, December 2011.

Jonah Lehrer, "The Reason We Reason," *Wired*, May 4, 2011, https://www.wired.com/2011/05/the-sad-reason-we-reason/.

"CDC Study: Flu Vaccine Saved 40,000 Lives During 9 Year Period," Centers for Disease Control and Prevention, March 30, 2015, https://www.cdc.gov/flu/news/flu-vaccine-saved-lives.htm.

Thomas Gilovich, *How We Know What Isn't So: The Fallibility of Human Reason in Everyday Life*, reprint edition (New York: Free Press, 1993).

Ben Tappin, Leslie Van Der Leer, and Ryan McKay, "The Heart Trumps the Head: Desirability Bias in Political Belief Revision," *Journal of Experimental Psychology: General* vol. 146, no. 8, August 2017, https://www.ncbi.nlm.nih.gov/pubmed/28557511.

P. C. Wason, "Reasoning about a Rule," *Quarterly Journal of Experimental Psychology* vol. 20, no. 3, 1968, pp. 273–81.

13—Appeal to Antiquity

Team Register, "Indian Courts 'Rule Astrology Is a Science,'" *The Register*, February 7, 2011. Accessed May 25, 2017. http://www.theregister.co.uk/2011/02/07/telegraph_astrology_bunkum/.

Thomas Paine, *The Age of Reason; Being an Investigation of True and Fabulous Theology* (Paris: Barrois, 1794).

Steven Novella, "Acupuncture Doesn't Work," *Science-Based Medicine*, June 19, 2013. Accessed May 25, 2017. https://sciencebasedmedicine.org/acupuncture-doesnt-work/.

14—Appeal to Nature

Food and Agriculture Organization of the United Nations. "Why Is Organic Food More Expensive Than Conventional Food?" Accessed May 25, 2017. http://www.fao.org/organicag/oa-faq/oa-faq5/en/.

James Randi, from his lecture presented at Yale University, October 1999.

David Hume, *A Treatise of Human Nature* (London: White-Hart, 1739), section 3.1.1.

G. E. Moore, *Principia Ethica*, revised edition (Cambridge, UK: Cambridge University Press, 1903).

Orac, "'Natural' doesn't necessarily mean better," *Respectful Insolence*, May 12, 2015. Accessed May 25, 2017. http://scienceblogs.com/insolence/2015/05/12/the-naturalistic-fallacy-on-parade/.

U.S. Food and Drug Administration, "What Is the Meaning of 'Natural' on the Label of Food"? Accessed May 25, 2017. http://www.fda.gov/AboutFDA/Transparency/Basics/ucm214868.htm.

15—Fundamental Attribution Error

Robert Todd Carroll, *The Skeptic's Dictionary: A Collection of Strange Beliefs, Amusing Deceptions, and Dangerous Delusions* (Hoboken, NJ: John Wiley, 2003).

16—Anomaly Hunting

Brian L. Keeley, "Of Conspiracy Theories," *Journal of Philosophy* vol. 96, no. 3, March 1999, pp. 109–26.

Karl Tate, "Space Radiation Threat to Astronauts Explained (Infographic)," Space.com, May 30, 2013, http://www.space.com/21353-space-radiation-mars-mission-threat.html.

Bill Kaysing and Randy Reid, *We Never Went to the Moon: America's Thirty Billion Dollar Swindle* (Pomeroy, WA: Health Research Books, 1997).

17—Data Mining

"You a Gemini? Drive Carefully and Get Insurance," Reuters, February 11, 2002.

Naomi Kim, "Had a Car Crash? It's All in the Stars, Study Says," Reuters, December 14, 2006, https://uk.reuters.com/article/oukoe-uk-astrology-driving/had-a-car-crash-its-all-in-the-stars-study-says-idUKNCD15739420061214.

18—Coincidence

Gina Kolata, "1-in-a-Trillion Coincidence, You Say? Not Really, Experts Find," *New York Times*, February 27, 1990.

J. A. Paulos, "Coincidences," *Skeptical Inquirer* vol. 15, no. 4, summer 1991, pp. 382–85.

Thomas L. Griffiths and Joshua B. Tenenbaum, "Randomness and Coincidences: Reconciling Intuition and Probability Theory," *Proceedings of the Annual Meeting of the Cognitive Science Society*, vol. 23, http://web.mit.edu/cocosci/Papers/random.pdf.

Donald Saucier and Scott Fluke, "Research Project Offers Insight into Superstitious Behavior," Kansas State University, September 2, 2010, http://www.k-state.edu/media/newsreleases/sept10/superstition90210.html.

The Monty Hall Problem: Ruma Falk, "A Closer Look at the Probabilities of the Notorious Three Prisoners," *Cognition* vol. 43, no. 3, pp. 197–223, doi: 10.1016/ 0010-0277(92)90012-7.

Steve Selvin, "A Problem in Probability (letter to the editor)," *American Statistician* vol. 29, no. 1, February 1975, https://www.jstor.org/stable/2683689?seq=1 #page_scan_tab_contents.

W. T. Herbranson and J. Schroeder, "Are Birds Smarter Than Mathematicians? Pigeons (Columba livia) Perform Optimally on a Version of the Monty Hall Dilemma," *Journal of Comparative Psychology* vol. 124, no. 1, pp. 1–13, doi: 10.1037/a0017703.

SCIENCE AND PSEUDOSCIENCE

Daryl J. Bem, "Feeling the Future: Experimental Evidence for Anomalous Retroactive Influences on Cognition and Affect," *Journal of Personality and Social Psychology* vol. 100, no. 3, 2011, pp. 407–425.

19—Methodological Naturalism and Its Critics

"The Wedge Strategy: Center for the Renewal of Science & Culture," The Discovery Institute, 1998.

20—Postmodernism

Dave Holmes et al., "Deconstructing the Evidence-Based Discourse in Health Sciences: Truth, Power and Fascism," *International Journal of Evidence-Based Healthcare* vol. 4, 2006, pp. 180–86.

Thomas Kuhn, *The Structure of Scientific Revolutions* (Chicago: University of Chicago Press, 1962).

E. O. Wilson, *Consilience: The Unity of Knowledge* (New York: Knopf, 1998).

21—Occam's Razor

Isaac Newton, "Newton's Principia: Rules of Reasoning in Natural Philosophy," Trans. A. Motte, 1729.

22—Pseudoscience and the Demarcation Problem

Larry Arnold, *Ablaze: The Mysterious Fires of Spontaneous Human Combustion* (New York: M. Evans, 1995).

Pew Research Center, "Public Praises Science; Scientists Fault Public, Media," 2009, http://www.people-press.org/2009/07/09/public-praises-science-scientists -fault-public-media.

Hannah Osborne, "Head Transplants: Sergio Canavero Says First Patient Will Be Chinese National, Not Valery Spiridonov," *Newsweek*, April 28, 2017, http:// www.newsweek.com/head-transplant-sergio-canavero-valery-spiridonov -china-2017-591772.

Harriet Hall, "Cranial Manipulation and Tooth Fairy Science," *Science-Based Medicine*, August 27, 2013, https://sciencebasedmedicine.org/cranial-manipulation-and-tooth-fairy-science/.

Richard P. Feynman, "Cargo Cult Science," Caltech commencement address 1974, http://calteches.library.caltech.edu/51/2/CargoCult.htm.

Sergio Canavero, *Immortal: Why CONSCIOUSNESS is NOT in the BRAIN* (Create Space, 2014).

23—Denialism

Stephen J. Gould, "Evolution as Fact and Theory" in *Hen's Teeth and Horse's Toes: Further Reflections in Natural History* (New York: W. W. Norton, 1994).

John Cook et al., "Quantifying the Consensus on Anthropogenic Global Warming in the Scientific Literature," *Environmental Research Letters* vol. 8, no. 2, 2013.

24—P-Hacking and Other Research Foibles

Kenneth L. Cavanaugh and Michael C. Dillbeck, "Field Effects of Consciousness and Reduction in U.S. Urban Murder Rates: Evaluation of a Prospective Quasi-Experiment," *Journal of Health and Environmental Research* vol. 3, no. 3-1, May 2017, pp. 32–43.

Regina Nuzzo, "Scientific Method: Statistical Errors: P Values, the 'Gold Standard' of Statistical Validity, Are Not As Reliable As Many Scientists Assume," *Nature* vol. 506, February 12, 2014, pp. 15–52.

John P. A. Ioannidis, "Why Most Published Research Findings Are False," *PLOS Medicine* vol. 2, no. 8, August 2005, doi: 10.1371/journal.pmed.0020124.

Joseph P. Simmons, Leif D. Nelson, and Uri Simonsohn, "False-Positive Psychology: Undisclosed Flexibility in Data Collection and Analysis Allows Presenting Anything as Significant," *Psychological Science* vol. 22, no. 11, pp. 1359–66.

Monya Baker, "1,500 Scientists Lift the Lid on Reproducibility: Survey Sheds Light on the 'Crisis' Rocking Research," *Nature* vol. 533, no. 7604, May 25, 2016, corrected July 28, 2016, https://www.nature.com/news/1-500-scientists-lift-the-lid-on-reproducibility-1.19970.

Daniel Engber, "Daryl Bem Proved ESP Is Real: Which Means Science Is Broken," *Slate*, May 17, 2017, https://slate.com/health-and-science/2017/06/daryl-bem-proved-esp-is-real-showed-science-is-broken.html.

Eric Loken and Andrew Gelman, "Measurement Error and the Replication Crisis," *Science* vol. 355, no. 6325, February 2017, pp. 584–85, doi: 10.1126/science.aal3618.

Open Science Collaboration, "Estimating the Reproducibility of Psychological Science," *Science* vol. 349, no. 6251, August 28, 2015, doi: 10.1126/science.aac4716.

25—Conspiracy Theories

Michael J. Wood, Karen M. Douglas, and Robbie M. Sutton, "Dead and Alive: Beliefs in Contradictory Conspiracy Theories," *Social Psychological and Personal-*

ity Science vol. 3, no. 6, January 25, 2012, http://journals.sagepub.com/doi/abs/10.1177/1948550611434786.

David Robert Grimes, "On the Viability of Conspiratorial Beliefs," *PLOS One*, January 26, 2016, http://dx.doi.org/10.1371/journal.pone.0147905.

Dean Koontz, *Fear Nothing* (New York: Bantam, 1998).

Kathy Frankovic, "Belief in Conspiracies Largely Depends on Political Identity," YouGov, December 27, 2016.

Conspiracy Theory Poll Results, Public Policy Polling, April 02, 2013, http://www.publicpolicypolling.com/main/2013/04/conspiracy-theory-poll-results-.html.

Gina Kolata, "1-in-a-Trillion Coincidence, You Say? Not Really, Experts Find," *New York Times*, February 27, 1990, http://www.nytimes.com/1990/02/27/science/1-in-a-trillion-coincidence-you-say-not-really-experts-find.html.

Viren Swami and Rebecca Coles, "The Truth Is Out There," *The Psychologist* vol. 23, July 2010, pp. 560–63, http://thepsychologist.bps.org.uk/volume-23/edition-7/truth-out-there.

Richard Hofstadter, "The Paranoid Style in American Politics," in *The Paranoid Style in American Politics and Other Essays* edited by Richard Hofstadter (New York: Knopf, 1966) pp. 3–40.

Damaris Graeupner and Alin Comin, "The Dark Side of Meaning-Making: How Social Exclusion Leads to Superstitious Thinking," *Journal of Experimental Social Psychology* vol. 69, pp. 218–222, http://psycnet.apa.org/record/2016-50282-001.

"Alex Jones Gives 'Final' Unhinged Rant Defending His Sandy Hook Conspiracy Theories," Media Matters for America, November 18, 2016.

26—Witch Hunts

History.com staff, "Salem Witch Trials," History.com, 2011, http://www.history.com/topics/salem-witch-trials.

Jess Blumberg, "A Brief History of the Salem Witch Trials," Smithonian.com, October 2007.

Heinrich Kramer and Jacob Sprenger, *Malleus Maleficarum*, http://www.malleusmaleficarum.org/.

Douglas Walton, "The Witch Hunt as a Structure of Argumentation," *Argumentation* vol. 10, no. 3, August 1996, pp. 389–407.

Paul Achter, "McCarthyism," Britannica.com, https://www.britannica.com/topic/McCarthyism.

Tom Dart, "Texas Pair Released after Serving 21 Years for 'Satanic Abuse'," *Guardian*, December 5, 2013.

27—Placebo Effects

"Radium Cures," Museum of Quackery, http://www.museumofquackery.com/devices/radium.htm.

Richard Van Vleck, "The Electronic Reactions of Albert Abrams," *American Artifacts* no. 39, http://www.americanartifacts.com/smma/abrams/abrams.htm.

Scientific American staff, "Our Abrams investigation (Staff) series I-XI, and Our Abrams verdict," *Scientific American* Oct 1923–Sept 1924.

Michael E. Wechsler et al., "Active Albuterol or Placebo, Sham Acupuncture, or No Intervention in Asthma," *New England Journal of Medicine* 365, pp.119–26, July 14, 2011, doi: 10.1056/NEJMoa110331.

28—Anecdote

"Interview: Andrew Weil, M.D.," *Frontline*, Public Broadcasting Service, November 2003, http://www.pbs.org/wgbh/pages/frontline/shows/altmed/interviews/weil.html.

Barry L. Beyerstein, "Why Bogus Therapies Seem to Work," *Skeptical Inquirer* vol. 21.5, September/October 1997, https://www.csicop.org/si/show/why_bogus_therapies_seem_to_work.

ICONIC CAUTIONARY TALES FROM HISTORY
29—The "Clever Hans" Effect

For an outline of the Clever Hans story, see http://www.intropsych.com/ch08_animals/clever_hans.html.

30—The Hawthorne Effect

J. R. P. French, "Experiments in Field Settings," in L. Festinger and D. Katz, *Research Methods in the Behavioral Sciences* (New York: Holt, Rinehart & Winston, 1953), ch. 3, pp. 98–135.

Jim McCambridge, John Witton, and Diana R. Elbourne, "Systematic Review of the Hawthorne Effect: New Concepts Are Needed to Study Research Participation Effects," *Journal of Clinical Epidemiology* vol. 67, no. 3, March 2014, pp. 267–277, doi: 10.1016/j.jclinepi.2013.08.015.

Robert Rosenthal and Lenore Jacobson, "Teachers' Expectancies: Determinants of Pupils' IQ Gains," *Psychological Reports* vol. 19, no. 1, 1966, pp. 115–18.

31—Cold Reading

Ray Hyman, "Cold Reading: How to Convince Strangers That You Know All About Them." *Skeptical Inquirer* vol. 1, no. 2, 1977, pp. 18–37.

Bertram R. Forer, "The Fallacy of Personal Validation: A Classroom Demonstration of Gullibility," *Journal of Abnormal and Social Psychology* (American Psychological Association) vol. 44, no. 1, 1949, pp. 118–23.

33—Quantum Woo

Liema Davidovich et al., "Mesoscopic Quantum Coherences in Cavity QED: Preparation and Decoherence Monitoring Schemes," *Physical Review A* vol. 53, no. 1295, 1995.

Michel Raimond, M. Brune, and S. Haroche, "Manipulating Quantum Entanglement with Atoms and Photons in a Cavity," *Review of Modern Physics* vol. 73, no. 585, 2001.

Deepak Chopra, "The Illusion of Past, Present, Future," *Huffpost*, March 18, 2010, updated December 6, 2017, www.huffingtonpost.com/deepak-chopra/the-illusion -of-past-pres_b_326250.html.

Oakridge Associated Universities, "Radithor," 1999, https://www.orau.org/ptp/ collection/quackcures/radith.htm.

https://chem.libretexts.org/Core/Physical_and_Theoretical_Chemistry/Quan tum_Mechanics/02._Fundamental_Concepts_of_Quantum_Mechanics/ De_Broglie_Wavelength.

34—Homunculus Theory

Paracelsus, "Concerning the Nature of Things" in *Hermetic Chemistry* (London, James Elliott & Co., 1894).

John C. McLachlan, "Integrative Medicine and the Point of Credulity," *BMJ*, December 8, 2010, doi: https://doi.org/10.1136/bmj.c6979.

35—Intelligent Design

"Definition of Intelligent Design," intelligentdesign.org, http://www.intelligentde sign.org/whatisid.php.

Michael Behe, *Darwin's Black Box: The Biochemical Challenge to Evolution* (New York: Free Press, 1996).

Jonathan Wells, "Darwin of the Gaps—Review of *The Language of God: A Scientist Presents Evidence for Belief* by Francis S. Collins," Discovery Institute, March 26, 2008, www.discovery.org/a/4529.

Kenneth R. Miller, "The Flagellum Unspun: The Collapse of 'Irreducible Complexity,'" in *Debating Design: From Darwin to DNA*, edited by William A. Dembski and Michael Ruse (Cambridge, UK: Cambridge University Press, 2004), http://www.millerandlevine.com/km/evol/design2/article.html.

36—Vitalism and Dualism

S. C. Morris, J. E. Taplin, and S. A. Gelman, "Vitalism in Naive Biological Thinking," *Development Psychology* vol. 36, no. 5, September 2000, pp. 582–95.

S. Wilson, "Vitalistic Thinking in Adults," *British Journal of Psychology* vol. 104, no. 4, November 2013, pp. 512–24, doi: 10.1111/bjop.12004. Epub 2012 Oct 16.

Linda Rosa et al., "A Close Look at Therapeutic Touch," *JAMA* vol. 279, no. 13, April 1998, pp. 1005–10, doi:10.1001/jama.279.13.1005.

Thelma Moss, *The Body Electric: A Personal Journey into the Mysteries of Parapsychological Research, Bioenergy, and Kirlian Photography* (Los Angeles: J. P. Tarcher, 1979).

Chun Siong Soon et al., "Unconscious Determinants of Free Decisions in the Human Brain," *Nature Neuroscience* vol. 11, 2008, pp. 543–45, doi:10.1038/nn .2112.

Susan Blackmore, *Journal of Consciousness Studies* vol. 9, no. 5–6, pp. 17–28.

Jerry Fodor, *The Mind Doesn't Work That Way* (Cambridge, MA: MIT Press, 2001).

David J. Chalmers, *The Conscious Mind: In Search of a Fundamental Theory* (New York: Oxford University Press, 1996).

Daniel C. Dennett, *Consciousness Explained* (Boston: Little, Brown, 1991).

37—N-Rays

Mary Jo Nye, *Science in the Provinces: Scientific Communities and Provincial Leadership in France, 1860–1930* (Oakland: University of California Press, 1986).

Robert W. Wood, "The n-Rays," *Nature* vol. 70, September 29, 1904, pp. 530–31.

Ernie Tretkoff, "This Month in Physics History, September 1904: Robert Wood Debunks N-rays," *APS News* vol. 17, no. 8, August/September 2007, https://www.aps.org/publications/apsnews/200708/history.cfm.

Elisabeth Davenas et al., "Human Basophil Degranulation Triggered by Very Dilute Antiserum Against IgE," *Nature* vol. 333, June 30, 1988, 816–18.

John Maddox, James Randi, and Walter W. Stewart, "'High-Dilution' Experiments a Delusion," *Nature* vol. 334, July 28, 1988, pp. 287–90, doi: 10.1038/334287a0.

38—Positive Thinking

James Coyne and Howard Tennen, "Positive Psychology in Cancer Care: Bad Science, Exaggerated Claims, and Unproven Medicine," *Annals of Behavioral Medicine* vol. 39, no. 1, February 2010, 16–26.

Lien B. Pham and Shelley E. Taylor, "From Thought to Action: Effects of Process-Versus Outcome-Based Mental Simulations on Performance," *Personality and Social Psychology Bulletin* vol. 25, no. 2, February 1999, pp. 250–60, https://doi.org/10.1177/0146167299025002010.

James Coyne, Howard Tennen, and Adelita Ranchor, "Positive Psychology in Cancer Care: A Story Line Resistant to Evidence," *Annals of Behavioral Medicine* vol. 39, no. 1, February 2010, 35–42.

39—Pyramid Scheme

Aditi Jhaveri, "The Telltale Signs of a Pyramid Scheme," FTC.gov, May 13, 2014.

Jon Taylor, "MLMs Evaluated with 4 Red Flags of a Product-Based Pyramid Scheme," MLM-thetruth.com, 2016.

Jon Taylor, "Multilevel Marketing Primer," MLM-thetruth.com, 2016.

SECTION 2: ADVENTURES IN SKEPTICISM

Maurice de Kunder, "The Size of the World Wide Web (The Internet)," 2017, http://www.worldwidewebsize.com/.

40—Motivated Reasoning About Genetically Modified Organisms

Pew Research Center, "Major Gaps Between the Public, Scientists on Key Issues," July 2015, http://www.pewinternet.org/interactives/public-scientists-opinion-gap/.

Mark Lynas, "How I Got Converted to G.M.O. Food," *New York Times*, April 24, 2015.

William Saletan, "Unhealthy Fixation: The War Against Genetically Modified Organisms Is Full of Fearmongering, Errors, and Fraud. Labeling Them Will Not Make You Safer," *Slate*, July 15, 2015, http://www.slate.com/articles/health _and_science/science/2015/07/are_gmos_safe_yes_the_case_against_them_is _full_of_fraud_lies_and_errors.html.

Chitra Chandrasekaran and Esther Betrán, "Origins of New Genes and Pseudogenes," *Nature Education* vol. 1, no. 1, p. 181, https://www.nature.com/scitable/ topicpage/origins-of-new-genes-and-pseudogenes-835.

Tina Kyndta et al., "The Genome of Cultivated Sweet Potato Contains *Agrobacterium* T-DNAs with Expressed Genes: An Example of a Naturally Transgenic Food Crop," *PNAS* vol. 112, no. 18, pp. 5844–49, https://doi.org/10.1073/ pnas.1419685112.

Theresa Phillips, "Genetically Modified Organisms (GMOs): Transgenic Crops and Recombinant DNA Technology," *Nature Education* vol. 1, no. 1, p. 213, https://www.nature.com/scitable/topicpage/genetically-modified-organisms -gmos-transgenic-crops-and-732.

American Association for the Advancement of Science, "Statement by the AAAS Board of Directors on Labeling of Genetically Modified Foods," October 20, 2012, https://www.aaas.org/sites/default/files/AAAS_GM_statement.pdf.

"Modern Food Biotechnology, Human Health and Development: An Evidence-Based Study," Food Safety Department, World Health Organization, Geneva, 2005, http://www.who.int/foodsafety/publications/biotech/biotech_en.pdf.

EFSA GMO Panel Working Group on Animal Feeding Trials, "Safety and Nutritional Assessment of GM Plants and Derived Food and Feed: The Role of Animal Feeding Trials," *Food and Chemical Toxicology*, Supplement 1, March 2008, pp. S2– 70, doi: 10.1016/j.fct.2008.02.008.

A. L. Van Eenennaam and A. E. Young, "Prevalence and Impacts of Genetically Engineered Feedstuffs on Livestock Populations," *Journal of Animal Science* vol. 92, no. 10, October 2014, pp. 4255–78, https://doi.org/10.2527/jas.2014-8124.

"Elsevier Announces Article Retraction from Journal Food and Chemical Toxicology," Elsevier press release, November 28, 2013, https://www.elsevier.com/ about/press-releases/research-and-journals/elsevier-announces-article -retraction-from-journal-food-and-chemical-toxicology.

David Shukman, "Genetically-Modified Purple Tomatoes Heading for Shops," BBC News, January 24, 2014, http://www.bbc.com/news/science-environment -25885756.

Keith Kloor, "The GMO-Suicide Myth," *Issues in Science and Technology* vol. 30, no. 2, winter 2014, http://issues.org/30-2/keith/.

Glenn Davis Stone, "Field *versus* Farm in Warangal: Bt Cotton, Higher Yields, and Larger Questions," *World Development* vol. 39, no. 3, March 2011, pp. 387–98, https://doi.org/10.1016/j.worlddev.2010.09.008.

Gayathri Vaidyanathan, "Genetically Modified Cotton Gets High Marks in India: Engineered Plants Increased Yields and Profits Relative to Conventional Varieties,"

Nature, July 3, 2012, https://www.nature.com/news/genetically-modified-cotton
-gets-high-marks-in-india-1.10927.

Dan Charles, "Top Five Myths of Genetically Modified Seeds, Busted," *The Salt*,
NPR, October 18, 2012, https://www.npr.org/sections/thesalt/2012/10/18/1630
34053/top-five-myths-of-genetically-modified-seeds-busted.

Claudia Reinhardt and Bill Ganzel, "The Science of Hybrids," Living History Farm,
2003, http://www.livinghistoryfarm.org/farminginthe30s/crops_03.html.

"OSGATA et al. v. Monsanto," Organic Seed Growers and Trade Association,
http://www.osgata.org/osgata-et-al-v-monsanto/.

"Monsanto Canada Inc. v. Schmeiser," Judgments of the Supreme Court of Canada,
https://scc-csc.lexum.com/scc-csc/scc-csc/en/item/2147/index.do.

USDA, "Background: The Science of Seed," https://www.ers.usda.gov/webdocs/pub
lications/42517/13593_aib786c_1_.pdf?v=41055.

Steven MacMillan, "Monsanto's GMO Food and its Dark Connections to the
'Military Industrial Complex,'" *Global Research*, July 03, 2014, https://www.glob
alresearch.ca/monsantos-gmo-food-and-its-dark-connections-to-the-military
-industrial-complex/5389708.

GMWatch, "Golden Rice: Scientific Realities," 2014.

Guangwen Tang et al., "β-Carotene in Golden Rice Is as Good as β-Carotene in
Oil at Providing Vitamin A to Children," (Retracted) *American Journal of
Clinical Nutrition*, vol. 96, no. 3, September 2012, pp. 658–64, doi: 10.3945/ajcn
.111.030775.

Antonio Regalado, "As Patents Expire, Farmers Plant Generic GMOs: Monsanto
No Longer Controls One of the Biggest Innovations in the History of Agriculture," *MIT Technology Review*, July 30, 2015, https://www.technologyreview
.com/s/539746/as-patents-expire-farmers-plant-generic-gmos/.

"Do Seed Companies Control GM Crop Research?" *Scientific American*, August 1,
2009, https://www.scientificamerican.com/article/do-seed-companies-control-gm
-crop-research/.

Andrew Pollack, "Crop Scientists Say Biotechnology Seed Companies Are Thwart-
ing Research," *New York Times*, February 19, 2009, https://www.nytimes.com/
2009/02/20/business/20crop.html.

Nathanael Johnson, "Genetically Modified Seed Research: What's Locked and
What Isn't," *Grist*, August 5, 2013, https://grist.org/food/genetically-modified
-seed-research-whats-locked-and-what-isnt/.

Brian Hutchinson, "Ex-Greenpeace president says group's opposition to genetically-
modified Golden Rice costing thousands of lives," *National Post*, October 11,
2013, last updated January 25, 2015, http://nationalpost.com/news/canada/ex
-greenpeace-president-says-groups-opposition-to-genetically-modified
-golden-rice-costing-thousands-of-lives.

Myriam Charpentier and Giles Oldroyd, "How Close Are We to Nitrogen-Fixing
Cereals?" *Current Opinion in Plant Biology* vol. 13, no. 5, October 2010, pp. 556–
64, doi: 10.1016/j.pbi.2010.08.003.

Anindya Bandyopadhyay et al., "Enhanced Photosynthesis Rate in Genetically Engineered Indica Rice Expressing Pepc Gene Cloned from Maize,"
Plant Science vol. 172, no. 6, June 2007, pp. 1204–09, https://doi.org/10.1016/j
.plantsci.2007.02.016.

41—Dennis Lee and Free Energy

Dennis Lee, "NO EXHAUST Engine—Dennis Lee Shows Running Geet Engine
with NO EXHAUST Closed Loop," May 2002, https://www.youtube.com/
watch?v=dW0moXn9y9U.

"FTC Sues Promoters of Bogus Fuel Efficiency Device: Ads Appeared in Major
Magazines Promising to Turn Any Car Into a Hybrid," Federal Trade Commission press release, February 2, 2009, https://www.ftc.gov/news-events/press
-releases/2009/02/ftc-sues-promoters-bogus-fuel-efficiency-device.

42—Holly-woo

Michael Hiltzik, "Reporting on Quacks and Pseudoscience: The Problem for Journalists," *Los Angeles Times*, April 13, 2015.

Kristen Brown, "The Next Pseudoscience Health Craze Is All About Genetics,"
Gizmodo, February 2017.

Sarah Waldorf, "Physiognomy, The Beautiful Pseudoscience," *The Iris*, October 2012.

Tyler Szelinski, "Pseudoscience in Hollywood: How I Avoided Being Victimized
by Pseudoscience on Hollywood Boulevard," Odyssey Online, September 2016.

Allegra Ringo, "I Took My Dog to Pet Reiki," Vice.com, November 2014.

Mack Rawden, "Gwyneth Paltrow Says You Need to Steam Your Vagina Because
Pseudoscience," Cinemablend.com, 2015.

Fox News Editors, "Gwyneth Paltrow Doesn't Care If You Think Her Health
Advice Is Weird," Fox News Health, March 2017.

Julie Kelly, "Hollywood Celebrities Embrace Pseudoscience, Promote Anti-GMO
Movie 'Consumed,'" Genetic Literacy Project, June 2015.

Rachael Rettner, "Celeb Trend of 'IV Vitamins' Not a Good Idea," Live Science,
June 2012.

Steven Novella, "Iridology," *Science Based Medicine*, December 2011.

Jessica Goldstein, "Is Gwyneth Paltrow Wrong About Everything? This Researcher
Thinks So," thinkprogress.org, April 2016.

Timothy Caulfield, *Is Gwyneth Paltrow Wrong About Everything?* (New York: Viking,
2015).

43—The Singularity

Stanislaw Ulam, "Tribute to John von Neumann, 1903–1957," *Bulletin of the American Mathematical Society* vol. 64, no. 3, May 1958.

Irving John Good, "Speculations Concerning the First Ultraintelligent Machine,"
Advances in Computers vol. 6 (New York: Academic Press, 1965), pp. 31–88,
https://doi.org/10.1016/S0065-2458(08)60418-0.

Raymond Kurzweil, "The Law of Accelerating Returns," Kurzweilai.net, Kurzweil
.net/the-law-of-accelerating-returns.
Fortune Magazine, March 1, 2017.

44—The Warrens and Ghost Hunting

Frank Moran, "The True Story of the Perron Family, The Harrisville Haunting," *Horror Galore*. Accessed May 25, 2017. http://www.horrorgalore.com/
true-story/true-story-perron-family-harrisville-haunting.
Joe Nickell, "Enfield Poltergeist: Investigative Files," *Skeptical Inquirer* vol. 36, no.
4, July/August 2012. Accessed May 25, 2017. http://www.csicop.org/si/show/
enfield_poltergeist.
David Thomas, "Artificial Intelligence Investing Gets Ready for Prime Time,"
Forbes, October 25, 2017.

45—Loose Thinking About Loose Change

James B., "It Was A Military Plane," *Screw Loose Change Exposing the Lies, Distortions
and Myths of the 9-11 "Truthers*," May 2006, http://screwloosechange.blogspot
.com/2006/05/it-was-military-plane.html.
ABC News, "Air Traffic Controllers Recall Sept. 11," 911Review.com, October
2001, http://911review.com/cache/errors/pentagon/abcnews102401b.html.
James B., "It's Just a Mistake," *Screw Loose Change Exposing the Lies, Distortions
and Myths of the 9-11 "Truthers*," May 2006, http://screwloosechange.blogspot
.com/2006/05/its-just-mistake.html.
Griffin, "Danielle Obrien," 911Myths.com, July 2012, http://www.911myths.com/
index.php?title=Danielle_Obrien.

SECTION 3: SKEPTICISM AND THE MEDIA

46—Fake News

John Carroll, "Pseudo-Journalists Betray the Public Trust," *Los Angeles Times*, May
16, 2004.
Oxford Dictionaries, "Word of the Year 2016 is…" https://en.oxforddictionaries
.com/word-of-the-year/word-of-the-year-2016.
CNN Wire Staff, "Onion: We Just Fooled the Chinese Government," November 2012.
Sharyl Attkisson, "Astroturf and Manipulation of Media Messages," TEDx Talk, Uni-
versity of Nevada, February 2015, https://www.youtube.com/watch?v=-bYAQ
-ZZtEU.
David Gorski, "Anti-Vaccine Propaganda from Sharyl Attkisson of CBS News,"
Science-Based Medicine, April 4, 2011, http://sciencebasedmedicine.org/anti-vaccine
-propaganda-from-sharyl-attkisson-of-cbs-news-2/.

48—Science Journalism

Roy Caldwell et al., "Beware of False Balance: Are the Views of the Scientific Community Accurately Portrayed?" University of California Museum of Paleontol-
ogy, http://undsci.berkeley.edu/article/sciencetoolkit_04.

David Shiffman, "World's Leading Experts Say There's a Problem with False Balance in Conservation Journalism; Steve Disagrees," Southern Fried Science, July 2013.

Joe Romm, "False Balance Lives At The New York Times," Thinkprogress.org, March 2012.

Joe Romm, "The Washington Post Doubles Down on False Balance," Thinkprogress .org, March 2012.

Gary Schwitzer, "How Do US Journalists Cover Treatments, Tests, Products, and Procedures? An Evaluation of 500 Stories," PLoS Med vol. 5, no. 5, e95, 2008, https://doi.org/10.1371/journal.pmed.0050095.

Amélie, Yavchitz et al., "Misrepresentation of Randomized Controlled Trials in Press Releases and News Coverage: A Cohort Study," PLoS Med vol. 9, no. 9, e1001308, 2012, doi: 10.1371/journal.pmed.1001308.t004.

49—Enter the Matrix

Jaehoon Choe at al., "Transcranial Direct Current Stimulation Modulates Neuronal Activity and Learning in Pilot Training," Frontiers in Human Neuroscience, February 2016.

HRL Laboratories, "HRL Demonstrates the Potential to Enhance the Human Intellect's Existing Capacity to Learn New Skills," 2016, http://www.hrl.com/news/2016/02/10/hrl-demonstrates-the-potential-to-enhance-the-human -intellects-existing-capacity-to-learn-new-skills.

50—Microbiomania

Luke Ursell et al., "Defining the Human Microbiome," Nutrition Reviews vol. 70, no. Suppl 1, August 2012, pp. S38–44, doi: 10.1111/j.1753-4887.2012.00493.x.

Ron Sender, Shai Fuchs, and Ron Milo, "Are We Really Vastly Outnumbered? Revisiting the Ratio of Bacterial to Host Cells in Humans," Cell vol. 164, no. 3, January 28, 2016, pp. 337–40.

Ed Yong, I Contain Multitudes: The Microbes Within Us and a Grander View of Life (London: Bodley Head, 2016).

Emma Bryce, "Do Probiotics Cure Asthma? Don't Believe the Hype," Wired, January 2016, http://www.wired.co.uk/article/microbiome-gut-defender.

Jonathan A. Eisen, "Microbiomania and 'Overselling the Microbiome,'" The Tree of Life, 2015, https://phylogenomics.blogspot.com/p/blog-page.html.

Jonathan A. Eisen, "Today's Misleading Overselling the #microbiome—U. Chicago on Alzheimer's and Gut Microbes," The Tree of Life, July 2016.

Myles Minter et al., "Antibiotic-Induced Perturbations in Gut Microbial Diversity Influences Neuro-Inflammation and Amyloidosis in a Murine Model of Alzheimer's Disease," Scientific Reports vol. 6, no. 30028, 2016, doi: 10.1038/srep30028.

Press Release, "Antibiotics Weaken Alzheimer's Disease Progression through Changes in the Gut Microbiome," University of Chicago Medical Center, July 2016.

Olivia Lerche, "Antibiotics Could Be Used to PREVENT Alzheimer's Disease by Changing GUT Bacteria," *Daily Express*, July 21, 2016.

Laura Sanders, "Antibiotics Might Fight Alzheimer's Plaques," *Science News*, July 21, 2016, https://www.sciencenews.org/article/antibiotics-might-fight-alzheimer%E2%80%99s-plaques.

Taoufiq Harach et al., "Reduction of Abeta Amyloid Pathology in APPPS1 Transgenic Mice in the Absence of Gut Microbiota," *Scientific Reports* vol. 7, no. 41802, 2017, doi: 10.1038/srep41802.

Joseph Mercola, "How Your Gut Microbiome Influences Your Mental and Physical Health," Mercola.com, January 2016.

Joseph Mercola, "What Happens If You Stop Showering? Dr. Mercola Shares the Surprising Truth," Consciouslifenews.com.

Joseph Mercola, *Fat for Fuel* (Carlsbad, CA: Hay House, 2017).

Marilyn Malara, "Man Skips Shower for 12 Years, Uses Bacterial Spray to Keep Clean," upi.com, September 12, 2015.

Kavin Senapathy, "Deepak Chopra Says Bacteria Listen to Our Thoughts," *Forbes*, January 2016.

51—Reporting Epigenetics

Brad Crouch, "Darwin's Theory of Evolution Challenged by University of Adelaide Genetic Memory Research, Published in Journal Science," *Advertiser*, August 2014.

Michelle Lane, Rebecca L. Robker, and Sarah A. Robertson, "Parenting from Before Conception," *Science* vol. 345, no. 6198, August 15, 2014, pp. 756–60, doi: 10.1126/science.1254400.

SECTION 4: DEATH BY PSEUDOSCIENCE

52—Death by Naturopathy

Britt Marie Hermes, "Naturopathic Doctors Look Bad after California Woman Dies from Turmeric Injection," *Forbes*, March 2017.

Britt Marie Hermes, "Confirmed: Licensed Naturopathic Doctor Gave Lethal 'Turmeric' Injection," *Forbes*, April 2017.

Kim Kelly, http://myemail.constantcontact.com/IV-Curcumin-and-Talk-At-Bella-Sareena-Spa-in-Solana-Beach.html?soid=1105463481806&aid=_FmLhr6qCmw.

Kathryn M. Nelson et al, "The Essential Medicinal Chemistry of Curcumin," *Journal of Medical Chemistry* vol. 60, 2017, pp. 1620–37.

Stephen Barrett, "A Close Look at Naturopathy," Quackwatch, November 2013.

David Gorski, "An As Yet Unidentified 'Holistic' Practitioner Negligently Kills a Young Woman with IV Turmeric (Yes, Intravenous)," *Respectful Insolence*, March 2017.

Klara Rombauts, Liene Dhooghe, and CAM-Cancer Consortium, "Curcumin [online document]," May 7, 2014, http://www.cam-cancer.org/The-Summaries/

Herbal-products/Curcumin.https://nccih.nih.gov/health/turmeric/ataglance
.htm.

Julianna LeMieux, "A Naturopath's Human Experiment Ends in Death," American
Council on Science and Health, March 23, 2017.

Elaine Hannah, "What Are the Natural Treatments That Could Be Dangerous to
Health?" *Science World Report*, March 2017.

FDA, "Enforcement Activities by the FDA," March 2017, https://www.fda.gov/
drugs/guidancecomplianceregulatoryinformation/enforcementactivitiesbyfda/
ucm171057.htm.

House of Delegates Position Paper, "Definition of Naturopathic Medicine," Natu-
ropathic.org, 2011.

AMA, "State Law Chart: Naturopath Licensure and Scope of Practice," 2017,
https://www.ama-assn.org/sites/default/files/media-browser/specialty%20
group/arc/ama-chart-naturopath-scope-practice-2017.pdf.

Pharmacy Webinar 28 Clinical Applications of Medical Ozone, Naturopathic
.org, 2015, http://www.naturopathic.org/ev_calendar_day.asp?date=12/1/2015
&eventid=14.

NCCIH, "Turmeric," https://nccih.nih.gov/health/turmeric/ataglance.htm.

53—Exorcism: Medieval Beliefs Yield Medieval Results

"Mother Who Killed Two Children in 'Exorcism' Will Go to Mental Hospital," Associated Press, September 15, 2016. Accessed May 25, 2017. http://
www.cbsnews.com/news/mother-who-killed-two-children-in-exorcism-will
-go-to-mental-hospital/.

Anelisa Kubheka, "Exorcism Victim's Last Moments," *IOL*, March 23, 2012. Accessed
May 25, 2017. http://www.iol.co.za/news/crime-courts/exorcism-victims-last
-moments-1262802.

"'Satan's Schoolgirl': Tortured to Death with Exorcism," *Bizarrepedia*, created November 14, 2016. Accessed May 25, 2017. https://www.bizarrepedia
.com/anneliese-michel/.

Rachel Ray, "Leading US Exorcists Explain Huge Increase in Demand for the
Rite—and Priests to Carry Them Out," *The Telegraph*, September 26, 2016,
https://www.telegraph.co.uk/news/2016/09/26/leading-us-exorcists-explain
-huge-increase-in-demand-for-the-rit/.

Joe Nickell, "Exorcism! Driving Out the Nonsense," *Skeptical Inquirer* vol. 25.1, Janu-
ary/February 2001.

54—Death by Denial

"A Timeline of HIV and AIDS," HIV.gov, https://www.aids.gov/hiv-aids-basics/hiv
-aids-101/aids-timeline/.

Ricardo A. Franco and Michael S. Saag, "When to Start Antiretroviral Therapy:
As Soon as Possible," *BMC Medicine* vol. 11, no. 147, June 2013, doi: 10.1186/
1741-7015-11-147.

Roger Cohen, "Mbeki's Shame," *New York Times*, July 3, 2008, https://www.nytimes.com/2008/07/03/opinion/03cohen.html.

P. Chigwedere et al., "Estimating the lost benefits of antiretroviral drug use in South Africa," *Journal of Acquired Immune Deficiency Syndromes* vol. 49, no. 4, December 2008, pp. 410–15.

55—Suffer the Children

Patrick Begley, "'Slapping Therapy' Death Case Referred to Prosecutor," *Sydney Morning Herald*, December 11, 2015, http://www.smh.com.au/nsw/slapping-therapy-death-case-referred-to-prosecutor-20151208-gli59j.html.

Jean Mercer, Larry Sarner, and Linda Rosa, "Attachment Therapies: A Deadly Cure without a Disease? Review of *Attachment Therapy on Trial: The Torture and Death of Candace Newmaker*," *Scientific Review of Mental Health Practice* vol. 3, no. 1, Spring/Summer 2004, https://www.srmhp.org/0301/review-01.html.

Attachment.org.

Mayo Clinic, "Reactive Attachment Disorder," July 2017, http://www.mayoclinic.org/diseases-conditions/reactive-attachment-disorder/basics/definition/con-20032126.

Quackerywatch, http://www.quackerywatch.com/Attachment-therapy/index.html.

SECTION 5: CHANGING YOURSELF AND THE WORLD

56—Being Skeptical

Lisa-Christine Girard, Orla Doyle, and Richard E. Tremblay, "Breastfeeding, Cognitive and Noncognitive Development in Early Childhood: A Population Study." *Pediatrics*, March 2017, doi: 10.1542/peds.2016-1848.

Index